FLORA OF TROPICAL EAST AFRICA

LEGUMINOSAE (Part 2)

Subfamily CAESALPINIOIDEAE

DC., Prodr. 2: 473 (1825), as suborder *Caesalpinieae*

J. P. M. BRENAN*

Trees, shrubs, sometimes lianes, or rarely herbs, unarmed, or often armed in the tribe *Caesalpinieae*. Leaves usually pinnate, sometimes bipinnate (tribes *Dimorphandreae* and *Caesalpinieae*; this condition considered by Dormer, in Ann. Bot., n.s., 9: 141–153 (1945), to be more primitive than pinnate), rarely unifoliolate or simple. Inflorescences usually of spikes or panicles of racemes, rarely of spikes or capitate; racemes sometimes (by reduction of the main axis) represented by umbelliform fascicles. Flowers usually small to medium or large, rarely very small, usually ± irregular. Sepals usually imbricate, rarely valvate, rarely open from an early stage of bud, free or sometimes ± connate; rarely calyx entire in bud and splitting afterwards (tribe *Swartzieae*). Petals imbricate in bud, usually with the dorsal one within and overlapped by the adjacent lateral ones, free or sometimes connate below, usually 5, sometimes ± reduced, even to only 1 or altogether absent. Stamens usually 10 or fewer, rarely numerous, free or ± connate below. Anthers various, but lacking the apical gland often seen in *Mimosoïdeae*. Pollen-grains usually simple (see p. 2). Ovules anatropous. Pods various. Seeds generally without areoles (see p. 2), with an apical or subapical hilum; embryo with a generally straight radicle.

The subfamily *Caesalpinioïdeae* seems best placed taxonomically between the *Mimosoïdeae*, whose floral characters are in general relatively less advanced, and the relatively more advanced *Papilionoïdeae*.** Within the *Caesalpinioïdeae* the tribe *Dimorphandreae* shows a very close approach to the *Mimosoïdeae*, and it is hard to decide whether to include the tribe *Swartzieae* in the *Papilionoïdeae* or the *Caesalpinioïdeae*. I have placed this latter tribe in the *Caesalpinioïdeae* on the strength of *Baphiopsis*. Sections through the flower-bud of *B. parviflora* show that the posticous petal is within the adjacent ones (as in *Caesalpinioïdeae*) and not outside them (as in *Papilionoïdeae*). It is impossible to assess this character in *Swartzia* itself, where only one petal is developed.

The *Dimorphandreae* and the *Swartzieae* thus are on the borderlines between the different subfamilies, and their inclusion in the *Caesalpinioïdeae* depends on the borderlines being arbitrarily rather than naturally fixed. Perhaps there is no clear natural boundary in this group.

Although some authors consider the three subfamilies of *Leguminosae* as separate families, this really represents no more than a slight difference of opinion. If emphasis is laid on the borderline tribes mentioned above, then subfamily is the reasonable rank; if, on the other hand, these are discounted in favour of the numerically much larger mass of genera about whose position there is no room for doubt or difference of opinion, then the subfamilies are reasonably considered as families. The three groups, however, remain very much the same in content whatever taxonomic rank is assigned to them.

The limits assigned here to the *Caesalpinioïdeae* result in all the genera of *Leguminosae*

* With the account of *Brachystegia* by Mr. A. C. Hoyle (Department of Forestry, Oxford).

** According to Article 19 of the "International Code of Botanical Nomenclature" (1961) this group should, I suppose, be called *Leguminosoïdeae*. However, in view of the special provision made in Article 18, Note 3, for the use of *Papilionaceae* as a family name, I am taking the view that the failure to provide for the use of the same stem in the rank of subfamily was an oversight in drafting the Code.

with more than 10 stamens being either in the *Caesalpinioïdeae* or *Mimosoïdeae*, as well as all those genera with regular flowers (except only for *Cadia* Forsk. and its relatives).

The naturalness of the *Caesalpinioïdeae* as a group and its relationship with the other subfamilies of *Leguminosae* have been additionally illustrated by certain special investigations, both morphological and physiological, which are mentioned under the following headings:—

1. *Pollen.* According to Erdtman, Pollen Morphology and Plant Taxonomy, 1 (Angiosperms): 225–231 (1952), there are no striking differences between the grains in the three subfamilies, but while tetrads or polyads are common in the *Mimosoïdeae*, they do not occur in the *Caesalpinioïdeae* except for tetrads in various species of *Afzelia* (though not *A. quanzensis*).

Sister M. V. Fashbender, O.S.B., in Lloydia 22: 107–162 (1959), studied the pollen-grain morphology of the tribes *Amherstieae*, *Cynometreae* and *Sclerolobieae* (not African) and its taxonomic significance. Her results did not support Léonard's classification of the two former tribes in Mém. 8°, Classe Sci., Acad. Roy. Belg. 30(2) (1957), nor a suggestion by Dwyer that all three tribes should be amalgamated. This investigation laid special emphasis on the American genera, and it is obvious that a similar one for Africa would be most valuable.

2. *Seeds.* For comments on the distinctive characters of the seeds of *Caesalpinioïdeae*, see Corner in Phytomorphology 1: 117–150 (1951). His example of *Swartzia pinnata* is not to be taken as implying that the radicle in the seeds of this genus is always straight—in fact it may be either straight or curved. However it remains true that the radicle in *Caesalpinioïdeae* is usually straight except in some species of *Swartzia* (and possibly some related genera). This is also true of *Mimosoïdeae*, though in *Papilionoïdeae* the radicle is usually curved (though said to be straight in, e.g., the *Galegeae*).

In F.T.E.A. *Leguminosae—Mimosoïdeae*: 1 (1959), I discussed the remarkable areole on each face of the seed, known as the pleurogram, which occurs so commonly in that subfamily. At that time I had met with a pleurogram among the *Caesalpinioïdeae* only in *Cassia*, but I have since seen similar structures in *Burkea*, *Tamarindus* and *Paramacrolobium*, though in none of the other genera of this subfamily.

The areoles in *Tamarindus* and *Paramacrolobium* are closed, i.e. with continuous margins, and reflect a small but abrupt change in level or surface-marking of the testa. In these ways they are different from those of *Mimosoïdeae*, where there is a definite crack, at least in the superficial layer of the testa, and the areole is open (non-marginate) at the hilar end of the seed. The very well-marked areoles of *Cassia* are similar to those of *Mimosoïdeae*, but again, at any rate as far as I have observed them, they are normally closed. Perhaps only in *Burkea*, a genus decidedly on the borderline with *Mimosoïdeae*, is the pleurogram really comparable with those of the latter subfamily.

3. *Anatomy.* Some anatomical distinctions for the *Caesalpinioïdeae* are given by Metcalfe & Chalk, Anatomy of the Dicotyledons 1: 528–9 (1950), though they perhaps seem less definite than some of the gross morphological ones already mentioned.

Dormer, in Ann. Bot., n.s., 9: 141–153 (1945), stated that " open " (i.e., without anastomoses) vascular systems in the stem are characteristic of the *Mimosoïdeae* and some *Papilionoïdeae*, but are not known in *Caesalpinioïdeae*, where " closed " systems are usual.

4. *Foliar pulvini and leuco-anthocyanins in the leaves.* According to Bate-Smith, in Proc. Linn. Soc. Lond. 169: 202–3 (1958), these characters appear to be correlated, and both are constantly present in the *Mimosoïdeae* and *Caesalpinioïdeae*, but present or absent (according to tribes) in the *Papilionoïdeae*. It is interesting that Dormer (in Ann. Bot., n.s., 9: 147 (1945)) considered that the presence of foliar pulvini was a relatively primitive character.

5. *Amyloids.* The occurrence of amyloids in plant seeds, substances causing a blue colour in the cell-walls of the cotyledons of some plant species when they are treated with a solution of iodine in potassium iodide, was studied by Kooiman in Acta Botanica Neerlandica 9: 208–219 (1960). He found that amyloid was strictly confined, in the *Caesalpinioïdeae*, to the tribes *Cynometreae*, *Amherstieae* and the non-African *Sclerolobieae*, though strangely lacking in the genera *Colophospermum*, *Gossweilerodendron* and *Oxystigma*. Although 126 species in 104 genera of *Papilionoïdeae* were tested for amyloid, it was found only in two, *Goodia* and *Mucuna*.

The arrangement of the native genera is intended to be natural, and follows the generally accepted grouping of the genera under a number of tribes. A conspectus of this system is given below:—

A. Leaves bipinnate (except in No. 9, *Stuhlmannia*, in *Caesalpinieae*):

1. Tribe **Dimorphandreae** *Benth.* in Hook., Journ. Bot. 2: 74 (1840). Unarmed. Flowers small, in elongate spikes or dense spike-like racemes,

often paniculately aggregated. Calyx-lobes ± connate below into a short tube (sometimes very short) extending beyond the hypanthium. Genera Nos. 1–2.

2. Tribe **Caesalpinieae** (*Eucaesalpinieae* Benth. in Hook., Journ. Bot. 2 : 72 (Mar. 1840) ; *Caesalpinieae* Endl., Gen. Pl. : 1310 (Oct. 1840)). Unarmed or armed. Flowers usually medium to large, sometimes small, in racemes or panicles of racemes. Sepals 5, free to the hypanthium. Genera Nos. 3–9.

B. Leaves simply pinnate, or sometimes simple or unifoliolate:

3. Tribe **Cassieae** *Bronn*, De Formis Pl. Legum. : 130 (1822) ; DC., Prodr. 2 : 478 (1825). Leaves normally simply pinnate. Bracteoles usually small or absent and caducous. Sepals distinct in bud, usually 5, free to base. Anthers characteristically firm in texture, often large and with comparatively short filaments, usually dehiscing by pores, which may be prolonged into short slits; sometimes slits extending to whole length of anther, anthers then basifixed. Genera Nos. 10–11.

4. Tribe **Cynometreae** *Benth.* in Hook., Journ. Bot. 2 : 74 (1840), emend. J. Léon. in Mém. 8°, Classe Sci., Acad. Roy. Belg. 30(2) : 54 (1957). Leaves simply pinnate, rarely unifoliolate or simple. Bracteoles small or large, usually caducous, not enclosing the flower-buds, or enclosing them, but then never valvate. Sepals distinct in bud, free to base. Anthers dorsifixed, dehiscing by slits. Genera Nos. 12–22.

5. Tribe **Amherstieae** *Benth.* in Hook., Journ. Bot. 2 : 73 (1840), emend. J. Léon. in Mém. 8°, Classe Sci., Acad. Roy. Belg. 30(2) : 163 (1957). Leaves simply pinnate, rarely unifoliolate. Bracteoles well-developed, enclosing the flower-buds, valvate, usually persistent. Sepals distinct in bud, free to base, or very small or absent. Anthers dorsifixed, dehiscing by slits. Genera Nos. 23–32.

6. Tribe **Cerceae** *Bronn*, De Formis Pl. Legum. : 131 (1822) (*Bauhinieae* Benth. in Hook., Journ. Bot. 2 : 74 (1840)). Leaves usually simple, bilobed or entire, or sometimes with 2 separate leaflets. Calyx gamosepalous above the hypanthium, campanulate or tubular, shortly toothed or lobed, sometimes more deeply divided into valvate lobes. Stamens 10 or fewer. Anthers dorsifixed, opening by longitudinal slits, rarely by pores. Genera Nos. 33–36.

7. Tribe **Swartzieae** *DC.*, Prodr. 2 : 422 (1825). Leaves simply pinnate, rarely unifoliolate. Calyx entire in bud, closed, not divided into sepals, becoming variously lobed or split as the flower opens. Stamens 9–many (always numerous in East Africa). Genera Nos. 37–40.

In the tribes *Cynometreae* and *Amherstieae*, the genera are arranged according to the brilliant revision by J. Léonard in Mém. 8°, Classe Sci., Acad. Roy. Belg. 30(2) (1957).

Dialium and *Cordyla* are often considered, with some reason, to be anomalous genera. Nevertheless I consider that Bentham was correct in placing them in the *Cassieae* and *Swartzieae* respectively, as he did in the Genera Plantarum (1865), and that neither genus is as anomalous as one might assume from a study only of the African genera in these tribes.

KEY TO GENERA BASED MAINLY ON VEGETATIVE AND FLORAL CHARACTERS*

1. Leaves simple or unifoliolate 2
   Leaves compound 7
2. Leaves obtusely pointed to acute or acuminate at apex, not emarginate 3
   Leaves emarginate at apex, or bilobed 5
3. Flowers very large (hypanthium in East African representatives 8–13 cm. long, and comprising most of the apparent

* I have had most valuable help and advice, in making both this key and the one for fruiting specimens (p. 10), from Mr. A. C. Hoyle, Department of Forestry, Oxford.

"pedicel", and petals 9–11 cm. long); petals 5; fertile stamens 10; ovary long-stipitate . . . . 33. **Gigasiphon**

Flowers much smaller (hypanthium not elongate, and petals 0·6–1 cm. long); petals 5–6; fertile stamens 10–41; ovary sessile to stipitate 4

4. Fertile stamens 10; sepals (4–)5, imbricate in bud; petals 5; ovary stipitate . . 15. **Zenkerella**

Fertile stamens 13–41; calyx entire in bud, not divided into sepals, later opening by a single slit, or into 2–3 lobes; petals 6; ovary sessile or nearly so . . . . 37. **Baphiopsis**

5. Plants with trailing or climbing mostly herbaceous stems arising from a large underground tuber; tendrils usually present; fertile stamens 2, accompanied by staminodes; calyx with the 2 upper sepals partly or completely fused, the rest free. 36. **Tylosema**

Plants growing as shrubs or small trees, in Flora area never climbing; tendrils absent 6

6. Calyx turbinate, with 4–5 short broad lobes; flowers normally dioecious and unisexual; ♂ flowers with 10 fertile stamens; ♀ flowers with the stigma sessile on the ovary . . . . 34. **Piliostigma**

Calyx spathaceous (splitting to the base down one side only), with the sepals fused or sometimes partly separated (then long and narrow); flowers ⚥; fertile stamens 1–10; style elongate . . . . . 35. **Bauhinia**

7. Leaves bipinnate 8

Leaves simply pinnate 20

8. Plant unarmed; shrubs or trees, not climbing 9

Plant armed with prickles or spines 15

9. Leaflets alternate; flowers small (petals 2–5 mm. long, white to cream or pale green); sepals open from a very early stage, leaving the petals covering the flower until anthesis 10

Leaflets opposite; flowers usually medium to large (petals 9–55 mm. long, usually yellow or white, sometimes red); sepals valvate or imbricate 11

10. Flowers pedicellate, with narrow petals which are ± densely pubescent, at least on margins; anthers 0·5–0·75 mm. long; stigma minute, cup-shaped-punctiform on a narrowly conical style; stamen-filaments glabrous or hairy; pods dehiscent, (by abortion 1–)2–11-seeded; seeds not areolate; hairs on vegetative buds, petiole-bases, young branchlets and ovary grey to yellowish or (rarely, and not in East Africa) rusty . . . . 1. **Erythrophleum**

Flowers sessile, with broad strongly imbricate petals which are glabrous or almost so; anthers 1·5–2 mm. long; style very short, ending in a funnel-shaped stigma slit down one side; stamen-filaments glabrous; pods indehiscent, 1-seeded; seeds areolate on each face; hairs on vegetative buds, etc., conspicuously rusty-red . . . 2. **Burkea**

11. Stigma peltate and abruptly enlarged; pods with 2 woody valves recurving elastically on opening, the outside of each with a conspicuous longitudinal groove down centre, or concave; inflorescences conspicuously rusty-tomentose . . . 4. **Bussea**

Stigma truncate or oblique or funnel-shaped, not wider than style, or gradually enlarged upwards but not abruptly peltate; pods dehiscing or not, but valves not recurving elastically nor grooved down centre outside 12

12. Ovary and style densely glandular and also pubescent; some peltate reddish glands scattered over surface of leaflets . . 5. **Caesalpinia** (spp. 2, 3)

Ovary and style pubescent or glabrous, but eglandular; leaflets without glands on surface 13

13. Flowers rather small, with petals 1–1·1 cm. long; pods indehiscent, with papery brown closely venose valves . . . 8. **Parkinsonia** (sp. 2)

Flowers larger, with petals 1·5–5·5 cm. long; pods dehiscent, with coriaceous or woody valves 14

14. Sepals valvate; trees 2·5–18 m. high . . 3. **Delonix**

Sepals imbricate; shrub 1–6 m. high . . 5. **Caesalpinia** (sp. 1)

15. Prickles scattered along stems 16

Prickles or spines paired, at nodes only 18

16. Petals 3 mm. long; pods samaroid, with a basal seed-containing portion with a single seed, and a distal wing bounded on one side by a prolongation of the upper suture, the whole resembling a samara of *Acer* . . . . 7. **Pterolobium**

Petals 6·5–25 mm. long; pod unwinged, or winged all along upper side of suture, and then (in East Africa) 2–4-seeded 17

17. Pod winged along upper suture; petals 6·5–9 mm. long, yellow; pedicels 4–11 mm. long 6. **Mezoneuron**

Pod unwinged; petals 9–25 mm. long, variously coloured; pedicels 2–75 mm. long . . 5. **Caesalpinia** (spp. 1, 4, 5, 7–10)

18. Anthers pubescent . . . . . . 5. **Caesalpinia** (sp. 6)

Anthers glabrous 19

19. Flowers large, often scarlet, with petals 15–25 mm. long and stamen-filaments 50–65 mm. long; pod dehiscent, with hard non-venose valves; leaflets 5–13 pairs per pinna, 4–15 mm. wide . . . . . . 5. **Caesalpinia** (sp. 1)

    Flowers rather small, yellow, with petals 6–8 mm. long and stamen-filaments ± 5–7 mm. long; pod indehiscent, with thin brown closely venose valves*; leaflets 3–6 pairs per pinna, 1·25–3·5 mm. wide 8. **Parkinsonia** (sp. 1)

20. Calyx closed, entire and undivided in bud, becoming divided into 2–5 lobes or irregularly torn as the flower opens; bracteoles very small and caducous, or absent; stamens always numerous (12–126) 21

    Calyx (when present) clearly divided into lobes or separate sepals in bud; calyx sometimes very small or 0, but then bracteoles large, enclosing the flower-bud, and usually persistent; stamens 10 or fewer (very rarely up to 20**, but then (in East Africa) bracteoles large, as described above) 23

21. Petals 1, large; leaflets without pellucid glands; stamen-filaments free or nearly so; disc and hypanthium absent (filaments inserted round base of gynophore or ovary) . . . . 38. **Swartzia**

    Petals 0; leaflets with circular or elongate pellucid glands; stamen-filaments slightly connate at base; disc or hypanthium obviously present 22

22. Flower-buds subglobose, becoming flattened or concave beneath as the calyx splits into (2–)3 unequal lobes down to the edge of a disc; stamens 12–18, united at base in one whorl round the well-marked margin of the disc; connective non-glandular . . . . 40. **Mildbraediodendron**

    Flower-buds turbinate, with the calyx dividing into 3–5 lobes only as far as the edge of the campanulate hypanthium; stamens 23–126, their filaments usually in several series, gradually confluent at base with the non-margined hypanthium; connective gland-tipped . . . . . 39. **Cordyla**

23. Bracteoles paired, valvate throughout, well-developed, completely enclosing the flower-bud, usually persistent 24

* If petals only 3 mm. long and pod conspicuously winged distally, compare genus 7, *Pterolobium*.

** In genera 23, *Isoberlinia*, 26, *Englerodendron* and 30, *Brachystegia*.

Bracteoles non-valvate, often caducous, or absent, usually not enclosing the flower-bud, but, if so, then one or both margins of one bracteole overlapping the other at least at the base 33

24. Bracteoles enclosing the young flower-bud but soon caducous, exposing the bud enclosed by the calyx; stamen- and staminode-filaments all connate to about half-way in a band; fertile stamens 3, alternating with 5 sterile teeth or short filaments; larger petals 3, subequal, gold with red veins . . . . . 27. **Tamarindus**

Bracteoles enclosing the flower-bud and persistent below the open flower, but sometimes caducous with the petal(s) and before the anthers in 32, *Cryptosepalum* 25

25. Bracteoles petaloid, white or pink; fertile stamens 3 or rarely up to 6(–8), sometimes with staminodes, all free from each other, with the 1(–3) petals on the edge of a small but obvious, almost naked hypanthium ("receptacle"); sepals minute or absent. . . . . . 32. **Cryptosepalum**

Bracteoles not petaloid, often thick and coriaceous; stamens (8–)9–18(–20), usually all fertile but, if not, then at least some united at the base and/or sepals well-developed 26

26. Leaflets all distinctly petiolulate; petiolules not obviously twisted; base of leaflet often symmetric or nearly so 27

Leaflets (at least the distal pair) sessile or rarely very shortly petiolulate and then petiolules obviously twisted as viewed from above; base of leaflet usually asymmetric with the proximal side decurrent almost or quite to the upper side of the leaf-rhachis and often twisted over the pulvinulus 30

27. Leaflets (in native species) with a very obvious, coppery sheen beneath; fertile stamens 5 (in East Africa); one petal much larger than the others (in East Africa) . . . . . . 28. **Anthonotha**

Leaflets without any metallic sheen beneath. 28

28. Petals 5, very unequal (in native species), one very large with a long, narrow basal part (subequal to the 5 narrow sepals and 4 smaller petals) and with a conspicuous, broad limb above; stamens with 9 shortly connate and 1 free . . . . . 24. **Berlinia**

Petals 5–7, equal or subequal at least in length, expanding from near the base; stamens 10–14, all free 29

29. Stamens 10, subequal, all long-exserted and fertile; sepals 5; petals 5, subequal but usually with 1–3 broader than the others . . . 23. **Isoberlinia**
Stamens with 6–7 long-exserted and fertile, alternating with 6–7 shorter and thinner, usually sterile, included ones; sepals 6(? –7); petals 6–7, subequal . . . 26. **Englerodendron**

30. Perianth 0 or of 1–7(–11) parts, usually 4–7 all sepaloid and of similar form, grading inwards from broader to narrower, or 1–3(–4) very small; if 5–11 and two distinct forms present, then the inner (petals ?) much narrower than the outer (sepals ?), even when (rarely) longer; stipellar expansions or wings often present on the upper side of the leaf-rhachis . 30. **Brachystegia**
Perianth clearly differentiated into 1–7 obvious petals, usually longer and broader* than the 4–7 sepals, the latter very rarely reduced or absent; leaf-rhachis without expansions or wings 31

31. Petals bluish-mauve; fertile stamens 3(–5), much longer and thicker than the (4–)6 staminodes; sepals 4, large, all obviously united at base in a persistent cup . . 29. **Paramacrolobium**
Petals white; fertile stamens 8–10, subequal; staminodes 0; sepals 0 or 1–5, sometimes minute, free or at least not united in an obvious cup 32

32. Stamens with 9 filaments shortly connate below and 1 free; sepals 5, well-developed, subequal at least in length; petals variable, usually 5, well-developed, subequal or 1 much larger; leaflets (except in *J. unijugata*) normally 4 or more, with midrib not marginal . . . . . 25. **Julbernardia**
Stamens 8–10, all free; sepals 0 or 1–5 small and unequal; petals 1, large, alone, or with 1–4 others rudimentary; leaflets (in East Africa) 2 only, with their principal nerve marginal . . . . 31. **Monopetalanthus**

33. Anthers opening by terminal or basal pores or short slits, usually basifixed; flowers arranged spirally along the inflorescence-axes; petals 5, all well-developed; glands often (by no means always) present on petiole or leaf-rhachis . . . . . 10. **Cassia**
Anthers opening by slits as long as the anther, dorsifixed (except in 11, *Dialium*); conspicuous glands not present on petiole or rhachis; other characters various 34

34. Petals 0 35
Petals 1–5, well-developed (but sometimes very quickly caducous, or shorter than the sepals, notably in 19, *Daniellia*) 37

* But sometimes scarcely different in texture, or not consistently broader, in 25, *Julbernardia*.

35. Stamens (in East Africa) 2; anthers basifixed; disc (in native species) much wider than ovary; pellucid gland-dots on leaflets few or absent . . . . . . 11. **Dialium**

Stamens (8–)10(–12); anthers dorsifixed; disc small; leaflets (in East Africa) with numerous pellucid gland-dots 36

36. Sepals 5(–6); leaflets (in East Africa) mostly 5–7, alternate . . . . . . 21. **Oxystigma**

Sepals 4; leaflets 2, opposite . . . . 22. **Guibourtia**

37. Flowers rich yellow, with 5 subequal petals; vegetative buds naked, i.e. without bud-scales. . . . . 9. **Stuhlmannia**

Flowers white, cream-coloured, pink or red (upper petal in 13, *Baikiaea*, pale yellow, others white or cream). 38

38. Flowers arranged in 2 rows (distichously) along the inflorescence-axes; petals 5, well-developed, subequal in length 39

Flowers arranged spirally in more than 2 ranks along the inflorescence-axes 40

39. Leaflets with numerous pellucid gland-dots, but without any marginal swelling near base, rounded or obtuse or emarginate at apex; petals comparatively small, ± 2 cm. long . . . . 12. **Tessmannia**

Leaflets without gland-dots, but with a small ± marked swelling near the posticous margin of each leaflet close to the base; lamina subacuminate to obtusely pointed at apex; petals (where known) large, 6 cm. long or more . . . . . 13. **Baikiaea**

40. Petals (4–)5, equal or nearly so 41

Petals 1–3 large, the others absent or very reduced 45

41. Leaflets with numerous pellucid gland-dots, in one pair per leaf; bracteoles large, enclosing the flower-buds . . . . 20. **Trachylobium**

Leaflets without pellucid gland-dots, in 1–many pairs per leaf; bracteoles small, not enclosing the flower buds 42

42. Leaflets opposite; inflorescence paniculate . 14. **Cynometra** (all species except *C. brachyrrhachis* and *C. filifera*)

Leaflets alternate (at least the upper ones); inflorescence racemose 43

43. Leaflets 8–12; young branchlets densely pilose 14. **Cynometra** (*C. filifera**)

Leaflets (in East Africa) 3–5 44

44. Axis of raceme 15–20 mm. long; pedicels 5–21 mm. long . . . . . . 16. **Scorodophloeus**

* The true position of this species (known to me from description only) is doubtful. It should probably not be kept in *Cynometra*.

Axis of raceme 4–10 mm. long; pedicels 4–7 mm. long . . . . . . 14. **Cynometra** (*C. brachyrrhachis**)

45. Fertile stamens 3–7(–9); staminodes 2–7; translucent gland-dots absent, though a small dot-like gland is present on proximal side of leaflet base either on lower surface in angle between midrib and margin, or on margin itself 46
    Fertile stamens 10; leaflets with numerous pellucid gland-dots 47
46. Fertile stamens 7(–9); leaflets in 2–6(–7) pairs; seeds with a conspicuous brightly coloured aril . . . . 17. **Afzelia**
    Fertile stamens 3; leaflets in (1–)2(–3) pairs; seeds without an aril . . . . . 18. **Intsia**
47. Leaflets in several pairs; petals (in East Africa) 1 large and 4 very small and inconspicuous; pods dehiscent . . . . 19. **Daniellia**
    Leaflets in one pair; petals (2–)3 large and 2(–3) very small; pods indehiscent . 20. **Trachylobium**

KEY TO GENERA BASED MAINLY ON VEGETATIVE AND POD CHARACTERS

This key has been made so that identification of genera should be easier when, as so often happens, leaves and pods are available, but no flowers. The key is at many points artificial, and the characters used must not necessarily be taken to apply to the genera throughout their ranges, but only as they are represented in the native and naturalized flora of East Africa.

1. Leaves simple or unifoliolate 2
   Leaves compound 7
2. Leaves obtusely pointed to acute or acuminate at apex, not emarginate 3
   Leaves emarginate at apex, or bilobed 5
3. Pod large, up to ± 30 × 6·8 cm.,** irregularly elliptic-oblong, up to ± 10-seeded, indehiscent, breaking irregularly . . . 33. **Gigasiphon**
   Pod (where known) much smaller, 3–6 × 1·5–4 cm., boat-shaped in outline, 1(–2)-seeded, indehiscent or dehiscent 4
4. Pod clearly stipitate, dehiscent along lower and partially at least along upper suture. 15. **Zenkerella**
   Pod not or scarcely stipitate, thick, black, apparently indehiscent. . . . . 37. **Baphiopsis**
5. Plants with trailing or climbing mostly herbaceous stems arising from a large underground tuber; tendrils usually present; pod woody, dehiscent or indehiscent, (1–)2-seeded . . . . . . 36. **Tylosema**
   Plants growing as shrubs or small trees, in East Africa never climbing; tendrils absent; pod normally with more than 2 (and often many) seeds, oblong to linear 6

* This inadequately known species, of which I have seen no material, is keyed out individually here, as it may well be at present in the wrong genus, and may perhaps be a *Scorodophloeus*.

** Material very limited, range of size hence uncertain.

6. Pod indehiscent; leaves rusty-puberulous or -pubescent beneath . . . . . 34. **Piliostigma**
   Pod dehiscent; leaves without rust-coloured indumentum beneath . . . . . 35. **Bauhinia**
7. Leaves bipinnate 8
   Leaves simply pinnate 17
8. Plant unarmed; shrubs or trees, not climbing 9
   Plant armed with prickles or spines 14
9. Leaflets alternate 10
   Leaflets opposite 11
10. Pod dehiscent, oblong or oblong-elliptic, woody, (by abortion 1–)2–11-seeded; seeds without areoles; hairs on vegetative buds and young shoots (in East Africa) not rusty-red . . . . . . 1. **Erythrophleum**
    Pod indehiscent, elliptic, coriaceous, 1-seeded; seeds areolate on each face; hairs on vegetative buds and young shoots conspicuously rusty-red . . . . . 2. **Burkea**
11. Pod indehiscent, linear-oblong, 0·9–1 cm. wide; valves papery, closely and ± longitudinally veined . . . . . 8. **Parkinsonia** (sp. 2)
    Pod dehiscent, 1·2–3·7 cm. wide; valves woody or coriaceous, not veined as above 12
12. Valves of pod longitudinally grooved down the centre, or concave, curving outwards from base or recurving on dehiscence but not twisted, thick, rigidly woody . . . 4. **Bussea**
    Valves of pod flat, neither longitudinally grooved nor concave in centre, flat or sometimes twisted on dehiscence, but not recurving, and not so thickly woody 13
13. Pod 13–25 cm. long, with a ± central apical beak (i.e. equidistant from lower and upper sutures) . . . . . . 3. **Delonix**
    Pod shorter, 7–12 cm. long, with an apical beak much nearer to and almost a continuation of the upper suture . . . 5. **Caesalpinia** (sp. 1*)
14. Pod with a basal seed-containing portion with a single seed, and a distal wing bounded on one side by a prolongation of the upper suture, the whole resembling a samara of *Acer* . . . . 7. **Pterolobium**
    Pod unwinged, or winged all along upper side of upper suture, and then (in East Africa) 2–4-seeded 15
15. Pod winged along upper suture . . . . 6. **Mezoneuron**
    Pod unwinged 16
16. Valves of pod stiffly papery, closely and longitudinally or obliquely veined; pods indehiscent, 1·1–2·5 cm. wide . . . 8. **Parkinsonia** (sp. 1)

* Pods unknown in spp. 2 and 3, which are likewise unarmed.

Valves of pod woody to coriaceous, hard, not veined as above, armed or unarmed; pods indehiscent or dehiscent, 1·5–6·5 cm. wide . . . 5. **Caesalpinia** (spp. 1, 4, 7–10*)

17. Petiole and/or rhachis of leaves with one or more conspicuous solitary sessile or sometimes projecting or stalked glands on the centre line of upper side . . . 10. **Cassia** (spp. 15–48)

Petiole and rhachis without conspicuous solitary glands, but pairs of glands or stipellar expansions sometimes present . . . 18

18. Seeds with a conspicuous areole (sometimes small) on each face . . . 19

Seeds without any areole on face . . . 21

19. Pod sausage-like, not much compressed, 3–13 cm. long, indehiscent, closely scurfy outside . . . 27. **Tamarindus**

Pod much compressed . . . 20

20. Valves of pod thin, papery to subcoriaceous, 1–2·6 cm. wide; seeds rather small, 6–9 × 3·5–5 mm.; inflorescence racemose; stipules free, small to large; shrubs . . . 10. **Cassia** (spp. 9–14)

Valves of pod woody, 2·5–6 cm. wide; seeds large, 13–25 × 10–20 mm.; inflorescence a compact corymbose panicle; each pair of stipules connate into a persistent intrapetiolar scale; trees . . . 29. **Paramacrolobium**

21. Pod with a conspicuous crenate-margined wing running longitudinally along the middle of each valve . . . 10. **Cassia** (sp. 8)

Pod unwinged or rarely (21, *Oxystigma*) with an entire marginal wing; sometimes a narrow wing-like ridge present running along the upper suture . . . 22

22. Pod dehiscent . . . 23

Pod indehiscent . . . 42

23. Leaflets (in East Africa) in 1 pair only, each with its main nerve running along the inner (i.e. distal) margin; valves of pod each with 1(–2) very prominent longitudinal nerves running from stipe to style . . . 31. **Monopetalanthus**

Leaflets rarely in 1 pair and then main nerve not marginal; pod usually without prominent longitudinal nerves . . . 24

24. Leaflets with a very obvious metallic coppery sheen beneath . . . 28. **Anthonotha**

Leaflets without any metallic coppery sheen beneath . . . 25

25. Pod dehiscing into 2 valves, the coriaceous endocarp of which curls up and separates from the stiffly papery to thinly woody exocarp, to the distal end of which the

* Ripe pods unknown in spp. 5 and 6.

solitary seed remains attached during dispersal; leaflets with pellucid gland-dots . . . . 19. **Daniellia**

Pod with valves whose exocarp and endocarp remain attached to each other; pellucid gland-dots present or absent. 26

26. Valves of pod remaining flat after separation 27

Valves of pod becoming spirally twisted after separation 29

27. Seeds with a conspicuous yellow to red basal aril; valves of pod thickly woody, with a deeply partitioned white inner layer 17. **Afzelia**

Seeds without any aril, though sometimes with some detachable rusty scurf; valves of pod thinly woody or coriaceous, not deeply partitioned within 28

28. Petiolules twisted; leaflets asymmetric at base; pod glabrous, 10–28 cm. long* . 18. **Intsia**

Petiolules not twisted; leaflets equal-sided at base; pod densely brown-tomentellous, 4–12·5 cm. long . . . . . . 26. **Englerodendron**

29. Upper suture of pod flattened or concave, with an obvious projecting ridge or wing along each side; ridges or wings spreading, suberect, or revolute 30

Upper suture of pod without any or with only very slight wing-like ridges as described above 38

30. Leaflets alternate (at least the upper ones); inflorescence racemose . . . . . 16. **Scorodophloeus****

Leaflets opposite 31

31. Petiolules distinct, 3–7(–9) mm. long, not twisted . . . . 24. **Berlinia*****

Petiolules 0–3 mm. long, usually twisted 32

32. Pod ± densely brown-tomentose when ripe; leaflets usually fringed with whitish pubescence on margins . . . . 25. **Julbernardia** (spp. 3–4)

Pod glabrous or nearly so when ripe 33

33. Leaflets in 1 pair . . . . . . 25. **Julbernardia** (sp. 2)

Leaflets in 2–many pairs 34

34. Suffrutices with racemes . . . . . 32. **Cryptosepalum**

Shrubs or trees 1·5–21 m. high, with racemes or panicles 35

35. Stipules broad, foliaceous, shortly connate through the leaf-axil, persistent (at least on sterile branchlets), with or without a

* If leaflets mostly in 3–4 pairs and upper suture of pod markedly ridged as described in section 29 of this key, then compare genus 29, *Paramacrolobium*.

** It is possible that *Cynometra filifera* (p. 119) and *C. brachyrrhachis* (p. 114), whose pods are unknown, but which are said to have alternate leaflets and racemose inflorescences, may prove to belong to *Scorodophloeus*. The distinctions are given in the other key to genera on pp. 9–10.

*** If midrib and lateral nerves of leaflets are pubescent beneath and the plant is from the Usambara Mts. or Morogoro District in Tanganyika, then compare *Isoberlinia scheffleri* (p. 139).

much smaller basal auricle; leaflets normally in 2–3 pairs . . . . . 25. **Julbernardia** (sp. 1)

Stipules usually linear to filiform, free or connate, with or without a subfoliaceous basal auricle (sometimes persistent alone); stipules proper rarely foliaceous and then free and normally caducous if leaflets in fewer than 4 pairs 36

36. Inflorescence paniculate* . . . . . 30. **Brachystegia**
    Inflorescence racemose 37
37. Leaflets in 2–5 pairs; leaf-rhachis without stipellar expansions or wings . . . 32. **Cryptosepalum**
    Leaflets in more than 5 pairs and/or leaf-rhachis with stipellar expansions or short wings below or between the leaflets of a pair . . . . 30. **Brachystegia**
38. Leaflets alternate, usually with a small ± marked swelling on the posticous margin of each leaflet close to the base; scars in 2 ranks on infructescence-axis . . . 13. **Baikiaea**
    Leaflets opposite, without basal swellings as described above; scars spirally arranged on the infructescence-axes 39
39. Petiolules absent or very short, at most up to 3 mm. long; pods (where known) short, up to 9 cm. long 40
    Petiolules usually well-developed, 3–9 mm. long; pods large, up to 35 cm. long 41
40. Petiole (0·6–)1–1·7 cm. long**; inflorescence simply racemose; flowers yellow; leaflets 3–5 pairs; axillary buds (or developing shoots) usually accompanied by additional buds in the same vertical line . 9. **Stuhlmannia**
    Petiole normally 0·1–1 cm. long, up to 1·8 cm. only in *Cynometra engleri* which has only 1 pair of leaflets per leaf; inflorescence usually paniculate; flowers white; axillary buds solitary . . . . . 14. **Cynometra**
41. Pod short, 4–12·5 cm. long . . . . 26. **Englerodendron**
    Pod large, 15–35 cm. long . . . . . 23. **Isoberlinia**
42. Pod flattened or markedly compressed 43
    Pod not compressed, round or nearly so in section 46
43. Leaflets always in one pair, ovate-falcate, with pellucid gland-dots; pod obovate-elliptic, 3–3·5 × 2–2·3 cm. . . . . . 22. **Guibourtia**
    Leaflets 4–26, or in one pair only on occasional reduced leaves, and then without pellucid dots 44
44. Leaflets without pellucid gland-dots, opposite or subopposite; pod 10–28 cm. long, not winged . . . . 18. **Intsia**

* If pods dehisce, initially at least, along upper suture only, and the valves are not much twisted, compare genus 29, *Paramacrolobium*.

** If petiole is up to 2 cm. long, and the plant has conspicuously foliaceous stipules and a paniculate inflorescence, compare *Julbernardia magnistipulata* (p. 146).

Leaflets with pellucid gland-dots, alternate to opposite; pod (where known) shorter than above 45

45. Pod with a rigid proximal wing, obovate-elliptic, 4–6 × 2·5–4 cm.; leaflets ± acuminate at apex . . . . 21. **Oxystigma**

Pod unwinged (not known in East African species, however!); leaflets rounded to obtuse and often emarginate . . . 12. **Tessmannia**

46. Leaves all with a single pair of leaflets, with numerous pellucid gland-dots; pod 2·5–5 × 1·5–3 cm., resinous-warted . . . 20. **Trachylobium**

Leaves with several to many (always more than 2) leaflets 47

47. Pod ellipsoid or spherical, less than twice as long as wide 48

Pod elongate, sausage-shaped or cylindrical, sometimes irregularly constricted, more than twice as long as wide 50

48. Pod small, 1·1–1·8 × 1–1·3 cm., with a hard red to red-brown exocarp and a pulpy mesocarp red to brown-orange at least when dry; inflorescences of terminal and lateral panicles . . . . 11. **Dialium**

Pod large, 2·5–8 × 2·5–6·5 cm., yellow or green outside, containing 1–6 large seeds embedded in pulp; inflorescences of lateral racemes; leaflets with pellucid dots or streaks 49

49. The two genera keyed out here, 39, *Cordyla* and 40, *Mildbraediodendron*, are best separated by floral characters (see p. 6). *Mildbraediodendron* is a tall tree of lowland rain-forest in Uganda, with more pointed leaflets than in *Cordyla*. The only forest species of *Cordyla*, *C. africana*, is not found in Uganda, although another species of drier habitats is.

50. Leaflets asymmetric at base, with their proximal side sessile on the rhachis; pod closely covered with brownish scurf . 27. **Tamarindus**

Leaflets distinctly petiolulate; pod usually black to blackish-brown and glabrous or nearly so; sometimes pod densely tomentellous, but not brownish-scurfy 51

51. Leaves ending in a pair of leaflets; leaflets in 4–12 pairs, opposite, or some alternate . 10. **Cassia** (spp. 1–7)

Leaves ending in a single leaflet; leaflets all alternate, (3–)7–11(–13) . . . . 38. **Swartzia**

The *Caesalpinioïdeae* account for a considerable number of exotic species planted in East Africa. Most of them are mentioned under their appropriate genera; *Bauhinia*, *Caesalpinia* and *Cassia* are perhaps the most prominent, but others are to be found under *Cynometra*, *Delonix*, *Mezoneuron* and *Parkinsonia*.

In addition to these, however, there are several genera only occurring as exotics in the area, which are briefly dealt with below. In order to assist naming, they are artificially grouped according to obvious foliage-characters.

A. Leaflets conspicuously crenate or crenulate-denticulate on margins in upper part; leaves simply pinnate and bipinnate mixed on the same shoot:

Gleditsia amorphoïdes (Griseb.) Taub., a native of South America, is planted in Kenya (*G. Williams* G. 389 !). It is a tree with small flowers in lateral spiciform racemes and pods 6–8 cm. long.

Gleditsia triacanthos L., native of North America, is recorded in Dale, Introd. Trees Uganda: 43 (1953). It is very similar to the last and, like that, has branched spines; its pods, however, are 20–40 cm. long.

B. Leaflets entire on margins; leaves normally all simply pinnate:

Brownea latifolia Jacq., the Mountain Rose or Cooper Hoop, native of Venezuela and Trinidad, is grown in Uganda and Tanganyika (T.T.C.L.: 94 (1949)). It is a small evergreen tree with very acuminate leaflets and reddish-pink flowers in dense bracteolate heads about 10–12 cm. across. Other species are also on record: *B. grandiceps* Jacq. on Zanzibar (U.O.P.Z.: 155 (1949)) and *B. ariza* Benth. and *B. rosa-de-monte* Berg. in Uganda (Dale, Introd. Trees Uganda: 14 (1953)). These species are all South American and similar. Identifications in this genus are so unsatisfactory at present that it seems advisable, in the absence of specimens, to do no more than record the names.

Ceratonia siliqua L., the Carob Tree or Locust Bean, native of the Mediterranean region, is grown in Uganda (Dale, Introd. Trees Uganda: 22 (1953)), Kenya (*Dyson* 21 !), and Tanganyika (T.T.C.L.: 99 (1949)). It is an evergreen tree with 2–8 pairs of coriaceous leaflets, small greenish ♂ ♀ or ⚥ flowers with no petals and 5 exserted stamens (in the ♂ and ⚥) in usually short racemes which are axillary terminal or clustered on the older wood, and thick coriaceous pulpy indehiscent pods 12–30 cm. long and 2 cm. thick.

Haematoxylon campechianum L., the Logwood, native of Central America, the West Indies and ? French Guiana, is cultivated in Uganda, Kenya and Tanganyika (Dale, Introd. Trees Uganda: 44 (1953); T.T.C.L.: 103 (1949)). A smallish tree, with closely veined obovate-emarginate or subtruncate leaflets (sometimes the lowest pair replaced by a pair of 1–2-jugate pinnae), rather small yellow flowers in axillary racemes, and rather small flattened elliptic-oblong pods which open down the middle of the valves.

Hymenaea courbaril L., native of tropical America, is cultivated in Uganda (Dale, Introd. Trees Uganda: 45 (1953)) and Kenya (*G. Williams* 429 !). A tree with 2 leaflets per leaf, very similar to *Trachylobium* (p. 132), but with non-clawed petals and smooth fruits.

Lysidice rhodostegia Hance, native of China, is cultivated for ornament in Uganda (Dale, Introd. Trees Uganda: 49 (1953)) and Tanganyika (T.T.C.L.: 105 (1949)). Small tree. Leaflets ± acuminate, in 3–4 pairs. Flowers pink to purple, with conspicuous pink bracts, in panicles.

Pterogyne nitens Tul., native of South America, is grown in Uganda and Kenya. Small tree. Leaflets mostly alternate. Flowers small, yellow, in racemes shorter than the leaves, with 5 petals and 10 stamens. Pods like those of *Pterolobium* (p. 41), with a single seed at proximal end and a distal parallel-veined falcate wing.

Saraca indica L., native of tropical Asia, is grown in Uganda (Dale, Introd. Trees Uganda: 64 (1953)), Tanganyika, (*Fernie* (EA !)), and Zanzibar (U.O.P.Z.: 442 (1949)). A small evergreen tree, with 3–6 pairs of oblong-lanceolate acuminate leaflets and short corymbose panicles of orange to deep pink or scarlet flowers with a narrow elongate hypanthium, four conspicuous sepals and exserted stamens.

Schotia brachypetala Sond., native from Rhodesia to South Africa, is grown in Kenya (*G. Williams* G. 357 !) and Uganda (*Dale* U807 !—see Dale, Introd. Trees Uganda: 65 (1953), where it is wrongly called *S. latifolia* Jacq.). Tree. Leaflets 4–7 pairs, rounded to emarginate at apex. Flowers deep red, in congested panicles. Pods oblong, containing seeds with a large yellow aril.

Schotia sp. Small evergreen tree. Leaflets mostly in 4 pairs, apiculate, mostly up to 2·5(–3) cm. long. Flowers small, white (stamens flesh-pink) in congested inflorescences. Cultivated at Nairobi (*Gardner* 1191 in *C.M.* 16887 !, *Greenway* 8748 !). The foliage is like *S. capitata* Bolle, but the flowers are smaller and paler, resembling those of *S. latifolia* Jacq. This plant has not been precisely matched yet.

C. Leaflets entire on margins; leaves all bipinnate:

Acrocarpus fraxinifolius Arn., native from India to Thailand, is grown in Uganda (Dale, Introd. Trees Uganda: 4 (1953)) and Tanganyika (T.T.C.L.: 87 (1949)). Unarmed, evergreen tree, with smooth pale grey bark. Leaves large; leaflets ovate-lanceolate, acuminate. Flowers usually nodding, in erect dense racemes. Sepals and petals green. Stamens 5, exserted, orange-red. Pods linear to linear-oblong, 2-valved, several- to many-seeded, narrowly winged along upper suture.

Peltophorum (Vogel) Walp. is a genus of trees rather closely related to *Bussea* (p. 25), with similar yellow flowers and a peltate stigma, but with a flattened indehiscent pod winged along both sutures. Three species of this genus are cultivated in East Africa:—

P. africanum Sond., native of southern tropical and South Africa, with 10–28 pairs of leaflets per pinna which are mucronate at apex and about 0·3–1(–1·2) cm. long and 0·1–0·45 cm. wide, and with pedicels at flowering time about 1–1½ times as long as the calyx. It is grown in Kenya and Tanganyika (see T.T.C.L.: 105 (1949) and Wilczek in F.C.B. **3**: 262 (1952)).

P. dasyrhachis (Miq.) Bak., native of tropical Asia, with foliage as in *P. pterocarpum* (see below), subulately branched stipules, lateral unbranched racemes, and pedicels ± 3–4 times as long as calyx. Grown in Uganda and Tanganyika (see Dale, Introd. Trees Uganda: 53 (1953) and T.T.C.L.: 106 (1949)).

P. pterocarpum (DC.) K. Heyne (*P. ferrugineum* (Decne.) Benth., *P. inerme* (Roxb.) Naves, *P. roxburghii* (G. Don) Degener), native of tropical Asia and Australia, with 10–19 pairs of leaflets per pinna which are emarginate and without a mucro at apex and ± 0·8–3 cm. long and 0·35–0·9 cm. wide, small unbranched stipules, racemes aggregated into a terminal panicle, and pedicels at flowering time at most as long as the calyx. Grown in Uganda, Kenya, Tanganyika and Zanzibar. See Dale, Introd. Trees Uganda: 54 (1953), T.T.C.L.: 106 (1949), U.O.P.Z.: 403 (1949), and Wilczek in F.C.B. **3**: 263 (1952).

Schizolobium parahybum (Vell.) Blake (*S. excelsum* Vogel), native of Brazil (and possibly elsewhere in tropical America), is grown in Uganda, Tanganyika and Zanzibar (see Dale, Introd. Trees Uganda: 65 (1953), T.T.C.L.: 106 (1949), and U.O.P.Z.: 443 (1949)). Unarmed tree with large leaves, yellow flowers in paniculate racemes, a minute stigma, and compressed oblong-cuneate pods dehiscing into two halves and liberating the single apical seed, which remains enclosed in the similar-shaped papery envelope of endocarp resembling a wing.

Wagatea spicata (Dalz.) Wight, native of India, is cultivated in Zanzibar (U.O.P.Z.: 486 (1949)). A very prickly shrub with ovate-elliptic to ovate-oblong leaflets, and fairly small red flowers in long spikes or spiciform racemes.

1. ERYTHROPHLEUM

R. Br. in Denham, Clapperton & Oudney, Trav. N. & Centr. Afr., Journ. Excurs.: 235 (1826); G. Don, Gen. Syst. 2: 424 (1832)*

Unarmed trees. Leaves alternate, bipinnate; stipules very small, quickly falling; specialized glands restricted to petiole and rhachis absent; leaflets alternate, eglandular, petiolulate. Inflorescences of pedunculate spike-like racemes usually aggregated into panicles; bracts very small, falling as or before the flowers open. Flowers ⚥. Calyx-lobes 5, ± connate below or almost free, slightly imbricate but open from an early stage. Petals 5, free, ± imbricate, equal, ± pubescent or tomentose, oblanceolate, narrowed towards base. Stamens 10, often alternately longer and shorter; filaments glabrous or hairy; anthers dorsifixed, 0·5–0·75 mm. long, dehiscing by longitudinal slits; connective not projecting beyond anther. Ovary stipitate, tomentose or densely pubescent, containing several ovules, tapering into a narrowly conical style; stigma minute, punctiform, cup-shaped and minutely ciliolate. Pods stipitate, woody, flattened, straight or slightly curved, oblong or oblong-elliptic, dehiscent, (by abortion 1–)2–11-seeded. Seeds not areolate, with endosperm, arranged transversely in the pod.

A genus of perhaps ten species in tropical Africa, Madagascar, Asia and Australia.

Leaflets at apex with ± pronounced obtuse acumen; calyx-lobes distinctly connate below; stamen-filaments glabrous or almost so; pods dehiscing, at least initially, along one suture only, 5–11-seeded; stipe of pod often lateral; seeds 4–7 mm. thick, with thick endosperm . . . . 1. *E. suaveolens*

Leaflets obtuse or rounded but not acuminate at apex; calyx-lobes free almost to base; stamen-filaments pubescent or tomentose to near apex; pods dehiscing simultaneously along both sutures, (1–)2–5-seeded; stipe of pod central; seeds 3–4 mm. thick, with thin endosperm . . . . 2. *E. africanum*

1. **E. suaveolens** (*Guill. & Perr.*) *Brenan* in Taxon 9: 194 (1960). Type: Gambia [" Senegal "], Albreda, on the R. Gambia, *Perrottet* (P, holo.). Specimen from " Senegal ", *Perrottet* 291 (BM !) may be an isotype.

Tree 9–30 m. high, with rough bark. Young branchlets slightly puberulous to glabrous. Leaves: petiole with rhachis 11–35 cm. long; pinnae 2–4 pairs, 9–19 cm. long; leaflets 7–14 per pinna, ovate, ovate-elliptic, or rarely lanceolate, usually asymmetric, mostly 2·7–9 cm. long and 1·3–5·3 cm. wide, obtusely ± acuminate at apex, glabrous or sometimes with spreading pubescence on midrib beneath; petiolules 3–5 mm. long. Racemes 3–8(–11) cm. long, shortly pubescent or puberulous (including the flowers). Flowers yellowish-white to greenish-yellow. Calyx-lobes 1–1·5 mm. long, distinctly connate below. Petals 2–3 mm. long, 0·5 mm. wide. Stamen-filaments glabrous or nearly so. Pods often slightly curved, dehiscing, at least initially, along one suture only, 6–11-seeded, 8–17 cm. long, 3–5·3 cm. wide, at apex broadly rounded; stipe of pod often lateral. Seeds brown, oblong-ellipsoid, 14–17 mm. long, 10–12 mm. wide, 4–7 mm. thick, with thick endosperm. Fig. 1.

UGANDA. W. Nile District: West Madi, Amua, *Eggeling* 909 in *F.H.* 1257 !; Bunyoro District: Budongo Forest, *Fraser* in *Eggeling* 3812 !; Masaka District: Sese Is., Bugala, Lutoboka Forest, Nov. 1931, *Eggeling* in *F.H.* 278 !

* See Taxon 9: 194 (1960) & 13: 181 (1964) for a discussion concerning the valid publication of this name.

FIG. 1. *ERYTHROPHLEUM SUAVEOLENS*—**1,** branchlet with inflorescences and mature leaf, × $\frac{2}{3}$; **2,** lower side of leaflet, showing venation, × 1; **3,** flower, × 6; **4,** stamen, × 6; **5,** ovary, showing insertion of stipe on receptacle, × 6; **6,** stigma, × 8; **7,** pod, × $\frac{2}{3}$; **8,** end-on view of dehisced pod, × $\frac{1}{2}$; **9,** seed, × $\frac{2}{3}$. 1, 3–6, from *Elliot* 1493; 2, from *Verdcourt* 2129; 7, from *Verdcourt* 2129, *S. Paulo* 149; 8, from *S. Paulo* 149; 9, from *Lebrun* 4224, *Eyles* 7917.

KENYA. Tana River District: N. of Witu, Mambosasa Forest Station, 29 Jan. 1958, *Verdcourt* 2129!; ? Lamu District: Witu district, *C. W. Elliot* 1493!

TANGANYIKA. Lushoto District: Sigi, Oct. 1934, *Greenway* 4927!; Kigoma District: Mkuti R., Sept. 1956, *Procter* 514!; Morogoro District: Turiani, 3 June 1933, *B. D. Burtt* 4748!

ZANZIBAR. Zanzibar I., Masingini ridge, 1 Feb. 1929, *Greenway* 1294! & 23 Feb. 1930, *Vaughan* 1266!; Pemba I., Madunga, 29 July 1929, *Vaughan* 419!

DISTR. **U**1, 2, 3 (I.T.U., ed. 2: 66), 4; **K**7; **T**3, 4, 6; **Z**; **P**; widespread in tropical Africa from Senegal to the Sudan Republic and southwards to Zambia, Rhodesia and Mozambique

HAB. Riverine and lowland rain-forest; from near sea-level to 1110 m.

SYN. *Fillaea suaveolens* Guill. & Perr. in Guill., Perr. & A. Rich., Fl. Seneg. Tent., t. 55 (? June–July 1832) & 242 (Oct. 1832)

Erythrophleum guineënse G. Don, Gen. Syst. 2: 424 (Oct. 1832); L.T.A.: 779 (1930); T.S.K.: 64 (1936); T.T.C.L.: 103 (1949); U.O.P.Z.: 245, fig. (1949); Wilczek in F.C.B. 3: 243, t. 18 (1952); I.T.U., ed. 2: 66 (1952); F.W.T.A., ed. 2, 1: 483, fig. 154/D (1958); K.T.S.: 104 (1961); F.F.N.R.: 124, fig. 20/C (1962). Type: from "Sierra Leone and other parts of Guinea" (Don, l.c.). Specimen from Sierra Leone, *G. Don* (BM, ? holo.!)

2. **E. africanum** (*Benth.*) *Harms* in F.R. 12: 298 (1913); L.T.A.: 777 (1930); T.T.C.L.: 103 (1949); Wilczek in F.C.B. 3: 244 (1952); Torre & Hillcoat in C.F.A. 2: 252 (1956); F.W.T.A., ed. 2, 1: 483 (1958); F.F.N.R.: 124, fig. 20/B (1962). Types: Angola, Huila, Mumpula, *Welwitsch* 591 (BM, isosyn.!) & Pungo Andongo, Calundo, *Welwitsch* 573 (BM, K, isosyn.!)

Tree 4–12 m. high. Leaves: petiole with rhachis (5–)7–23 cm. long; pinnae (2–)3–5 pairs, (4·5–)7–17 cm. long; leaflets (8–)10–17 per pinna, narrowly elliptic to elliptic, or with rhombic or ovate tendency, usually somewhat asymmetric, 1·5–6(–8) cm. long, 0·9–3·3(–4·6) cm. wide, obtuse or sometimes rounded and ± emarginate at apex, not acuminate; petiolules 1·5–4 mm. long. Racemes 3·5–9 cm. long, densely pubescent or tomentose (including the flowers). Flowers cream to greenish-yellow. Calyx-lobes 1·5–2·5 mm. long, free almost to base. Petals 3–4 mm. long, 0·75–1 mm. wide. Stamen-filaments pubescent or tomentose to near apex. Pods dehiscing simultaneously along both sutures, (1–)2–5-seeded, 5–17 cm. long, (2–)2·7–4·9 cm. wide, at apex rounded or tapering to an obtuse or acute point; stipe of pod central. Seeds brown, rotund to suborbicular, 10–14 mm. long, 8–12 mm. wide, 3–4 mm. thick, with thin endosperm.

TANGANYIKA. Tabora, 14 Sept. 1938, *Lindeman* 708!; Morogoro Fuel Reserve, Nov. 1954, *Semsei* 1874!; Songea District: E. of R. Mtandazi W. of Gumbiro, 9 May 1956, *Milne-Redhead & Taylor* 10137!

DISTR. **T**4, 6, 8; widespread from Senegal to the Sudan Republic, southwards to Zambia, Rhodesia and South West Africa; strangely absent from Kenya, Uganda and other territories to the north-east

HAB. Deciduous woodland (*Brachystegia*); ± 600–1370 m.

SYN. *Gleditsia africana* Benth. (as "*Gleditschia*") in Trans. Linn. Soc. 25: 304 (1865)

NOTE. A distinctly variable species. In Tanganyika the branchlets and leaves are normally, at least when young, densely pubescent or tomentose. *Semsei* 1874 and *Milne-Redhead & Taylor* 10137 (cited above) are, however, aberrant in having the leaflets glabrous or almost so, slight pubescence on the leaf-rhachides, and puberulous branchlets. They would, no doubt, be referable to *E. africanum* forma *glabrissimum* De Wild. or forma *subglabrum* De Wild., Contrib. Fl. Katanga, Suppl. 1: 28 (1927) (see also L.T.A.: 778 (1930)), which seem scarcely distinct from one another and to represent no more than a glabrescent variant of little importance and doubtfully worth naming. Similar trees occur sporadically also in Mozambique, Zambia and Rhodesia. One of the syntypes of *E. africanum*, *Welwitsch* 591, is densely pubescent, while the other, *Welwitsch* 573, is subglabrous.

E. africanum var. *stenocarpum* Harms in N.B.G.B. 13: 414 (1936); T.T.C.L.: 103 (1949), based on *Schlieben* 6536 (BM, iso.!) from Tanganyika, Lindi District, Lake Lutamba, 23 May 1935, characterized especially by having comparatively elongate and narrow pods 12–20 cm. long and 2–3 cm. wide, falls within what I consider to be the normal range of variation of *E. africanum*, although representing an extreme of pod-size. I do not consider it to be worth retention as a taxonomic entity.

2. BURKEA

Benth. in Hook., Ic. Pl. 6, t. 593–4 (1843)

Unarmed trees with rusty-tomentose young shoots. Leaves alternate, bipinnate; stipules very small, quickly falling; specialized glands restricted to petiole and rhachis absent; leaflets alternate, eglandular. Inflorescences of elongate spikes often paniculately aggregated; bracts very small, persistent until the flowers open. Flowers ⚥. Calyx-lobes 5, slightly imbricate, but almost open from an early stage, connate below into a short tube. Petals 5, imbricate, equal, glabrous, without claws. Stamens 10; filaments glabrous; anthers dorsifixed, 1·5–2 mm. long, dehiscing by longitudinal slits; connective shortly projecting beyond the anther. Ovary subsessile, densely rusty-tomentose, 1–2-ovulate; style very short, ending in a funnel-shaped stigma slit down one side. Pods stipitate, coriaceous, flattened, elliptic, indehiscent, one-seeded. Seeds areolate on each face (as in most *Mimosoïdeae*), with endosperm.

A genus of a single known species.

The characteristic areole on each face of the seed is very similar to those so frequent in *Mimosoïdeae*. Except for the genera mentioned in the discussion on areoles on p. 2 they do not otherwise occur in the *Caesalpinioïdeae* of East Africa.

B. africana *Hook.*, Ic. Pl. 6, t. 593–4 (1843); Oliv. in F.T.A. 2: 320 (1871); L.T.A.: 776 (1930); T.T.C.L.: 94 (1949); Wilczek in F.C.B. 3: 238 (1952); I.T.U., ed. 2: 58 (1952); Torre & Hillcoat in C.F.A. 2: 250 (1956); F.F.N.R.: 118, fig. 20/A (1962). Type: South Africa, Transvaal, Magaliesberg, *Burke* 274 (K, holo.!)

Tree 4–20 m. high. Bark grey to blackish, fissured and scaly. Branchlets often rather thick. Leaves silvery when young; petiole with rhachis 7–32 cm. long; pinnae (1–)2–5(–7) pairs,(3·5–)6–20(–28) cm. long; leaflets 6–15, elliptic or sometimes ovate-elliptic or ± obovate, mostly 1·5–7·5 cm. long, 0·7–4·2 cm. wide, obtuse or rounded and somewhat emarginate at apex, ± asymmetric at base, appressed-puberulous on both sides (or rarely pubescent —see note below); petiolule 2–5 mm. long. Flowers white or pale green, in pendulous spikes 6–27 cm. long. Sepals rounded at apex, erose and ciliate on margins. Petals obovate-oblong, ± 4–5 mm. long and 2–2·5 mm. wide, rounded at apex, ultimately recurving. Pods elliptic or narrowly elliptic, 4–6·5 cm. long, 2–3 cm. wide, brown. Seeds elliptic, compressed, ± 9–12 mm. long and 7–8 mm. wide, brown; areoles ± 7–8 mm. long and 3·5–4 mm. wide. Fig. 2, p. 22.

UGANDA. W. Nile District: W. Madi, " Assua " [? Amua], Mar. 1935, *Eggeling* 1776 in *F.H.* 1630! & Aringa, Mt. Kei Central Forest Reserve, 25 Feb. 1955, *Dale* U864!
TANGANYIKA. Mwanza District: near Mwanza, Nyegezi, *Rounce* 237!; Morogoro District: Mtibwa Forest Reserve, Nov. 1954, *Semsei* 1862!; Lindi District: without precise locality, *Gillman* 1543!
DISTR. **U**1; **T**1, 4, 6–8; widespread in tropical Africa, except for the forest regions, and extending southwards to South West Africa and the Transvaal
HAB. Deciduous woodland; 270–1300 m.

NOTE. *Carmichael* 950, from Tanganyika, Mpanda District, Kuuragwa R. Forest Reserve, is exceptional in having the leaves and inflorescence-axes densely covered with coarse tomentose pubescence. The pubescence on the inflorescence-axes, the rhachides of the leaves and the midribs of the leaflets is rusty-red; on the rest of the leaflet-surface grey or slightly rusty. For the present this plant is best accepted as an extreme variant, though more information is needed about its occurrence. Some specimens from Zambia and Angola show a close approach to this sort of indumentum.

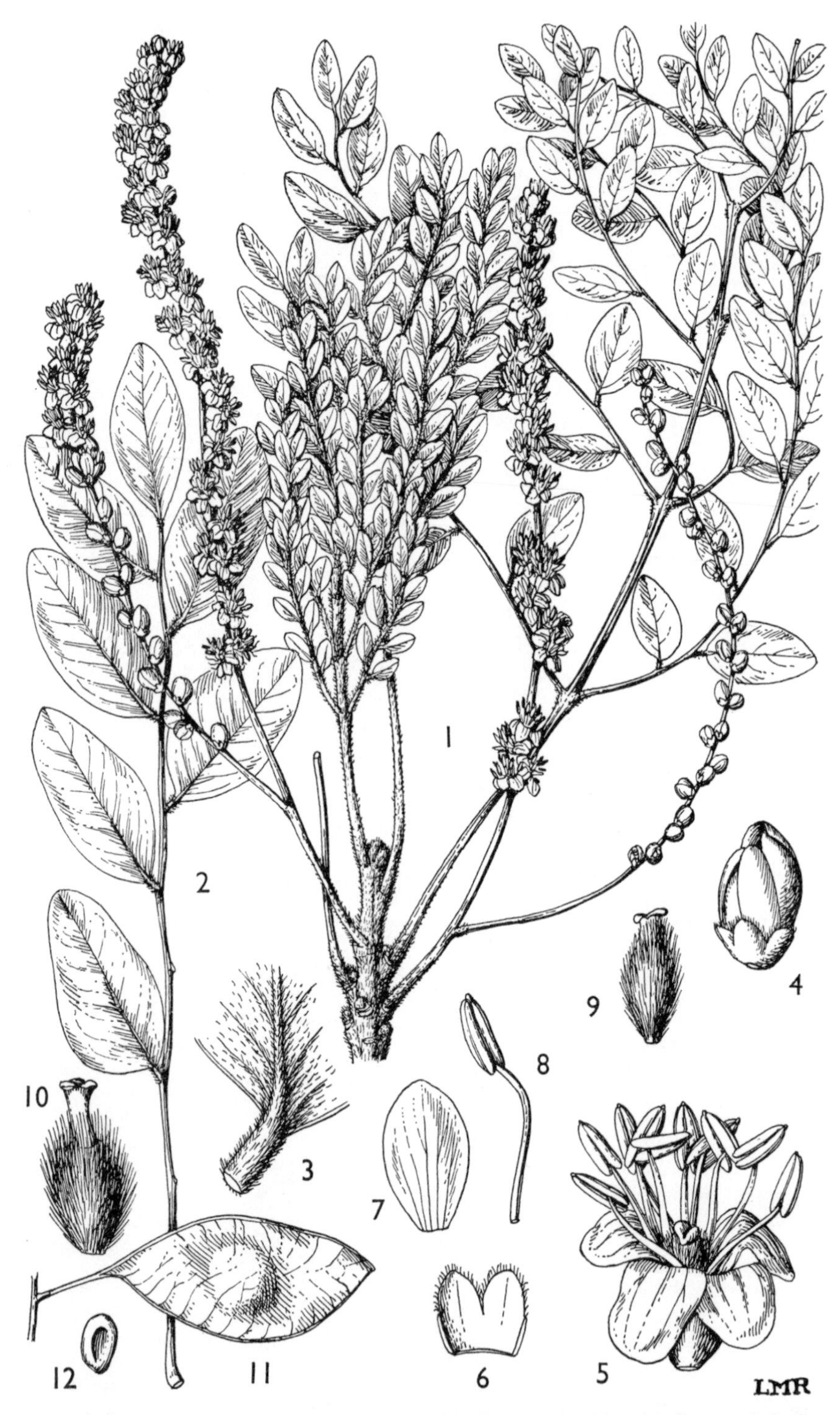

FIG. 2. ***BURKEA AFRICANA***—**1**, part of branchlet with inflorescences and immature leaves, × $\frac{2}{3}$; **2**, pinna of mature leaf, × $\frac{2}{3}$; **3**, basal part of mature leaflet, lower surface, × 3; **4**, flower-bud showing imbricate petals, × 4; **5**, flower, × 4; **6**, two calyx-segments, × 4; **7**, petal, × 4; **8**, single stamen enlarged, × 4; **9**, **10**, ovary × 6; **11**, pod, showing single seed, × $\frac{2}{3}$; **12**, seed, showing areole, × $\frac{2}{3}$. 1, 4–9, from *Semsei* 1862; 2, 3, from *Eggeling* 5776; 10, from *Aylmer* 27/17; 11, 12, from *Gillman* 1543.

3. DELONIX

Raf., Fl. Tellur. 2: 92 (1836)

Unarmed trees. Leaves bipinnate; stipules (in native species) small, subulate, quickly caducous, or (in *D. regia*) small but leafy and forked at base into two pinnate divisions; specialized glands absent from petiole and leaf-rhachis, although clustered brown cylindrical possibly glandular hairs may be present at the insertions of the pinnae; leaflets opposite, eglandular. Inflorescences of short axillary corymbose racemes aggregated near ends of branchlets. Bracts small, inconspicuous, falling while the flower-buds are still young (in native species), or persistent (in *D. regia*). Flowers large, ⚥. Sepals 5, valvate. Petals 5, conspicuously clawed, obovate to transversely elliptic or reniform, equal or subequal except that the upper posticous one is somewhat different from the rest in shape and colour. Stamens 10, fertile, exserted; filaments pubescent or tomentose below, eglandular; anthers dorsifixed, dehiscing by longitudinal slits. Ovary shortly stipitate, pubescent to glabrous; ovules numerous; style ± as long as the stamens, glabrous, or pubescent below, not or only somewhat enlarged near apex into a transverse, ciliate-margined, but not peltate stigma. Pods linear-oblong, flattened, many-seeded, dehiscing into two woody or coriaceous valves. Seeds transverse, hard, flattened and oblong-elliptic (in native species), or oblong-subcylindrical (*D. regia*); endosperm present.

A genus of three species, from eastern Africa through Arabia to India, also in Madagascar. Five species from the latter island, recently described under *Poinciana* by Viguier, may prove to be correctly placed under *Delonix*.

D. regia (Hook.) Raf., the well-known Flamboyant or Flame Tree, a very rare native of Madagascar, with magnificent scarlet flowers, is a favourite in cultivation in East Africa as it is throughout the tropics.

Leaflets 1·25–4 mm. wide; ovary pubescent or tomentose all over; petals ± rounded in outline or shortly pointed at apex . . . . 1. *D. elata*
Leaflets 4·5–20 mm. wide; ovary glabrous or nearly so except for some pubescence on margins and stipe; a usually larger more spreading tree than *D. elata*, with papery-peeling bark;* petals, especially the upper one, ± distinctly tapering and pointed at apex . . . . . . 2. *D. baccal*

1. **D. elata** (*L.*) *Gamble*, Fl. Madras 1(3): 396 (1919); L.T.A.: 624 (1930); B. D. Burtt, Field Key Savannah Gen. & Sp. Tang. Terr.: 30 (1939); T.T.C.L.: 102 (1949); I.T.U., ed. 2: 65 (1952); Roti-Michelozzi in Webbia 13: 195 (1957); K.T.S.: 103, t. 9 (1961); F.F.N.R.: 433 (1962). Type: locality uncertain, *Herb. Linnaeus* 529.3, 529.4 (LINN, syn. !)

Tree 2·5–15 m. high, with rounded-spreading crown; bark rather smooth, buff or grey. Leaves with 2–12 pairs of pinnae; leaflets (8–)11–25 pairs, linear-oblong, 4–17 mm. long, 1·25–4 mm. wide, usually rounded at apex, appressed-puberulous on both surfaces. Inflorescences of 5–20-flowered (or more) racemes, puberulous to densely pubescent (including the calyces outside). Petals 1·6–3·8 cm. long, 1·8–4·2 cm. wide; the upper one smaller than the rest, pale yellow; the remainder white; later all turning apricot; all with their lamina ± rounded in outline or shortly pointed at apex and irregularly erose-lacerate and crisped on margins. Stamen-filaments (3–)6–11 cm. long, red. Ovary silky to pubescent or tomentose all over outside. Pods 13–25 cm. long, 2·1–3·7 cm. wide. Fig. 3/1–5, p. 24.

**D. elata* may sometimes perhaps have similar bark: *J. Wilson* 189 (EA !), from Uganda, Karamoja District, Lokitonyala, is said to have had " bark smooth, shiny, peeling in papery strips ".

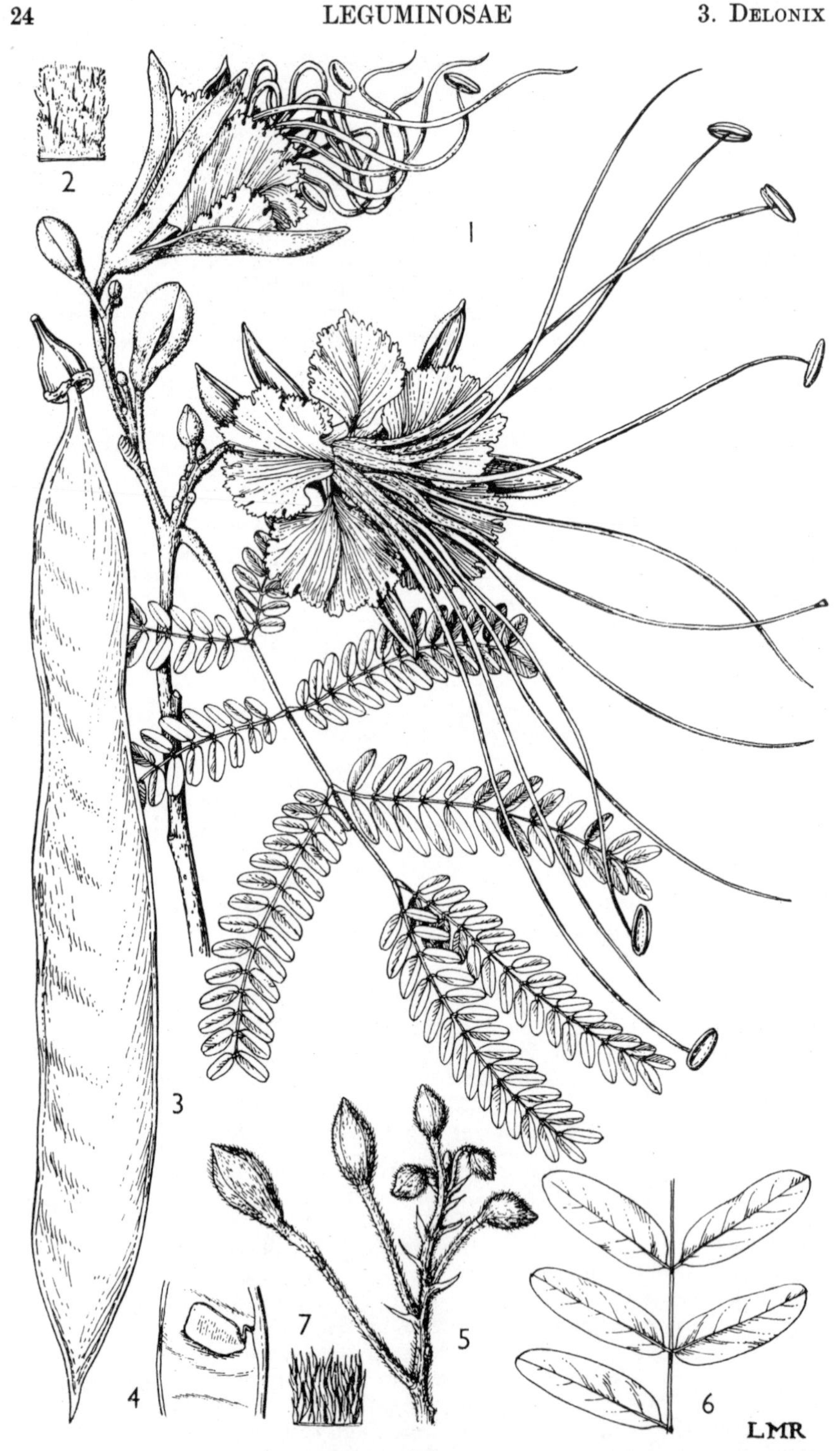

FIG. 3. ***DELONIX ELATA*—1,** part of flowering branch, × $\frac{2}{3}$; **2,** part of outside of sepal, to show appressed indumentum, × 10; **3,** pod, × $\frac{2}{3}$; **4,** seed attached to part of valve of pod, × $\frac{2}{3}$; **5,** young inflorescence of variant in NW. and central Tanganyika, to show spreading pubescence, × $\frac{2}{3}$. *D. BACCAL*—**6,** leaflets × $\frac{2}{3}$; **7,** part of outside of sepal, to show indumentum, × 10. 1, from *Newbould* 3446, *Wilson* 189, *Padwa* 294; 2, 3, from *Newbould* 3446; 4, from *Wilson* 189; 5, from *B. D. Burtt* 794; 6, 7, from *Gillett* 13278.

UGANDA. Karamoja District: Lokitonyala, *Wilson*, 189!
KENYA. Northern Frontier Province: Wajir, Jan. 1949, *Dale* K719!; Turkana District: Lokitaung, 15 June 1953, *Padwa* 294!; Masai District: 19 km. Magadi–Nairobi, 30 Mar. 1956, *Verdcourt* 1465!
TANGANYIKA. Lushoto District: Umba Steppe, Kivingo, 2 Jan. 1930, *Greenway* 2005!; Kondoa, 11 Feb. 1928, *B. D. Burtt* 794!; Iringa District: 80 km. N. of Iringa, near Izazi, 3 Feb. 1962, *Polhill & Paulo* 1325!
DISTR. **U**1; **K**1, 2, 4, 6, 7; **T**1–5, 7; in the eastern Congo Republic, and northwards from the Flora area to Egypt; also in Arabia and extending to India
HAB. Deciduous thickets and bushland; 430–1400 m., perhaps higher

SYN. *Poinciana elata* L., Cent. II. Pl.: 16 (1756)

NOTE. This attractive-flowered tree is known in India as the White Gul Mohur, and there is a very good colour plate of it in Blatter & Millard, " Some Beautiful Indian Trees," ed. 2, t. XI (1954).

Delonix elata is a decidedly variable species. In East Africa there are two main variants, both showing correlation with geography. In Kenya and NE. Tanganyika (**T**2 & 3) the pinnae are normally comparatively few, (3–)5–9 pairs per leaf, and the inflorescence (including the outside of the calyces) is minutely appressed-puberulous. In NW. and central Tanganyika (**T**1, 4, 5 & 7) the pinnae are more numerous, (8–)10–11 pairs per leaf, and the inflorescence is clothed with dense spreading pubescence; this variant occurs elsewhere only in the Congo Republic and (almost certainly as an introduction) in Zambia.

At first sight there seems a strong case for recognizing these as subspecies or varieties. There are, however, good reasons for not doing so, at least not at present. In NE. tropical Africa many specimens resemble in their foliage the first variant described above, but their inflorescences have spreading pubescence, and in India the foliage is again similar but the indumentum on the inflorescence may be either spreading or appressed. This means that the second variant cannot be safely distinguished from the first by the indumentum alone. The distinctive foliage of the second variant can in the same way be a not always reliable distinction. *Gillett* 13030, from Kenya, Northern Frontier Province, Dandu, 4 May 1952, was collected well within the geographical area of the first variant, of which it has the indumentum, but the leaves (of which 10 are present) have 8–11 pairs of pinnae.

2. **D. baccal** (*Chiov.*) *Bak.f.*, L.T.A.: 624 (1930); Roti-Michelozzi in Webbia 13: 200, fig. 8/A (1957); K.T.S.: 103 (1961). Types: Somali Republic (S.), Benadir, near Hele Scid (di Marda), *Paoli* 836 (FI, syn. !) & along the Giuba [Juba] R., near Biobahal, *Paoli* 852 (FI, syn. !)

Tree 6–18 m. high, with umbrella-shaped crown; bark smooth, grey, with the outer surface peeling off and papery. Leaves with (1–, *fide* Roti-Michelozzi)2–5 pairs of pinnae; leaflets (4–)6–15 pairs, elliptic-oblong, 1–3·5 cm. long, 0·45–2 cm. wide, rounded at apex, minutely appressed-puberulous on both surfaces. Inflorescences of 4–9-flowered racemes, minutely and thinly puberulous (including the calyces outside). Petals: the upper one 4·2–5·5 cm. long, 1·8–2·2 cm. wide, oblanceolate or obovate, bright yellow; the lateral and lower ones ± 2·8–3·5 cm. long and 1·5–2·2 cm. wide, with a more rounded or somewhat rhombic lamina, white; all, especially the upper one, ± distinctly tapering and pointed at apex. Stamen-filaments 3–4 cm. long. Ovary glabrous or nearly so, except for some pubescence on stipe and margins. Pods 15–24 cm. long, 2·5–3·5 cm. wide. Fig. 3/6, 7.

KENYA. Northern Frontier Province: Lag Ola, 45 km. W. of Ramu on the Banessa road, 23 May 1952, *Gillett* 13278! & about 23 km. from Moyale on the Marsabit road, 12 Aug. 1952, *Gillett* 13717! & Malka Murri, 2 Oct. 1955, *J. Adamson* 79!
DISTR. **K**1; Somali Republic, and most probably also in southern Ethiopia
HAB. Deciduous bushland (*Commiphora-Acacia*); 450–1080 m.

SYN. *Poinciana baccal* Chiov. in Ann. Bot., Roma 13: 389 (1915)

4. **BUSSEA**

Harms in E.J. 33: 159 (1902)

Shrubs or more usually trees, unarmed and always with brown or rusty indumentum on young shoots and also inflorescences. Leaves usually opposite or subopposite, but varying sometimes to alternate (although not in East Africa), bipinnate; stipules small, subulate; specialized glands restricted to petiole and rhachis absent; leaflets opposite (always so in East Africa), or alternate, eglandular. Inflorescences of terminal panicles with the flowers densely and racemosely arranged along the branches; rarely and casually (in *B. massaiensis*) the inflorescence reduced to a simple raceme; bracts small, falling as or before the flowers open. Flowers ⚥. Sepals 5, imbricate, the inner ones with hyaline erose margins. Petals 5, the upper one smaller than the rest, all with rusty indumentum in middle and on claw. Stamens 10; filaments rusty-tomentose at base, eglandular; anthers dorsifixed, dehiscing by longitudinal slits. Ovary free, subsessile, densely brown-tomentose; ovules 2–3; style ± clothed like the ovary; stigma enlarged and peltate. Pods erect, linear-oblanceolate, ± compressed, not winged, elastically dehiscing into 2 recurving woody valves each of which has a conspicuous longitudinal groove down the middle or (in *B. eggelingii*) is concave. Seeds 1–3, longitudinal; funicle short; endosperm absent.

Five species in tropical Africa and one from Madagascar.

Pinnae (1–)2(–3) pairs and leaflets 5–8 pairs (in East Africa, but see note under the species); leaflets ± appressed-pubescent beneath; pods up to 12 cm. long and 2·2 cm. wide; valves deeply grooved down middle . . . . 1. *B. massaiensis*

Pinnae 4–6 pairs and leaflets (at least of the upper pinnae) 8–11 pairs; leaflets glabrous or almost so beneath; pods up to 15 cm. long and 2·6 cm. wide; valves concave but not grooved . . . . 2. *B. eggelingii*

1. **B. massaiensis** (*Taub.*) *Harms* in E.J. 33: 159 (1902); V.E. 3 (1): 512, fig. 274 (1915); L.T.A.: 617 (1930); B. D. Burtt, Field Key Savannah Gen. & Sp. Tang. Terr.: 30 (1939); T.T.C.L.: 94 (1949); Verdc. in K.B. 12: 350 (1957). Type: Tanganyika, Dodoma District, Saranda [Salanda], *Fischer* 226 (B, holo. †)

Shrub or tree 2–12 m. high, with smooth grey bark (T.T.C.L.). Leaves: rhachides with spreading rusty or brown pubescence; pinnae (1–)2–5 pairs; leaflets 5–10(–11) pairs per pinna, oblong-elliptic to elliptic, (0·7–)1–3·6(–4) cm. long, 0·4–1·9 cm. wide, unequal-sided at base, obtuse to emarginate at apex, ± pubescent beneath. Inflorescence-axes with a very dense coarse rusty tomentum; pedicels 0·4–1 cm. long. Outside of sepals clothed like the inflorescence-axes. Petals yellow, crinkled, obovate, the four lower and lateral ones 12–30 mm. long, 9–21 mm. wide, the fifth upper one 9–10 mm. long, 7–8 mm. wide. Stamen-filaments 7–9 mm. long. Pods 7–12 cm. long, 1·2–2·2 cm. wide; valves with a conspicuous longitudinal groove down middle. Seeds 1·6–2·2 cm. long, 0·9–1·2 cm. wide. Fig. 4.

TANGANYIKA. Singida District: near Swagaswaga, 10 Jan. 1928, *B. D. Burtt* 887 !; Kondoa District: 45 km. S. of Kondoa, 23 Jan. 1962, *Polhill & Paulo* 1241 !; Dodoma District: Manyoni, 1 Dec. 1926, *B. D. Burtt* 534 ! & Dodoma, 13 Dec. 1925, *Peter* 33146 !

DISTR. T4, 5; the typical plant not known elsewhere, but subsp. *rhodesica* in Zambia
HAB. Thickets, deciduous bushland, deciduous woodland; 1070–1370 m.

SYN. *Peltophorum massaiense* Taub. in P.O.A. C: 202 (1895)

NOTE. In East Africa only typical *B. massaiensis* (subsp. *massaiensis*) occurs, with (1–)2(–3) pairs of pinnae, 5–8 pairs of leaflets per pinna, and the lower and lateral petals 15–20 × 10–14 mm. Subsp. *rhodesica* Brenan in K.B. 17: 197 (1963), which should be looked for in the Ufipa District of Tanganyika, has 3–5 pairs of pinnae, 8–10(–11) pairs of leaflets (at least on the upper pinnae), and the lower and lateral

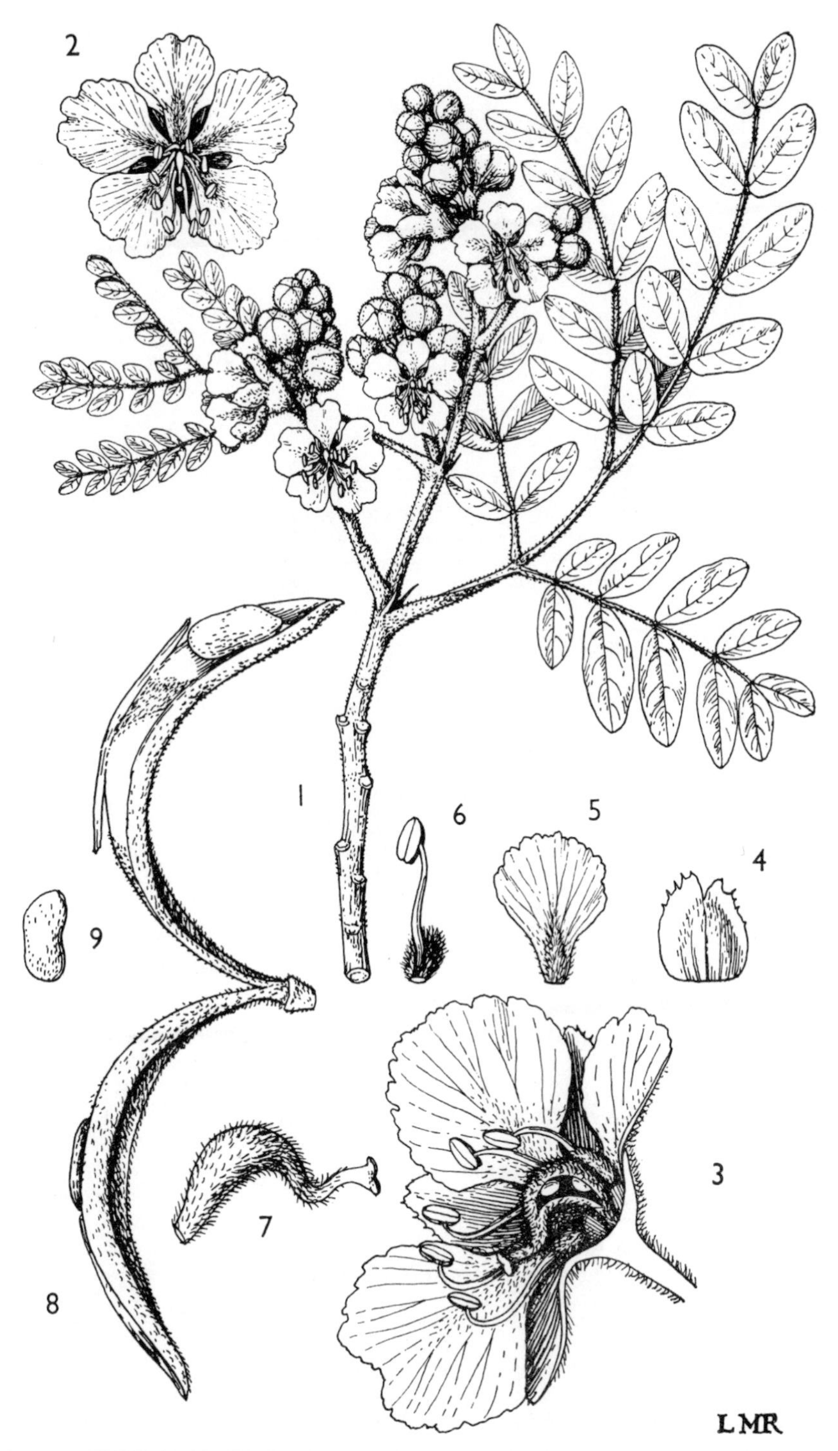

FIG. 4. *BUSSEA MASSAIENSIS* subsp. *MASSAIENSIS*—**1**, part of flowering branch, × 2/3; **2**, flower, front view, × 1; **3**, flower, longitudinal section, × 3; **4**, sepal, seen from outside, × 2; **5**, lateral petal, seen from inside, × 1; **6**, stamen, × 2; **7**, ovary, × 3; **8**, open pod, × 2/3; **9**, seed, × 2/3. 1–7, from *B. D. Burtt* 887; 8, 9, from *B. D. Burtt* 4438.

petals 12–15 × 9–12 mm. This is what is referred to as *B. massaiensis* in F.F.N.R.: 118, fig. 20/I (1962).

2. **B. eggelingii** *Verdc.* in K.B. 12: 350 (1957). Type: Tanganyika, Lindi District, *Eggeling* 6396 (EA, holo.!, FHO, K, iso.!)

Very similar in appearance to *B. massaiensis* subsp. *massaiensis*, differing as follows: Leaf-rhachides with appressed dark pubescence; pinnae 4–6 pairs; leaflets 5–11 pairs per pinna, but always 8–11 pairs on each of the upper pinnae, glabrous or almost so beneath. Inflorescence-axes (and outside of sepals) with a very dense ± appressed blackish-rusty indumentum. Petals: 4 lower and lateral ones 21–27 mm. long, 14·5–20 mm. wide, the fifth upper one 14–17 mm. long, 9–10 mm. wide. Pods 12–16 cm. long, 2–2·8 cm. wide; valves flattened or concave, but not conspicuously grooved.

TANGANYIKA. Lindi District: Rondo Plateau, Mchinjiri, Nov. 1951, *Eggeling* 6396! & Mar. 1952, *Semsei* 723!
DISTR. **T8**; not known elsewhere
HAB. " On the margin of closed forest "; ± 820 m.

NOTE. In view of the fact that *B. massaiensis* subsp. *rhodesica* is somewhat intermediate in its characters between typical *B. massaiensis* and *B. eggelingii*, I retain the latter as a species only with some hesitation, and mainly because of the distinctive pods, which may not be quite mature. Those of *B. massaiensis* subsp. *rhodesica* are similar to those of subsp. *massaiensis*.

5. CAESALPINIA

L., Sp. Pl.: 380 (1753) & Gen. Pl., ed. 5: 178 (1754) (as " *Caesalpina* ")

Shrubs, erect or more often scrambling or climbing, or trees, armed with spines or prickles or sometimes unarmed. Leaves bipinnate, very rarely (and not in East Africa) reduced to scales (" switch-plant " habit); stipules various, minute to conspicuously leafy; specialized glands restricted to petiole and rhachis absent, although a general glandular indumentum may be sometimes present; leaflets opposite, rarely alternate, glandular or sometimes eglandular. Inflorescences of terminal, sometimes falsely lateral, or terminal and axillary racemes or panicles; rarely racemes much reduced, the flowers thus single or very few together; bracts usually quickly falling. Flowers ⚥, or ♂ ⚥ in *C. bonduc* and *C. volkensii* among native species. Sepals 5, imbricate, sometimes very narrowly so, or almost valvate; the lower sepal often cucullate at apex and clasping the others. Petals 5, subequal except for the usually somewhat modified upper one, which usually has a rather smaller lamina and more pronounced claw. Stamens 10, fertile, rarely and casually with 1 ± abortive; filaments alternately longer and shorter, all pubescent or villous and often glandular below; anthers dorsifixed, dehiscing by longitudinal slits. Ovary free, subsessile or shortly stipitate, usually with 2 to ± 10 ovules, glabrous, pubescent or glandular; style glabrous, or clothed below like the ovary; stigma truncate or oblique, ciliolate or glabrous. Pods very variable, usually flat or ± compressed, very rarely cylindrical, not winged, indehiscent or dehiscent and 2-valved, hard and woody or thick and pulpy. Seeds transverse or nearly so, hard; funicle short; endosperm present or absent.

About 150–200 species throughout the tropics, but most numerous in the New World.

When the characteristic pods are not present, *Mezoneuron angolense* is liable to be taken for a *Caesalpinia*. The differences are given under *M. angolense* on p. 40.

Several exotic species of *Caesalpinia* are in cultivation in East Africa. The following key is provided to distinguish them:—

KEY TO EXOTIC SPECIES

Stamens scarlet, long-exserted, with filaments 6–12·5 cm. long; shrubs ± 1–6 m. high:
- Inflorescence, including outside of sepals, glabrous (only stamen-filaments hairy); sepals entire or almost so; petals scarlet, red and yellow, or sometimes all yellow; plant with, or sometimes without scattered prickles . . . . *C. pulcherrima* (L.) Sw. (see p. 31)
- Inflorescence, including outside of sepals, pubescent and copiously glandular-hairy; sepal-margins lacerate towards apex; petals yellow; plant unarmed . . . . *C. gilliesii* (Hook.) Dietr. (Native of South America)

Stamens not scarlet, not or shortly exserted, with filaments ± 0·5–1·5 cm. long; shrubs or trees; flowers yellow or yellowish:
- Lower sepal fimbriate-pectinate, the others ± erose or lacerate; pods indehiscent, often reddish; prickly tree to ± 6 m. high . . . . *C. spinosa* (Mol.) O. Kuntze (*C. pectinata* Cav., *C. tinctoria* (H.B.K.) Benth.) (Native of South America)
- Lower sepal, as are the others, entire or almost so:
  - Plants unarmed; flowers fragrant:
    - Leaflets oblong-linear, in many pairs; pod coiled or twisted; tree to 9 m. high, with small flowers in compact panicles . . . . *C. coriaria* (Jacq.) Willd. (Native of tropical America)
    - Leaflets elliptic, in few (± 2–5) pairs; pod straight; small tree to 6 m. high . . . *C. paucijuga* Oliv. (Native of the West Indies)*
  - Plants prickly:
    - Midrib arising from lower corner of leaflet at base, running ± diagonally, the leaflets thus very asymmetric at base; straggling shrub or small tree . . . . *C. sappan* L. (Native of India and Burma)
    - Midrib at base of leaflet arising subcentrally, the base thus ± symmetric:
      - Pedicels of flowers widely spreading, almost horizontal or slightly deflexed; pods thick, not or scarcely beaked, indehiscent; branchlets and racemes subglabrous; climbing shrub . . . . *C. digyna* Rottl. (Native of India, Ceylon and Malaya)**

* I have seen no material of this from the Flora area. It is said to be cultivated in Uganda at the Entebbe Gardens, see Dale, Introd. Trees Uganda: 14 (1953).

** No material seen from the Flora area. Cultivated in Zanzibar, according to U.O.P.Z.: 159 (1949).

Pedicels of flowers ascending at an acute angle from the main axis; pods strongly beaked, ultimately dehiscent; branchlets and racemes usually ± densely puberulous or pubescent; shrub 2·5–10 m. high . *C. decapetala* (Roth) Alston (see p. 36)

KEY TO NATIVE AND NATURALIZED SPECIES

Stamen-filaments 5–6·5 cm. long, long-exserted, scarlet; petals scarlet to orange-red or yellow; plant glabrous except for stamen-filaments . . . 1. *C. pulcherrima*

Stamen-filaments up to ± 1·5 cm. long, yellowish to red; petals yellow, white or pink; plant usually ± pubescent, sometimes glabrous except for stamen-filaments:

- Plant unarmed; ovary and lower part of style densely clothed with brown glands and pubescence; trees:
  - Pinnae 1–3 pairs; leaflets (2–)3–4 pairs, 1·2–8·2 cm. long, 0·6–3 cm. wide . . . . 2. *C. dalei*
  - Pinnae 6–10 pairs; leaflets 8–12 pairs, 0·5–0·8 cm. long, 0·25–0·4 cm. wide . . . . 3. *C. insolita*
- Plant armed with prickles; ovary and style glabrous to pubescent or setulose, eglandular; erect or climbing shrubs, rarely shrubby trees:
  - Pod not prickly, smooth except sometimes for pubescence; ovary and style glabrous, or pubescent in 8, *C. decapetala*, but not setulose; petals pink, white, yellow or greenish-yellow:
    - Anthers glabrous or subglabrous; prickles scattered over stems:
      - Pedicels 0·2–0·7 cm. long; ovary glabrous:
        - Petals pink (? rarely yellow); bracts 3–6 mm. long, usually falling off while flower-buds are still young . . . 4. *C. trothae*
        - Petals white; bracts ± 9 mm. long, persistent until the flower-buds open . 5. *C. sp. A*
      - Pedicels 1–2·4 cm. long; ovary pubescent or glabrous:
        - Stem with downwardly hooked prickles; leaflets 15–25 pairs per pinna, 0·1–0·35 cm. wide; bracts about 0·5 mm. long; ovary glabrous; pod glabrous, not beaked, with thickened upper suture . . . . 7. *C. welwitschiana*
        - Stem with usually straight spreading prickles; leaflets 8–12 pairs per pinna, 0·3–0·8 cm. wide; bracts 5–8 mm. long; ovary pubescent; pod shortly pubescent, beaked . . . . . 8. *C. decapetala*
    - Anthers densely woolly-pubescent; scattered prickles absent, but short stipular spines present in pairs at nodes . . . . 6. *C. erianthera*

Pod densely spreading-prickly outside; ovary densely setulose; lower part of style pubescent; petals yellow or greenish-yellow:
Leaflets acuminate or subacuminate at apex; stipules minute, not leafy; sepals 10–14 mm. long; petals 16 mm. long; seeds brown; growing inland at about 90–1770 m. alt. . 9. *C. volkensii*
Leaflets rounded or emarginate at apex, usually not or scarcely acuminate; stipules conspicuous, leafy; sepals 6·5–8 mm. long; petals 12–13 mm. long; seeds leaden-grey; growing on or near sea-shores . . . 10. *C. bonduc*

1. **C. pulcherrima** (*L.*) *Sw.*, Observ. Bot. Pl. Ind. Occ.: 166 (1791); Oliv., F.T.A. 2: 262 (1871); L.T.A.: 616 (1930); T.T.C.L.: 95 (1949); U.O.P.Z.: 160 (1949); Wilczek in F.C.B. 3: 254 (1952); Torre & Hillcoat in C.F.A. 2: 172 (1956); Roti-Michelozzi in Webbia 13: 214 (1957); F.F.N.R.: 119 (1962). Type: in *Herb. Linnaeus* 529.1 (LINN, syn. !)

Shrub 1–6 m. high, quite glabrous except for the stamen-filaments. Spines absent, or short, here and there in pairs at nodes, rarely sparsely scattered. Leaves: petiole with rhachis 8–30 cm. long, sometimes armed with small prickles or spinulose stipels at insertions of pinnae and leaflets; pinnae 3–10 pairs, 1·8–8·5 cm. long; leaflets 5–11(–13) pairs per pinna, oblong-elliptic to elliptic, 5–18(–28) mm. long, 4–9(–15) mm. wide, rounded to emarginate at apex. Racemes terminal or with axillary ones in addition, up to ± 37 cm. long. Bracts lanceolate or linear-lanceolate, 2·5–7·5 mm. long, falling when the flower-buds are very young. Flowers scarlet, red and yellow, orange-red, or yellow (var. *flava* L. H. Bailey). Pedicels 2·5–7·5 cm. long in flower, longer in fruit. Sepals 7–14 mm. long, the lower one 14–17 mm. long and hooded towards apex. Petals 15–25 mm. long, long-clawed, with a lamina ± 10–20 mm. wide broadly rounded at apex and erose-crisped on margin; upper petal with smaller lamina. Stamen-filaments scarlet, long-exserted, 5–6·5 cm. long; anthers glabrous. Pods asymmetrically oblanceolate-oblong, 7–12 cm. long, 1·8–2·2 cm. wide, compressed, dehiscent, unarmed, purple to blackish-brown. Seeds obovate, subtruncate at apex, 9–10 mm. long, 7–8 mm. wide, somewhat compressed, brown.

KENYA. Machakos District: Makindu R., 12 Apr. 1902, *Kassner* 571 !; Mombasa, on mainland, July 1876, *Hildebrandt* 2005b !; Lamu District: Witu, *F. Thomas* 156 !
DISTR. **K**4, 7; probably native of tropical America (often, but on too slender evidence, alleged to be Asiatic), but cultivated in most parts of the tropics and frequently becoming naturalized
HAB. Usually a relic of cultivation; near sea-level to ± 2000 m.

SYN. *Poinciana pulcherrima* L., Sp. Pl.: 380 (1753); Howes in K.B. 1: 78 (1947)

NOTE. I am obliged to Dr. B. Verdcourt for advice on the status of this beautiful plant, which seems at any rate persistent in **K**4 and 7. It is of course a favourite in gardens wherever possible, under the name " Barbados Pride ", and it may easily become established more often than the above account suggests.

2. **C. dalei** *Brenan & Gillett* in K.B. 17: 198 (1963). Type: Kenya, Kwale District, *Dale* in *F.H.* 3572 (K, holo. !, BM, EA, iso. !)

Unarmed spreading buttressed tree, with smoothish grey bark. Young stems densely clothed with a short velvety dark brown indumentum composed of glands and hairs mixed, soon glabrescent and grey. Leaves: petiole with rhachis 1·5–10 cm. long; stipels absent; pinnae 1–3 pairs, 1–8 cm. long; leaflets (2–)3–4 pairs per pinna, elliptic or narrowly elliptic, slightly obliquely rhombic, 1·2–8·2 cm. long, 0·6–3 cm. wide, obtuse at apex, glabrous except for inconspicuous pubescence on midrib beneath and for sessile reddish peltate glands scattered on lower surface; venation raised and reticulate on both surfaces. Racemes terminal on main or lateral branches, 2–8 cm. long, simple, very densely clothed like the young stems. Bracts 1 mm. long. Pedicels 4–11 mm. long. Sepals 5·5–6 mm. long, brown-velvety especially outside. Petals bright yellow, the upper one with brown spots, narrowly obovate-spathulate, 9–10 mm. long, 3–4 mm. wide; the upper petal bent upwards at apex of claw, but without a transverse projection on inner side. Anthers glabrous. Ovary and lower part of style densely covered with brown glands and pubescence. Pods and seeds unknown.

KENYA. Kwale District: Mwachi, Sept. 1936, *Dale* in *F.H.* 3572 & in *C.M.* 9273!
DISTR. **K**7; not known elsewhere
HAB. "In a wooded gulley"; 150 m.

SYN. *Caesalpinia* sp. sensu Dale, Woody Vegetation Coast Prov. Kenya (Imperial Forestry Institute Paper 18): 21 (1939); K.T.S.: 100 (1961)

NOTE. This species needs re-collecting, particularly in fruit. It is exceedingly distinct, and readily separated from all the other native species except *C. insolita* by being an unarmed tree and by the dense glands on the young stems and on the inflorescences.

3. **C. insolita** (*Harms*) *Brenan & Gillett* in K.B. 17: 200 (1963). Type: Tanganyika, Lindi District, *Schlieben* 5682 (B, holo.†, BM, BR, K, LISC, iso.!)

Tree 16–20 m. high, unarmed. Young stems densely clothed with a short velvety dark brown indumentum composed of glands and hairs mixed, soon glabrescent and grey. Leaves: petiole with rhachis 6–11 cm. long; stipels absent; pinnae 6–10 pairs, 2–5 cm. long; leaflets 8–12 pairs per pinna, elliptic-subrhombic, asymmetric, 0·5–0·8 cm. long, 0·25–0·4 cm. wide, obtuse or rounded at apex, glabrous or nearly so or with a few sessile reddish peltate glands scattered over lower surface; venation raised and reticulate on both surfaces. Racemes terminal on main or lateral branches, 5–11 cm. long, simple, densely clothed like the young stems. Bracts 1 mm. long. Pedicels 6–13 mm. long. Sepals 5–6 mm. long, brown-velvety especially outside. Petals yellow, narrowly obovate-spathulate, 10–11 mm. long, 3–4 mm. wide; the upper one more or less as in *C. dalei*. Anthers glabrous. Ovary and lower part of style densely clothed with brown glands and pubescence. Pods and seeds unknown.

TANGANYIKA. Lindi District: Lake Lutamba, 4 Dec. 1934, *Schlieben* 5682!
DISTR. **T**8; not known elsewhere
HAB. Woodland; ± 240 m.

SYN. *Hoffmanseggia insolita* Harms in N.B.G.B. 13: 416 (1936); T.T.C.L.: 103 (1949)

4. **C. trothae** *Harms* in E.J. 26: 277 (1899), as "*trothaei*"; L.T.A.: 615 (1930); T.S.K.: 60 (1936); T.T.C.L.: 95 (1949); Roti-Michelozzi in Webbia 13: 210 (1957); K.T.S.: 99 (1961). Type: Tanganyika, ? Dodoma District, Chumo Pass, *von Trotha* 186 (B, holo. †)

Shrub 0·3–3(–5) m. high, erect or sometimes somewhat climbing. Young stems glabrous to pubescent, armed with scattered straight and spreading or ± curved and deflexed prickles varying in length up to 9 mm. Leaves: petiole with rhachis 1·5–15 cm. long, often ± prickly; pinnae 3–17 pairs, 0·7–5·3(–8·5) cm. long; leaflets 4–33 pairs per pinna, narrowly oblong, or oblong-elliptic, or the uppermost slightly obovate, (2·5–)4–10(–13) mm. long, 1–4(–5) mm. wide, rounded at apex and usually mucronate, ± pubescent or puberulous on midrib and margins or becoming quite glabrous, margins usually minutely crenulate (? glandular, use lens of at least × 20). Racemes terminal and lateral, 2·5–15(–24) cm. long, simple. Bracts ovate- or elliptic-acuminate, 3–6 mm. long, usually falling off while flower-buds are still young. Flowers mauve-pink (or ? rarely yellow: see note below). Pedicels 2–7 mm. long in flower, to 15 mm. in fruit. Sepals 8–12 mm. long. Petals obovate, 9–20 mm. long, 4·5–15 mm. wide; upper one smaller than others, with a transverse usually bidentate wing-like projection on inner side at apex of claw. Anthers appearing glabrous but, under a × 20 lens, a sparse minute puberulence is visible. Ovary and style glabrous. Pods shortly boat-shaped or anvil-shaped, 1·5–5 cm. long, 1·5–2·2 cm. wide, compressed, narrowed to an acute beak directed forwards or upwards and continuing line of upper suture at apex. Seeds obovoid, scarcely compressed, 10–11 mm. long, 7–9 mm. wide, olive, or mottled olive and brown. Fig. 5, p. 34.

subsp. **trothae**

Young stems glabrous to pubescent. Pinnae (4–)6–17 pairs. Leaflets 9–33 pairs. Axis of raceme normally glabrous. Pedicels 2–6 mm. long in flower. Petals up to 20 mm. long and 15 mm. wide. Pods 3·7–5 cm. long. Fig. 5/1–10.

Kenya. Machakos District: 38 km. beyond Mtito Andei towards Mombasa, 4 Nov. 1956, *Verdcourt* 1588!; Teita District: near Tsavo Bridge, 3 Feb. 1953, *Bally* 8681 in *C.M.* 20522!; Kwale District: Kinango, Jan. 1930, *R. M. Graham* U.760 in *F.H.* 2239 & in *C.M.* 13918!

Tanganyika. Dodoma District: about 104 km. S. of Dodoma, 17 July 1956, *Milne-Redhead & Taylor* 11177!; Morogoro Fuel Reserve, Nov. 1954, *Semsei* 1879!; Iringa District: Lukose valley between Mahenge and Mdahira on the road to Kilosa, 23 Oct. 1936, *B. D. Burtt* 6302!

Distr. **K**4, 7; **T**5–7; not known elsewhere

Hab. Dry scrub with trees, ? deciduous bushland; 150–1070 m.

Note. In T.S.K.: 60 (1936), the flowers of *C. trothae* are said to be yellow; this probably originates in a field-note, "corolla yellow" with *Battiscombe* 276, from Kenya, Tana R. Either this is a mistake or else (and not impossibly) a rare and local colour-variant occurs; further evidence is needed to decide.

The Kenya specimens *Verdcourt* 1588 and *Bally* 8681, cited above, both have the inflorescence-axes puberulous. They are perhaps better regarded as intermediates between subsp. *trothae* and subsp. *erlangeri*; *Grenfell*, from Kenya, Teita Hills, is similar. See also the note under subsp. *erlangeri*.

subsp. **erlangeri** (*Harms*) *Brenan* in K.B. 17: 201 (1963). Type: Ethiopia, Galla Sidama, Borana, Tarro Cumbi, *Ellenbeck* 2071 (B, holo. †)

Young stems puberulous. Pinnae 3–6(–7) pairs. Leaflets 4–12 pairs. Axis of raceme puberulous. Pedicels 5–7 mm. long in flower. Petals up to ± 13 mm. long and 9 mm. wide. Pods 1·5–3·5 cm. long, 1–2-seeded. Fig. 5/11, 12.

Kenya. Northern Frontier Province: between Dandu and Gaddaduma, 18 May 1952, *Gillett* 13223! & Wajir, 3 Jan. 1955, *Hemming* 478!; Northern Frontier Province/Tana River District: Garissa, 26 Dec. 1942, *Bally* 1983 in *C.M.* 11619!

Distr. **K**1, ? 7; Ethiopia, Somali Republic (S.)

Hab. Dry scrub with trees; 100–800 m.

Syn. *C. erlangeri* Harms in E.J. 33: 160 (1902); V.E. 1: 181, fig. 155 (1910) & 3(1): 510, fig. 272 (1915); L.T.A.: 616 (1930); Roti-Michelozzi in Webbia 13: 209 (1957); K.T.S.: 99 (1961)

Note. *Verdcourt* 2099 from Kenya, Tana River District, 336 km. from Nairobi towards Garissa, is intermediate between subsp. *erlangeri* and subsp. *trothae*. The pinnae are in 3–9 parts, the leaflets in 7–19 pairs, the inflorescence-axes are puberulous, and the pedicels about 5 mm. long.

5. **C. sp. A**

Small bush. Stems grey-puberulous, here and there sparsely armed with scattered straight spreading prickles ± 3 mm. long. Stipules triangular, acute, ± 3·5 mm. long. Leaves: petiole with rhachis ± 1–1·5 cm. long, armed with paired prickles or spinulose stipels at insertions of pinnae; pinnae ± 3–4 pairs, 1–1·5 cm. long; leaflets 8–10 pairs per pinna, elliptic-

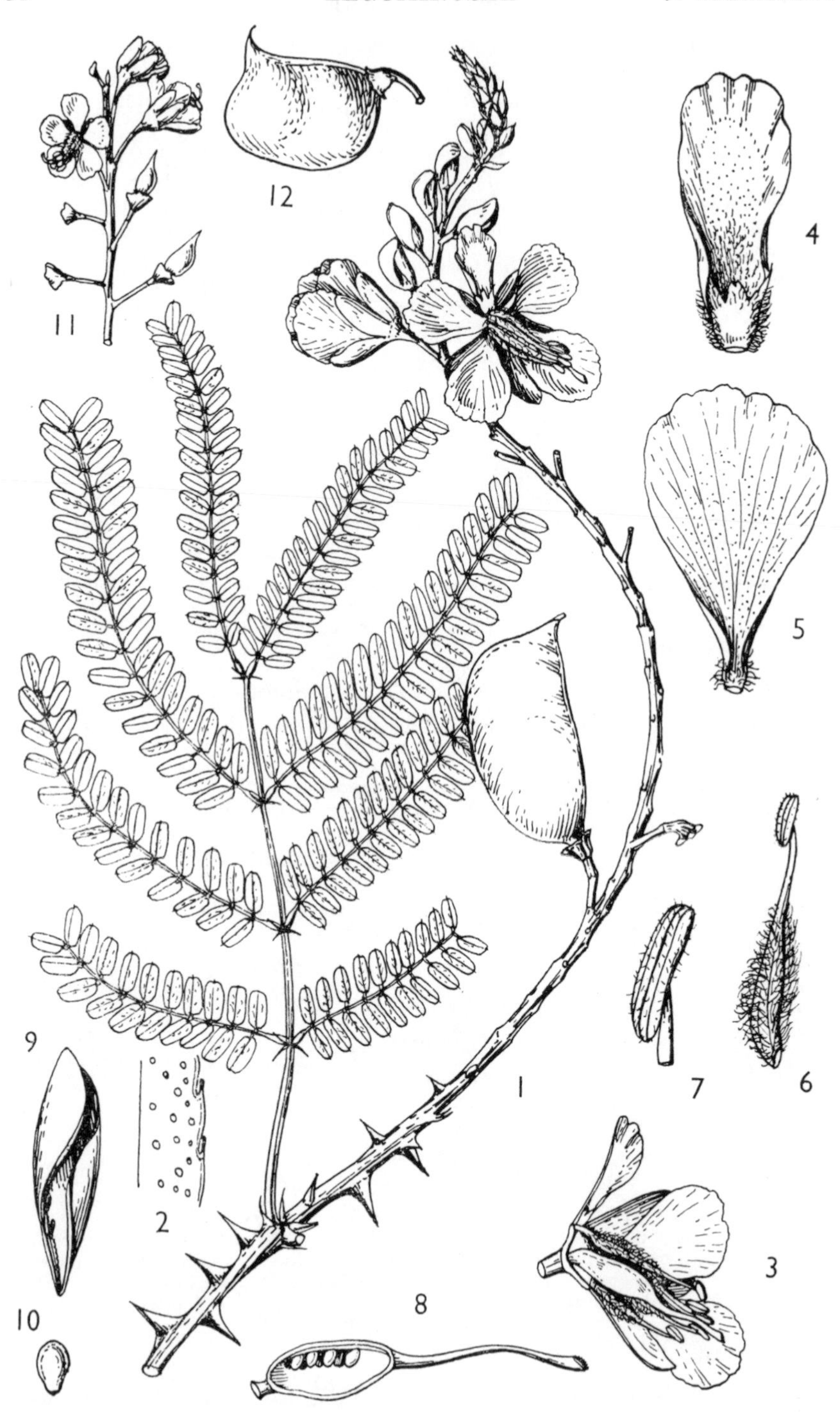

FIG. 5. *CAESALPINIA TROTHAE* subsp. *TROTHAE*—**1,** part of branch, showing inflorescence with flowers and pod, × $\frac{2}{3}$; **2,** marginal part of leaflet, seen from below, × 10; **3,** flower, longitudinal section, × 1; **4,** upper petal, × 2; **5,** lateral petal, × 2; **6,** stamen, × 2; **7,** anther and part of filament, × 6; **8,** ovary, with part of wall removed in order to expose ovules, × 2; **9,** valve of pod after dehiscence, × $\frac{2}{3}$; **10,** seed, × $\frac{2}{3}$. *C. TROTHAE* subsp. *ERLANGERI*—**11,** part of inflorescence, × $\frac{2}{3}$; **12,** pod, × $\frac{2}{3}$. 1–8, from *Milne-Redhead & Taylor* 11177; 9, 10, from *Ward* U27; 11, from *Gillett* 13223; 12, from *Hemming* 478.

oblong, ± 4–5 mm. long and 1·5–1·9 mm. wide, rounded at apex, puberulous on both surfaces. Racemes at ends of short lateral branches, simple, ± 2·5–7 cm. long. Bracts persistent until flower-buds open, ovate-elliptic, concavo-convex, ± 9 mm. long and 7 mm. wide, rounded and mucronate at apex. Pedicels ± 3 mm. long. Sepals 9–10 mm. long. Petals white, narrowly obovate, all ± 9–10 mm. long and 4–4·5 mm. wide. Anthers glabrous. Ovary and style glabrous. Pod and seeds unknown.

KENYA. Northern Frontier Province: Malka Murri, 2 Oct. 1955, *J. Adamson* 81!
DISTR. **K1**; not known elsewhere
HAB. Stony ground; 1010 m.

NOTE. As far as one can judge from a dried specimen, the rather conspicuous bracts seem not to have been green, but perhaps yellowish or white. The measurements in the above description, being taken from only a single gathering, are liable to alteration in the future. More material of this probably new species is desired.

6. **C. erianthera** *Chiov.*, Fl. Somala 1: 155, t. 16, fig. 2/a, b (1929); L.T.A.: 884 (1930); Roti-Michelozzi in Webbia 13: 211 (1957); K.T.S.: 99 (1961). Types: Somali Republic (S.), from Obbia to Uarandi, *Robecchi-Brichetti* 534 (FI, syn., K, fragments!) & between Attod and Dolobscio, *Puccioni & Stefanini* 450 (FI, syn.)

Shrub 0·3–2 m. high. Young stems glabrous to pubescent or tomentellous, armed here and there with recurved or sometimes spreading stipular spines 2–5 mm. long in pairs at the nodes; scattered prickles absent. Stipules on leafy shoots often foliaceous, rounded or reniform, 0·5–2 cm. in diameter. Leaves: petiole with rhachis 0·7–5·5(–7·7) cm. long, armed with minute prickles or spinulose stipels at insertions of pinnae; pinnae 2–6 pairs, 0·4–3·2 cm. long; leaflets (2–)3–5 pairs per pinna, elliptic, broadly elliptic or rotund, 2–9·5(–18) mm. long, 1·5–6(–13) mm. wide, glabrous to pubescent. Racemes lateral or terminating short lateral shoots, 1–15-flowered. Bracts caducous, 3–3·5 mm. long. Pedicels 5–17 mm. long. Sepals 5–6 mm. long (lower sepal up to 10 mm. long). Petals greenish-yellow, 5–6·5 mm. long, with a suborbicular or obovate-suborbicular lamina 3–6 mm. wide. Anthers densely woolly-pubescent. Ovary and style glabrous. Pod (only young ones seen) flattened, unarmed.

var. **pubescens** *Brenan* in K.B. 17: 203 (1963). Type: Kenya, Northern Frontier Province, *Gillett* 13274 (K, holo.!, EA, iso.!)

Stems, rhachides of leaves and pinnae, and both surfaces of leaflets shortly and sometimes densely pubescent, the stems being sometimes tomentellous.

KENYA. Northern Frontier Province: Banessa to Ramu, 23 May 1952, *Gillett* 13274!
DISTR. **K1**; Somali Republic and Arabia
HAB. *Acacia* open scrub on limestone; ? 760 m.

NOTE. *Gillett* 13274 is said to have greenish-yellow petals and ± red stamens and calyx. Typical var. *erianthera* has glabrous stems and leaves, and has not so far been found in East Africa, though occurring in the Somali Republic and Arabia.

7. **C. welwitschiana** (*Oliv.*) *Brenan* in K.B. 17: 203 (1963). Type: Angola, Cuanza Norte, Golungo Alto, *Welwitsch* 608 (LISU, holo., BM, K, iso.!)

Liane to 45 m. (*fide* F.C.B.). Stems subglabrous or with usually sparse pubescence when young, armed with scattered downwardly hooked prickles. Leaves: petiole with rhachis 21–36 cm. long, ± armed with paired deflexed prickles; pinnae 14–19 pairs, 3–6·8(–8, *fide* F.C.B.) cm. long; leaflets 15–25 pairs, narrowly oblong, 6–10 mm. long, 1–3·5 mm. wide, minutely appressed-puberulous beneath. Racemes terminal and often also axillary, 8–24(–36) cm. long. Bracts minute, scale-like, ± 0·5 mm. long. Pedicels 1–2·3 cm. long.

Sepals 6–12 mm. long, glabrous outside. Petals yellow, 11–13 mm. long, 7–8 mm. wide. Stamen-filaments 1·2–1·5 cm. long; anthers glabrous. Ovary glabrous. Pod shortly oblong-elliptic to elliptic, 2·5–6 cm. long, 2·2–3·1 cm. wide, straight, ± compressed, indehiscent, unarmed, glabrous, 1–3-seeded, with the upper suture thickened but not winged. Seeds globose or subglobose, 1·2–1·6(–2, *fide* F.C.B.) cm. in diameter, olive-brown.

UGANDA. Mengo District: Kyadondo, Kiwatule, May 1929, *Liebenberg* 832! & near Kampala, July 1931, *Staff of Herbarium*, Botanist, Dept. of Agriculture 2249!
TANGANYIKA. Kigoma District: Gombe Stream chimpanzee reserve, 1961, *Morris-Goodall*! & Kabogo Mts., Sept. 1963, *Azuma*!
DISTR. **U**4; ?**T**4; Cameroun Republic, Gabon, Congo Republic and Angola
HAB. Probably in lowland rain-forest; ± 800–1200 m.

SYN. *Mezoneuron welwitschianum* Oliv., F.T.A. 2: 261 (1871); L.T.A.: 613 (1930); Wilczek in F.C.B. 3: 260, t. 20 (1952); Torre & Hillcoat in C.F.A. 2: 171 (1956); F.F.N.R.: 126, figs. 20/K, 26 (1962)

8. **C. decapetala** (*Roth*) *Alston* in Trimen, Handb. Fl. Ceyl. 6 (suppl.): 89 (1931); T.T.C.L.: 94 (1949); Wilczek in F.C.B. 3: 253 (1952); Torre & Hillcoat in C.F.A. 2: 172 (1956); F.F.N.R.: 118, fig. 20/H (1962). Type: India, *Heyne* (whereabouts of holo. uncertain, ? K, iso.!)

Climbing or straggling bushy shrub 2·5–10 m. high. Stems when young ± densely clothed with short brownish pubescence or puberulence, rarely sparsely clothed or subglabrous, also armed with scattered usually straight spreading prickles up to 8(–10) mm. long. Stipules asymmetrically ovate, acuminate, wavy-margined, 4–20 mm. long, 2–8 mm. wide. Leaves: petiole with rhachis ± 12–47 cm. long, armed with downwardly hooked prickles often in pairs especially at insertions of pinnae; pinnae 4–10 pairs, 2·5–9·8 cm. long; leaflets 8–12 pairs per pinna, elliptic-oblong or slightly obovate, (0·8–)1–2(–2·7) cm. long, 0·3–0·8(–1·7) cm. wide, rounded at apex, pubescent or puberulous on both surfaces. Racemes at ends of main or lateral branches, simple, (10–)16–30 cm. or more long. Bracts very caducous, ovate-triangular to lanceolate, 5–8 mm. long. Pedicels 1·5–2·4(–4) cm. long. Sepals 9–10 mm. long. Petals yellow to yellowish-white, 10–15 mm. long, 8–15 mm. wide (the upper one, however, smaller, ± 8–11 mm. long and 5–6 mm. wide, with margins inflexed in middle). Anthers glabrous. Ovary pubescent. Pod oblong-elliptic, 6–9·5(–12) cm. long (excluding beak), (2–)2·4–2·7(–3·5) cm. wide, straight or slightly curved, compressed, ultimately dehiscent along upper suture, unarmed, shortly pubescent, brown, with a slender beak 0·6–3 cm. long arising near line of upper suture at apex. Seeds ellipsoid, 0·9–1 cm. long, 0·6–0·8 cm. wide, mottled brown and blackish-grey, sometimes all black.

UGANDA. Kigezi District: Kachwekano Farm, Jan. 1950, *Purseglove* 3223!; Mengo District: Wabusana, 28 July 1956, *Langdale-Brown* 2250!
KENYA. Elgon, Dec. 1930, *Lugard* 298!; Kiambu District: Limuru, 10 June 1918, *Snowden* 585!; Nairobi, Oct. 1943, *Bally* in *C.M.* 11778!
TANGANYIKA. Arusha District: Songoro Hill, 22 June 1955, *Willan* 243!; Lushoto District: Magamba Forest, 27 Oct. 1951, *Willan* 10!; Rungwe District: without more precise locality, 12 Mar. 1932, *R. M. Davies* D29!
DISTR. **U**2, 4; **K**?3, 4, 5; **T**1–3, 6, 7; originally from tropical and subtropical Asia, but now widely cultivated and often naturalized; in Africa from the Flora area southwards to Angola and South Africa
HAB. Clearings in lowland rain-forest, scattered-tree grassland, bushland; 880–2130 m.

SYN. ? *Reichardia decapetala* Roth, Nov. Pl. Sp. Ind. Or.: 212 (1821)
Caesalpinia sepiaria Roxb., Fl. Ind., ed. 2, 2: 360 (1832); L.T.A.: 615 (1930); T.S.K.: 60 (1936); Howes in K.B. 1: 63, fig. (1947). Type: India, *Roxburgh* (whereabouts of holo. uncertain, K (*Roxburgh* in *Wallich* 5834a) ? iso. or isosyn.!)

NOTE. *C. decapetala* has often been used for a hedging plant, and it may be expected to become naturalized in other parts of East Africa where the rainfall is comparatively high.

As might be expected, *C. decapetala* shows a wider range of variation in Asia, its native home, than in East Africa, where, for example, the indumentum on the stems is usually fairly dense. In *Battiscombe* 169 in *C.M.* 13917, from Kenya, S. Nyeri District, Wambugu's, and in *Battiscombe* 58, from Kenya, Kiambu District, Kikuyu country, however, the stems are subglabrous.

9. **C. volkensii** *Harms* in E.J. 45 : 304 (1910); L.T.A.: 615 (1930); T.T.C.L.: 95 (1949); K.T.S.: 99 (1961). Types: Tanganyika, Moshi District, Marangu, *Volkens* 1454 (B, syn. †) & Moshi, *Merker* 509 (B, syn. †, BM, drawing!) & Lushoto District, Handei, *Holst* 9123 (B, syn. †, K, isosyn., said to come from Kwa Mshusa!) & Amani, Monga, *Braun* 1549 (B, syn. †, EA, isosyn.!)

Climbing bush or liane. Stems shortly ± pubescent and also armed with scattered ± downwardly hooked or deflexed prickles ± 2–4 mm. long. Stipules of 2–3 minute subulate inconspicuous points ± 3 mm. long. Leaves: petiole with rhachis 14–±50 cm. long, armed as in *C. bonduc*; pinnae 3–6 pairs, 6·5–16 cm. long; leaflets 3–7 pairs per pinna, ovate to ovate-elliptic, becoming stiffer than in *C. bonduc*, (2–)3·2–8 cm. long, (1–)1·6–4·3 cm. wide, acuminate or subacuminate at the obtuse to subacute apex, glabrous or nearly so, except when very young and for some persistent pubescence on midrib. Racemes axillary, pedunculate, simple or with a few branches below. Bracts only shortly exceeding buds, ovate-acuminate, ascending or rarely somewhat spreading in upper part, 4–5(–7) mm. long. Pedicels 4–14 mm. long. Sepals 10–14 mm. long, 3·5–5·5 mm. wide. Petals yellow, ± 16 mm. long and 3·5–4·5 mm. wide. Anthers 1·5–1·75 mm. long. Pods broadly oblong- to obovate-elliptic, (6–)7–13 cm. long, 3·6–6·5 cm. wide, densely spreading-prickly outside. Seeds subglobose, hard, brown, 1·7–2·3 cm. in diameter; cuticle of testa finely and ± transversely cracked.

UGANDA. Busoga District: Bunya, Nov. 1937, *Webb* 11!; Mengo District: Entebbe, Bawanya, June 1923, *Maitland* 745! & Kajansi Forest, Entebbe road, Oct. 1935, *Chandler* 1425!

KENYA. ? S. Kavirondo District: Utende [? Butende], Sept. 1933, *Napier* 2945 in *C.M.* 5295!; Kericho District: Cheptuiyet, Belgut Location, 28 Aug. 1960, *Kerfoot* 2157!; Kwale District: Shimoni Forest, June 1962, *Birch* 62/146!

TANGANYIKA. Mbulu District: Ufiome Mt., Irago Hill, 21 Jan. 1928, *B. D. Burtt* 1230!; Tanga District: Bushiri Estate, 28 Dec. 1950, *Faulker* 745!; Morogoro District: Turiani, Nov. 1953, *Semsei* 1455!

DISTR. **U**3, 4; **K**4, 5, 7; **T**2, 3, 6; not known elsewhere

HAB. In or on margins of lowland rain-forest; perhaps also occurring in upland rain-forest at its lowest altitudes; 90–1770 m.

SYN. [*C. bonduc* sensu L.T.A.: 614 (1930), *non* (L.) Roxb.]
[*C. crista* sensu Chiov., Racc. Bot. Miss. Consol. Kenya: 38 (1935), *non* L.]
[*C. major* sensu T.T.C.L.: 95 (1949), *non* (Medic.) Dandy & Exell]

NOTE. *C. volkensii* is more closely related to *C. major* (Medic.) Dandy & Exell than it is to *C. bonduc* (L.) Roxb. *C. major* differs from *C. volkensii* in having linear-lanceolate bracts, smaller flowers, and (as far as Asiatic material is concerned) leaflets normally with appressed pubescence on lower surface.

10. **C. bonduc** (*L.*) *Roxb.*, Fl. Ind., ed. 2, 2: 362 (1832); Dandy & Exell in J.B. 76: 179 (1938); T.T.C.L.: 94 (1949); Wilczek in F.C.B. 3: 250 (1952); F.W.T.A., ed. 2, 1: 481, fig. 154/A (1955); Torre & Hillcoat in C.F.A. 2: 171 (1956); Roti-Michelozzi in Webbia 13: 204 (1957); K.T.S.: 99 (1961). Types: Ceylon, *Herb. Hermann* vol. 2, fol. 17 & vol. 3, fol. 35 (BM, syn.!)

Bush or shrubby tree, spreading or half-climbing, up to ± 5 m. high. Stems pubescent and also ± densely armed with spreading straight (or the larger ones slightly deflexed) prickles of various lengths. Stipules conspicuous, leafy, of 2–3 unequal-sized often asymmetric lobes resembling leaflets, each 0·3–2·5 cm. long, 0·2–3·6 cm. wide, mucronate and rounded to emarginate at apex. Leaves: petiole with rhachis to ± 50 cm. long, armed on lower side with recurved prickles often paired at the insertion of pinnae and there sometimes also with a straight prickle on upper side; pinnae 3–9 pairs, 4–15 cm. long; leaflets 6–9 pairs per pinna, ovate-oblong or elliptic-oblong, (0·8–)1·3–4·5(–5) cm. long, (0·5–)0·8–2·2(–2·5) cm. wide, obtuse, sometimes subacute, not or scarcely acuminate, pubescent to glabrous except for midrib and margins. Racemes axillary, pedunculate, simple or with 1–2 branches below. Bracts much longer than buds, linear-lanceolate, recurving, 0·8–1·6 cm. long. Pedicels 4–6 mm. long. Sepals 6·5–8 mm. long, 2–3 mm. wide. Petals yellow or greenish-yellow, 12–13 mm. long, 3 (or upper to 4) mm. wide. Anthers 1–1·25 mm. long. Pod ± oblong-elliptic, 4·5–7·6 cm. long, 3·5–4·5 cm. wide, densely spreading-prickly outside, dehiscent, with coriaceous valves. Seeds globose to subglobose, 1·5–2 cm. in diameter, hard, leaden-grey; cuticle of testa regularly transversely and finely cracked.

KENYA. Kilifi, *R. M. Graham* R 144 in *F.H.* 1562!; Lamu District: Witu, *F. Thomas* 199! & Lamu, Jan. 1943, *Bally* 2192 in *C.M.* 11617!

TANGANYIKA. Tanga District: Pangani, Mkwaja, 8 July 1951, *van Rensburg* 528! & Sawa, Kigiweni beach, 6 May 1956, *Faulkner* 1864!; Rufiji District: Mafia I., 3 Apr. 1933, *Wallace* 744!

ZANZIBAR. Zanzibar I., Mbweni, Feb. 1929, *Greenway* 1322! & Kiwengwa, 9 June ? 1952, *Oxtoby*!

DISTR. **K**7; **T**3, 6, 8; **Z**; widespread in the tropics of the Old and New Worlds

HAB. On and near sea-shores; 0–15 m.

SYN. *Guilandina bonduc* L., Sp. Pl.: 381 (1753)
Guilandina bonducella L., Sp. Pl., ed. 2: 545 (1762). Type: as for *C. bonduc*
Caesalpinia bonducella (L.) Fleming in As. Research. 11: 159 (1810); Oliv., F.T.A. 2: 262 (1871); P.O.A. C: 202 (1895)
[*C. crista* sensu L.T.A.: 614 (1930); T.S.K.: 60 (1936); U.O.P.Z.: 159 (1949), *non* L.]

NOTE. The very hard-shelled seeds will float well in the sea and have been shown to keep their power of germination unimpaired after as much as 2½ years afloat. In this way they may be carried very far by the currents of the sea, and have even been washed up, still alive, on British shores. The habitat of the plant in East Africa is therefore exactly where it might be expected. For further information, see Ridley, Dispersal of Plants Throughout the World: 282–3 (1930) (as *Guilandina bonducella* L.) and Guppy, Observations of a Naturalist in the Pacific 2: 183–97 (1906) (as *Caesalpinia bonducella* (L.) Fleming).

There is a specimen of *C. bonduc* at Kew, labelled: " Baringo, March/01 [1901]. Alt. 3400 ft. Comm. Sir H. H. Johnston 1901." This is the only evidence for the species occurring naturally inland in Kenya and at so high an altitude. However, I suspect mislabelling and have therefore neglected this specimen in giving the altitude and distribution of *C. bonduc*.

6. MEZONEURON

Desf. in Mém. Mus. Paris 4: 245, t. 10, 11 (1818) (as " *Mezonevron* ")

Climbing shrubs, usually armed with prickles on stem and leaves, rarely (not in East Africa) unarmed. Leaves bipinnate; stipules very small, soon falling; specialized glands restricted to petiole and rhachis absent; leaflets opposite or (but not in East Africa) alternate. Inflorescences of terminal and axillary, often paniculately aggregated racemes; bracts small or very small. Flowers ⚥. Sepals 5, imbricate, zygomorphic, the lower one cucullately embracing the others; all arising from a zygomorphic hypanthium. Petals 5, subequal or the upper one somewhat modified. Stamens 10, all fertile; filaments alternately longer and shorter, all pubescent or villous below, or (but not in East Africa) glabrous; anthers dorsifixed, dehiscing by longitudinal slits. Ovary free, sessile or shortly stipitate; style gradually enlarged near apex; stigma small, oblique, not peltate, ciliolate or glabrous. Pods flat, indehiscent, subcoriaceous, venose, longitudinally and ± broadly winged along the upper suture (wing 3–18 mm. wide). Seeds 1–9 per pod, ± transverse, compressed, without endosperm; funicle slender.

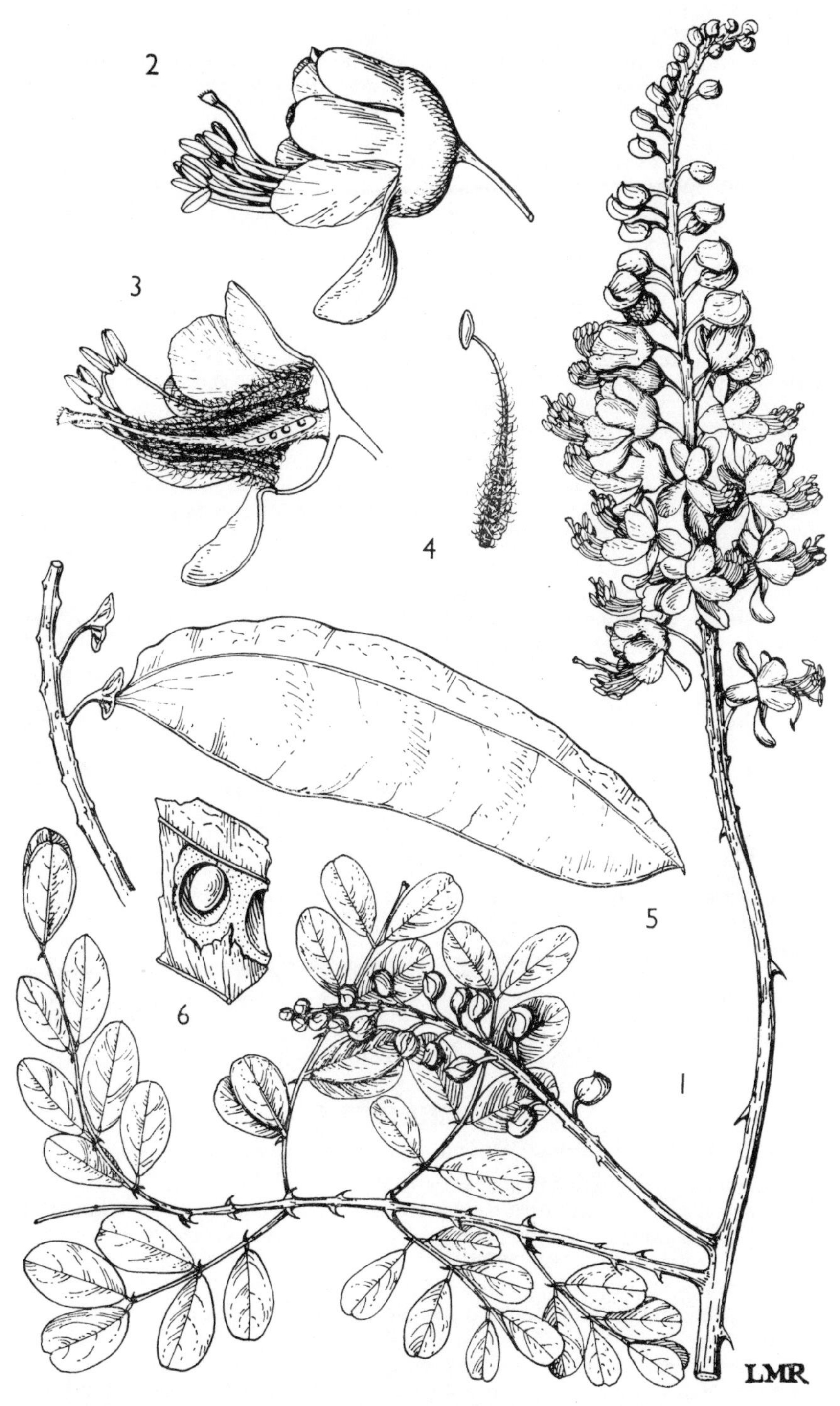

FIG. 6. ***MEZONEURON ANGOLENSE***—**1**, part of flowering branch, × $\frac{2}{3}$; **2**, flower, × 2; **3**, flower, longitudinal section, × 2; **4**, stamen, × 2; **5**, part of infructescence, with pod, × $\frac{2}{3}$; **6**, part of valve of opened pod, to show seed, × $\frac{2}{3}$. 1, 5–6, from *Purseglove* 911; 2–4, from *Purseglove* 2644.

About 20 species, mostly in tropical Asia and extending to Australia and Polynesia, 2 of them in Madagascar and 2 in tropical Africa.

NOTE. *M. cucullatum* (Roxb.) Wight & Arn., a tropical Asiatic species with rather large ovate to elliptic acuminate or subacuminate leaflets and one-seeded pods has been in cultivation in Tanganyika (*Grote* in *Herb. Amani* 6962 !).

M. angolense *Oliv.*, F.T.A. 2: 261 (1871); L.T.A.: 613 (1930); T.T.C.L.: 105 (1949); Wilczek in F.C.B. 3: 258 (1952); Torre & Hillcoat in C.F.A. 2: 170 (1956); F.F.N.R.: 125 (1962). Type: Angola, Pungo Andongo, *Welwitsch* 606 (LISU, lecto., BM, isolecto. !) & Golungo Alto, between Sange and Camilungo, *Welwitsch* 607 (LISU, syn., BM, K, isosyn. !)

Liane to 12 m. or more (to 20 m. *fide* F.C.B.). Stems armed with scattered downwardly hooked or deflexed prickles, which on the old stems become enlarged and raised on subconical-cylindrical corky bosses. Leaves: petiole with rhachis (6–)10–28 cm. long, the latter armed with hooked paired prickles; pinnae 4–10 pairs; leaflets 4–9 pairs, oblong-elliptic or the upper obovate-elliptic, 10–26(–35, *fide* F.C.B.) mm. long, 6–15(–19) mm. wide, rounded to emarginate at apex, glabrous except for some pubescence or puberulence especially on lower part of midrib, sometimes puberulous over the surface. Racemes 5–40 cm. long, simple or branched. Pedicels 4–11 mm. long. Sepals 6–9 mm. long. Petals yellow, ± 6·5–9 mm. long, 5·5–7·5 mm. wide, with a short claw and suborbicular lamina, pubescent near claw. Stamens declinate; filaments ± 1·3–1·5 cm. long, villous-pubescent below. Ovary densely tomentellous, or sometimes subglabrous in Tanganyika. Style ciliolate. Pods (7–)8·5–14·5 cm. long, (2·4–)2·8–4·3 cm. wide, 2–4-seeded. Seeds elliptic, smooth, ± 8–9 × 6–7 mm. Fig. 6, p. 39.

UGANDA. W. Nile District: near Niapea, banks of Nyagak R., May 1936, *Eggeling* 3012 !; Kigezi District: Kayonza, Apr. 1948, *Purseglove* 2644 !; Entebbe, 8 Aug. 1905, *Bagshawe* 720 !

TANGANYIKA. Lushoto District: E. Usambara Mts., Amani–Bomole, 21 Jan. 1915, *Peter* K129 !; Uzaramo District: Dar es Salaam, Mlalakwa [Mulalakuwa] R., 2 July 1939, *Vaughan* 2830 ! & Kimboza Forest Reserve, July 1952, *Semsei* 791 !

ZANZIBAR. Zanzibar I., Chukwani, 9 Sept. 1960, *Faulkner* 2706 !

DISTR. **U**1, 2, 4; **T**3, 4, 6; **Z**; Liberia, Cameroun Republic (*fide* F.C.B.), Congo Republic, Mozambique and Angola

HAB. Lowland rain-forest, riverine and swamp forest, coastal evergreen bushland; *Purseglove* 2644 from " savannah ", presumably as a relic of former forest; near sea-level to 1680 m.

NOTE. When in flower, and without the characteristic fruits, *M. angolense* is deceptively like a *Caesalpinia*, and confusion is likely. It may be separated from the native species of *Caesalpinia* by the following combination of characters: scattered deflexed prickles on stem; leaflets rounded or emarginate at apex; very small bracts; pedicels 0·4–1·1 cm. long; yellow petals; ovary usually tomentellous.

7. PTEROLOBIUM

Wight & Arn., Prodr. Fl. Ind. Or. 1: 283 (1834), *nom. conserv.*

Shrubs, normally climbing, armed with prickles on stem and leaves. Leaves bipinnate; stipules small, inconspicuous, soon falling, subulate or triangular-subulate; specialized glands restricted to petiole and rhachis absent; leaflets opposite. Inflorescences of terminal and axillary often paniculately aggregated racemes; bracts small, soon falling. Flowers ⚥. Sepals 5, imbricate, unequal, the lower one cucullately embracing the others; hypanthium cupular, regular. Petals 5, equal or almost so. Stamens 10, all fertile; filaments alternately rather longer and shorter, all pubescent below; anthers dorsifixed, dehiscing by longitudinal slits. Ovary free, very shortly stipitate; ovule 1, attached near top of ovary; style gradually enlarged near apex; stigma transverse, not peltate. Pods resembling the samara of a sycamore

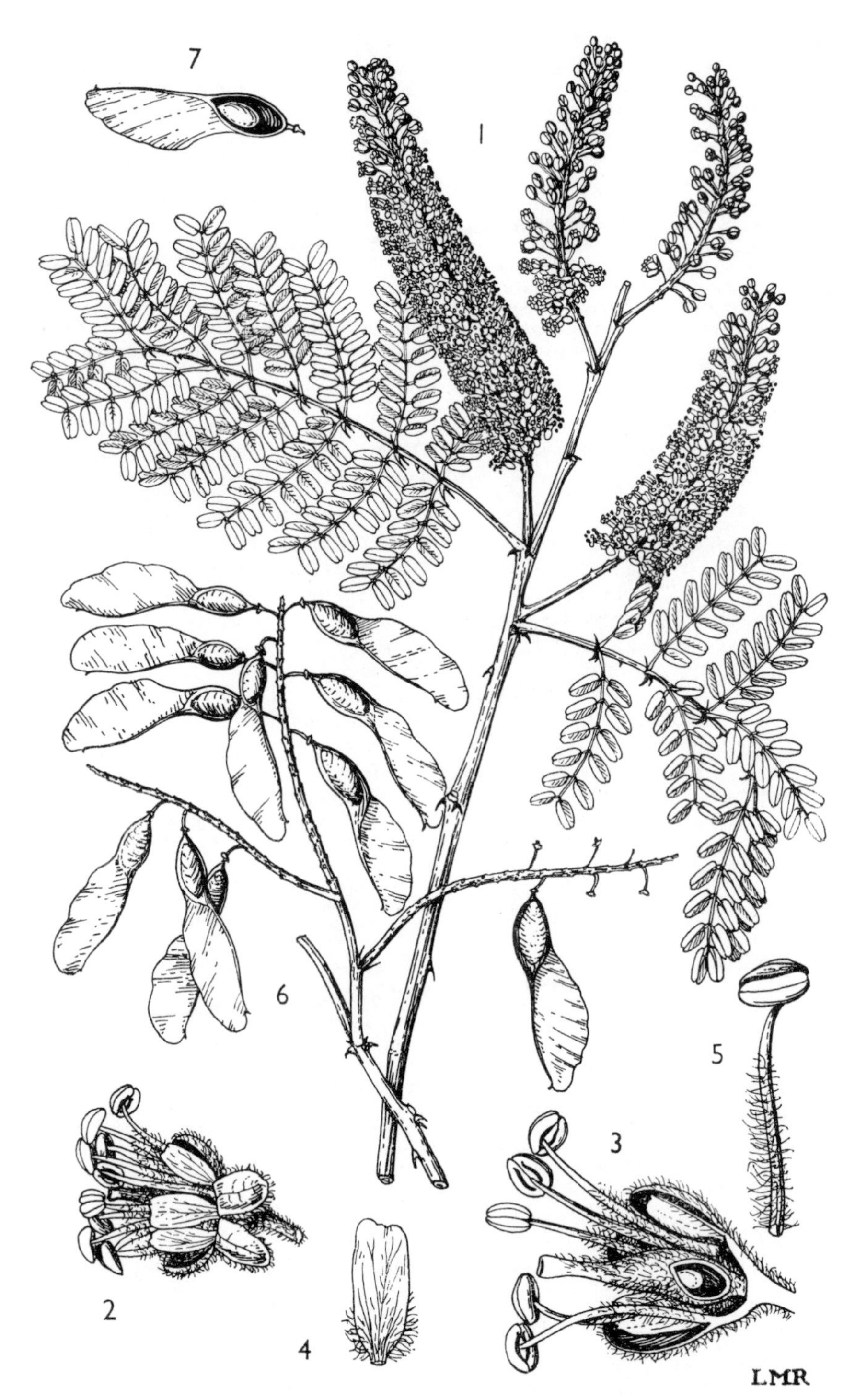

FIG. 7. *PTEROLOBIUM STELLATUM*—**1,** part of flowering branch, $\times \frac{2}{3}$; **2,** flower, $\times$ 4; **3,** flower, longitudinal section, $\times$ 6; **4,** petal, $\times$ 9; **5,** stamen, $\times$ 9; **6,** part of branch with mature pods, $\times \frac{2}{3}$; **7,** pod, with seed-bearing part opened, $\times \frac{2}{3}$. 1–5, from *Richards* 11275; 6, from *Eggeling* 3400; 7, from *Sandwith* 25.

(*Acer*), with a shortly stipitate basal seed-containing portion (with a single seed) whose upper suture is much prolonged beyond the seed-containing part of the pod and is broadly winged on its lower side, the wing usually becoming wider distally. Seed pendulous, ± compressed, without endosperm.

About 10 species, mostly in Asia, extending eastwards to the Philippines; only one species in Africa.

The attribution of the name *Pterolobium* is not beyond doubt. When originally published by Robert Brown in Salt, Abyss., app. lxiv (1814), there was no description, but " *Kantuffa* " Bruce was cited in synonymy. It is clear from the generic name that Brown had seen and appreciated the taxonomic importance of the pods—without pods the plant is not clearly separable from *Caesalpinia*. Bruce however did not describe and in fact never saw the pods of " *Kantuffa* ". Brown must therefore have based *Pterolobium* on other fruiting specimens, probably collected by Salt, which however were neither cited nor described. It seems preferable therefore to accept Wight & Arnott's description of *Pterolobium* as the first valid one.

P. stellatum (*Forsk.*) *Brenan* in Mem. N.Y. Bot. Gard. 8: 425 (1954); Roti-Michelozzi in Webbia 13: 181 (1957); K.T.S.: 94 (1961); F.F.N.R.: 128, fig. 20/J (1962). Type: Yemen, Kurma, *Forsskål* (C, lecto. !, K, photo. !)

Scrambling or climbing, rarely semi-erect shrub 2–15 m. high. Stems ± puberulous, at least when young, armed with reflexed prickles paired at nodes and often also with scattered ones between the nodes. Leaves: petiole with rhachis 6–16(–25) cm. long, the latter armed with mostly paired reflexed prickles, straight ascending ones also being often present on the upper side of the rhachis at the insertions of pinnae; pinnae 5–13 pairs; leaflets 7–15(–16, *fide* F.C.B.) pairs, narrowly oblong or elliptic-oblong, with the terminal ones ± obovate, 4–12 mm. long, 2–4·5(–5) mm. wide, rounded to emarginate at apex, ± puberulous or pubescent to almost glabrous. Racemes 5–13(–18) cm. long, puberulous or shortly pubescent, aggregated into panicles up to ± 35 cm. long and 22 cm. wide. Pedicels 3–5(–6) mm. long. Flowers sweetly scented, pale creamy-yellow except for the pale green calyx. Sepals ± 2–3 mm. long. Petals oblanceolate-oblong, ± 3 mm. long and 1·5 mm. wide, pubescent towards base. Stamen-filaments 4–5 mm. long. Ovary pubescent. Pods brick-red to scarlet, ultimately brown, 3–6 cm. long, thinly pubescent or puberulous to glabrous; wing 2·3–4·5 cm. long, 0·9–1·6 cm. wide at widest point. Seeds ovoid-ellipsoid, ± 11 mm. long, 6·5 mm. wide, olive. Fig. 7, p. 41.

UGANDA. W. Nile District: Aringa County, Midigo, Sept. 1937, *Eggeling* 3400 !; Karamoja District: Moroto, Oct. 1956, *Wilson* 288 !; Mengo District: near Najembe, Mabira Forest, 6 Mar. 1950, *Dawkins* 535 !
KENYA. Elgon, Oct.–Nov. 1930, *Lugard* 237 !; Baringo District: Ol Arabel ravine, July 1939, *Lewis* 222 !; 32 km. E. of Nairobi, 28 Aug. 1951, *Bogdan* 3213 !
TANGANYIKA. Mwanza District: Igalukiro, 22 Aug. 1951, *Tanner* 303 !; Arusha District: Engare Olmotoni, 8 Mar. 1954, *Matalu* in *F.H.* 3009 !; Lushoto District: between Soni and Mombo, 14 May 1956, *Willan* 290 !
DISTR. **U**1–4; **K**1–6; **T**1–7; widespread in eastern Africa, from Eritrea and the Sudan Republic southwards to the Transvaal; also in Arabia
HAB. Upland dry evergreen forest (? also on edges of upland rain-forest), riverine forest, deciduous woodland, sometimes on termite-mounds in grassland; 850–2290 m.

SYN. *Mimosa stellata* Forsk., Fl. Aegypt.-Arab.: cxxiii, 177 (1775); Vahl, Symb. Bot. 1: 81 (1790), *non M. stellata* Lour. (1790)
Cantuffa exosa J. F. Gmel., Syst. Nat. 2: 677 (1791). Type: Bruce, Travels 5: app. 49 (1790)
Pterolobium lacerans R. Br. in Salt, Abyss., app.: lxiv (1814), nomen ipse nudum sed cum syn. " *Kantuffa* " Bruce; Oliv., F.T.A. 2: 264 (1871). Type: Ethiopia, *Salt* (BM, syn. !)
P. exosum (J. F. Gmel.) Bak. f., L.T.A.: 621 (1930); T.T.C.L.: 106 (1949); Wilczek in F.C.B. 3: 256 (1952)

NOTE. This viciously armed plant varies little, except slightly in its indumentum and rather more in the size and shape of its ornamental pods.

8. PARKINSONIA

L., Sp. Pl.: 375 (1753) & Gen. Pl., ed. 5: 177 (1754); Brenan in K.B. 17: 203 (1963)

Peltophoropsis Chiov. in Ann. di Bot. 13: 385 (1915)

Shrubs or trees, not climbing, armed with spines or unarmed, eglandular. Leaves bipinnate; stipules various, minute and scale-like to conspicuous and spinescent; leaflets opposite or (*P. aculeata*) partly alternate, sometimes much reduced or absent. Inflorescences of axillary racemes which are sometimes corymbose and short; bracts minute and scale-like, falling quickly. Flowers ⚥. Sepals 5, valvate or subvalvate to very narrowly imbricate, or more widely so. Petals 5, subequal except for the usually somewhat modified upper one, which usually has a more pronounced claw than the others. Stamens 10; filaments alternately longer and shorter, all pubescent below; anthers dorsifixed, opening by longitudinal slits. Ovary free, stipitate, with 2–6 ovules, glabrous to (more usually) ± pubescent; style glabrous or clothed below like the ovary, often ± spirally twisted; stigma truncate, ciliolate or glabrous. Pods flat or turgid, ± constricted or not between the seeds, not winged, indehiscent, with usually papery or thinly coriaceous brown valves. Seeds usually ± oblique or longitudinal, hard; funicle usually rather long and slender; endosperm present.

About 14 species, mostly in the drier parts of North and South America, but one in South and South West Africa, and two species in East and North-east Africa.

The native species of *Parkinsonia* in East Africa are distinguished from the native species of *Caesalpinia* by the following combination of features: plant eglandular, unarmed or only with spinescent stipules; flowers yellow (probably but not certainly in species No. 2); pods indehiscent, with thin brown closely venose valves.

An exotic species, *P. aculeata* L., native of tropical and subtropical America, is cultivated in Uganda, Kenya and Tanganyika, and may become naturalized, although there is no evidence of it happening so far. It is a small tree up to 6 m. high, with smooth green bark; branchlets armed with spines (= modified leaf-rhachides or stipules); bipinnate leaves with 1–3 pairs of pinnae which are composed of elongate flattened green rhachides up to more than 39 cm. long bearing very small obovate-elliptic to obovate-oblong or oblong leaflets along the margins; yellow flowers in ± elongate racemes; pods usually ± elongate, pointed or beaked at apex, constricted into oblong or elliptic segments.

Plant armed with paired stipular spines; upper petal rhombic, ± 8 × 4·5 mm., not or scarcely clawed; other petals lanceolate, 6–6·5 × 1·5–2 mm.; pods 1·1–2·5 cm. wide . . . . 1. *P. scioana*
Plant unarmed; upper petal 11–12 × 6–6·5 mm., distinctly clawed; other petals elliptic or rhombic-elliptic, 10–12 × 4–6 mm.; pods 0·9–1·6 cm. wide . 2. *P. anacantha*

1. **P. scioana** (*Chiov.*) *Brenan* in K.B. 17: 209 (1963). Type: Ethiopia, Shoa, between Tadeccia Melca and Cioba, *Negri* 1358 (FI, holo.)

Shrub or small tree 0·6–5 m. high, branching from or near base. Young branchlets shortly pubescent, going grey or brown. Stipular spines paired at nodes, 2–5 mm. long, straight or hooked. Leaves up to 8 cm. long; pinnae 2–8 pairs, rather short; leaflets 3–6 pairs, mostly elliptic, 2–7 mm. long, 1·25–3·5 mm. wide, mostly rounded at apex, puberulous. Racemes 1–4(–10, *fide* Chiovenda) cm. long. Sepals 4–5 mm. long. Petals yellow, upper one rhombic, 8 mm. long, 4·5 mm. wide, not or scarcely clawed; the other petals lanceolate, 6–6·5 mm. long, 1·5–2 mm. wide, tapering towards apex, ± erose upwards. Pods narrowly elliptic to oblanceolate, rarely linear-oblong, 4·5–9·5 cm. long, 1·1–2·5 cm. wide, 1–4-seeded, acute or sometimes rounded and apiculate at apex, attenuate-stipitate at base; valves thin, stiffly papery,

FIG. 8. *PARKINSONIA SCIOANA*—**1,** part of leafy branch, × $\frac{2}{3}$; **2,** leaflet, lower surface, × 4; **3,** part of flowering branch, × $\frac{2}{3}$; **4,** bud, × 6; **5,** bud, cross-section, to show aestivation of sepals, × 6; **6,** flower, longitudinal section, × 4; **7,** upper petal, × 4; **8,** lateral petal, × 4; **9,** stamen, × 4; **10,** part of branch with mature pods, × $\frac{2}{3}$; **11,** seed, × 2. 1, 2, from *Gillett* 13463; 3–9, from *Ellis* 144; 10, 11, from *Hemming* 491.

closely and longitudinally or obliquely veined; upper suture ± thickened, and flattened transversely to the flat plane of the pod or somewhat channelled. Seeds somewhat compressed, obovoid-ellipsoid, 6–7 mm. long, 5–6 mm. wide, subtruncate at apex, mottled with olive and dark brown, smooth but not glossy. Fig. 8.

KENYA. Northern Frontier Province: Mandera, 24 May 1952, *Gillett* 13308! & Dandu, 21 June 1952, *Gillett* 13463! & 36 km. NE. of Wajir, 18 June 1955, *Hemming* 491!
DISTR. **K**1; Ethiopia and Somali Republic
HAB. Dry scrub with trees; 300–760 m.

SYN. *Peltophoropsis scioana* Chiov. in Ann. di Bot. 13: 385 (1915); Roti-Michelozzi in Webbia 13: 221, fig. 9 (1957); K.T.S.: 105 (1961)
Peltophorum scioanum (Chiov.) Bak. f., L.T.A.: 612 (1930)
Caesalpinia gillettii Hutch. & E. A. Bruce in K.B. 1941: 114 (1941). Type: Somali Republic (N.)/Ethiopia, " Boundary Pillar 93 " [45° 9′ E., 8° 37′ N.], *Gillett* 4157 (K, holo.!)

2. **P. anacantha** *Brenan* in K.B. 17: 209 (1963). Type: Kenya, Meru District, between Mt. Kinna and Garba Tula, *T. Adamson* in *E.A.H.* 12684 (K, holo.!)

Shrub up to 2·1 m. high or spreading tree, unarmed. Young branchlets shortly appressed-pubescent or puberulous, later glabrescent and turning grey-brown. Leaves with petiole 1·5–4·5 cm. long and rhachis 1·5–11·8 cm. long; pinnae 2–6 pairs; leaflets 4–9 pairs, oblong-elliptic, 3–10 mm. long, 1·5–5 mm. wide, rounded or slightly emarginate at apex, appressed-puberulous especially beneath. Racemes 2·5–10 cm. long. Sepals ± 7 mm. long. Petals yellow, the upper one 11–12 mm. long and 6–6·5 mm. wide, shortly (3 mm.) unguiculate at base; the other petals elliptic or rhombic-elliptic, 10–12 mm. long, 4–6 mm. wide, somewhat narrowed towards apex. Pods narrowly elliptic to linear-oblong, 6·5–11·5 cm. long, 0·9–1·6 cm. wide, 1–4-seeded, obtuse to subacute at apex, attenuate at base; valves thin, papery, closely finely and almost longitudinally veined; upper suture somewhat thickened and longitudinally channelled. Seeds somewhat compressed, ellipsoid, ± 9 mm. long and 5–6 mm. wide, brown, smooth.

KENYA. Northern Frontier Province: Lorian, end of Golana Gof, 19 Jan. 1943, *Bally* 2107! & 22 km. N. of Isiolo, 14 Oct. 1963, *Verdcourt* 3798!; Tana River District: Garissa road near Tula [Tola], 7 Dec. 1947, *L. C. Edwards* 168!
DISTR. **K**1, 4, 7; not known elsewhere
HAB. Semi-desert scrub with *Acacia mellifera*; 300–910 m.

NOTE. Easily separated from *P. scioana* by being unarmed and having larger petals.

9. STUHLMANNIA

Taub. in P.O.A. C: 201 (1895)

Unarmed tree. Leaves simply paripinnate; stipules apparently represented only by minute conical projections; conspicuous glands absent from petiole and rhachis; leaflets opposite to subopposite, eglandular. Inflorescences of terminal (sometimes falsely lateral) racemes; bracts small, quickly falling. Flowers ⚥. Calyx-segments 5, valvate in bud, quickly falling. Petals 5, subequal. Stamens 10; alternate filaments longer and shorter, all hispidulous below with bulbous-based spreading or deflexed hairs; anthers small, dorsifixed, dehiscing by longitudinal slits. Hypanthium cup-shaped. Ovary stipitate, with 2 ovules, with sessile glands, otherwise glabrous except for some puberulence on the stipe; style glandular, transversely truncate and not enlarged at apex, ciliate round the very small stigma. Pods flattened, not winged, elastically dehiscent into 2 valves, 1-seeded; valves hard and almost woody, smooth. Seeds transverse, on a short funicle, compressed; testa thin, crustaceous; endosperm absent.

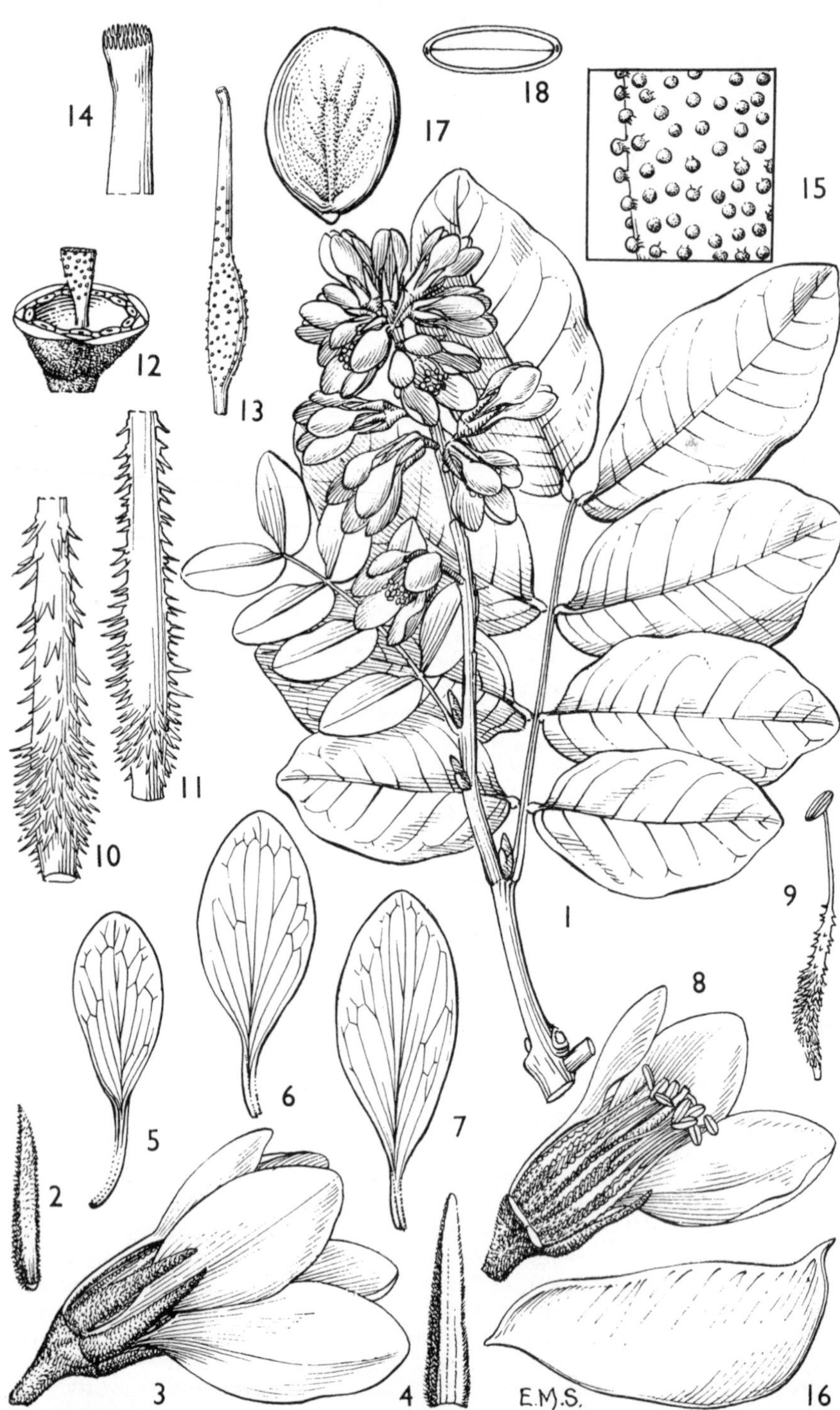

FIG. 9. *STUHLMANNIA MOAVI*—**1,** part of flowering branch, × $\frac{2}{3}$; **2,** flower-bract, × 6; **3,** flower, × 2; **4,** sepal, × 3; **5,** upper petal, × 2; **6,** lateral petal, × 2; **7,** lower petal, × 2; **8,** flower, with sepals and petals removed from one side to show arrangement of stamens, × 2; **9,** stamen, × 3; **10,** lower portion of stamen-filament, seen from inside the flower, × 6; **11,** same, seen from outside the flower, × 6; **12,** hypanthium after fall of calyx-segments, petals and stamens, × 3; **13,** gynoecium, × 3; **14,** stigma and apical part of style, × 12; **15,** part of outer surface of ovary, showing sessile glands, × 30; **16,** pod, × 1; **17,** seed, × 2; **18,** seed, transverse section, × 2. 1, from *Tanner* 3167; 2, 16–18, from *Tanner* 3724; 3–15, from *Tanner* 2467. Reproduced by permission of the Bentham-Moxon Trustees.

S. moavi is the only known species of the genus.

The very closely related *Cordeauxia edulis* Hemsl., from Ethiopia and Somali Republic, with conspicuous red glands on the foliage and stems, has been cultivated in Tanganyika.

S. moavi *Taub.* in P.O.A. C: 201 (1895); L.T.A.: 620 (1930); T.T.C.L.: 106 (1949). Types: Tanganyika, Tanga District, Pangani, *Stuhlmann* 467 & 616 (B, syn. †, BM, drawing of 616 !)

Tree 7·5–9 m. high, apparently evergreen. Bark rough or smooth. Young branchlets tomentellous to puberulous with minute white to brown hairs and with numerous other minute blackish-red opaque worm-like quite possibly glandular hairs intermixed. Leaves: petiole (0·6–)1–1·7 cm. long; rhachis (3·5–)6–12 cm. long, at first with a sparse indumentum like that of the branchlets, glabrescent; leaflets in 3–5(–6, *fide* Taubert) pairs, ovate-elliptic to elliptic, usually with a subrhombic tendency, 2·5–9(–12·5) cm. long, 1·3–5(–6·5) cm. wide, obtuse to rounded at apex, asymmetric at base, glabrous (unless on midrib when young), with nerves and venation prominent on both surfaces. Racemes ± 9–13 cm. long (to 20 cm. in fruit); axis, pedicels (± 0·6–1 cm. long, and to 1·3 cm. in fruit) and outside of calyx with a dense tomentellous brownish indumentum like that of the branchlets; calyx with some scattered round red sessile glands in addition. Petals yellow, unguiculate, obovate-spathulate, 1·6–2 cm. long, 6–9 mm. wide, obtuse or rounded at apex. Pods obliquely oblanceolate, ± 5 cm. long and 2 cm. wide, olive-brown, glabrous and rather glossy. Seeds broadly obovate-elliptic, ± 14 mm. long and 10 mm. wide, brown, glossy. Fig. 9.

TANGANYIKA. Pangani District: Mkwaja, Mkaramo Wachenya, 23 Nov. 1955, *Tanner* 2404 ! & Kumbamtoni, 25 Oct. 1955, *Tanner* 2467 ! & 12 Oct. 1956, *Tanner* 3167 ! & Madanga, Jasini, 10 Oct. 1957, *Tanner* 3724 !
DISTR. **T3**; not known elsewhere
HAB. Lowland dry evergreen forest and riverine forest; 15–150 m.

10. CASSIA

L., Sp. Pl.: 376 (1753) & Gen. Pl., ed. 5: 178 (1754)

Annual or perennial herbs, shrubs or trees, rarely scrambling or climbing, unarmed. Leaves simply paripinnate or rarely (not in East Africa) modified to simple phyllodes; stipules various; conspicuous glands often present on petiole and/or rhachis; leaflets 1 to many pairs. Inflorescences of racemes varying from elongate and many-flowered to short subumbellate and 1- to few-flowered; sometimes racemes sessile or almost so, so that the flowers appear axillary or lateral and solitary or fascicled; sometimes racemes aggregated into terminal panicles; bracts and bracteoles various. Flowers normally ⚥, very rarely ⚥ ♀, or ⚥ ♂. Sepals 5, imbricate. Petals 5, imbricate, the upper ones often somewhat smaller, yellow, or less commonly pink, red or white. Stamens usually 10, subequal, or of various sizes with the 2–3 lower longer; sometimes the 1–3 upper reduced and staminodial; rarely stamens only 4–5; anthers basifixed, or sometimes dorsifixed, dehiscing by terminal or basal pores or short slits. Ovary sessile or stipitate, with several to many ovules. Pod very variable, cylindrical to flat, rarely winged, indehiscent or dehiscent and 2-valved, woody to coriaceous or membranaceous, with or without septa between the seeds, sometimes pulpy, rarely longitudinally septate. Seeds transverse, sometimes oblique or longitudinal, usually ± horizontally or vertically compressed, smooth or punctate, sometimes areolate; endosperm present.

About 500–600 species, pantropical but most numerous in America.

NOTE. Unlike those preceding them, the species numbered here 27–48, all belonging to the Section *Chamaecrista* (Moench) DC., are critical and often closely related and taxonomically extremely difficult. Many of them are imperfectly known and often with very little material available for study. There can also be little doubt that in East Africa there are additional species not yet known from there. The keys and descriptions therefore must often be read with caution, and I hope also with charity towards their imperfections.

Those who have studied carefully these critical Cassias in the living condition agree about their distinctness, but not infrequently it happens that the species appear easier to separate when alive than in the herbarium. Much help could be given, however, by careful observation in the field and corresponding notes with collected specimens.

The following seem some of the most significant features for separating the species:

1. *The duration of the plant.* Some are decidedly perennial with a thickened rootstock from which herbaceous stems arise annually. Others are certainly annual, dying away at the end of the growing season. However, some of the latter may develop a certain woodiness near the base of the stem, and may evidently at times survive for more than a season by this means, although their habit seems essentially different from that of the truly perennial species with a woody rootstock.

2. *Habit.* The direction of the stems is important. Some are prostrate; others erect. The amount of branching is at times significant, although in general it appears rather variable.

3. *The petiolar gland.* This is of great importance in separating these closely related species in the herbarium, and the little evidence available suggests that it may be usefully employed in the field also. This gland, to see which clearly a hand-lens is usually desirable, may be sessile or shortly stalked; it may be very small to comparatively large, round or elongate, and cup-shaped to cushion-shaped, see fig. 10, p. 57. The colour may also be significant, but further information about this would be welcome.

4. *The leaf-rhachis.* In most of the species there is a channel running along the upper side of the leaf-rhachis whose sides may be somewhat raised or even very narrowly wing-like, but whose margins between the points of insertion of the individual leaflets have, when viewed in outline from the side, a generally flat and even line. Some species, however, including the very common and variable *C. mimosoïdes*, have a raised wing-like projection running along the upper side of the rhachis, whose outline seen from sideways is more or less distinctly crenate or dentate, the sinuses between the crenations or teeth corresponding to the points of insertion of the leaflets.

5. *Leaflets.* The size and number of the leaflets is also significant in separating the species. The position of the midrib, whether placed centrally in the leaflets or somewhat displaced to one side, or running along one margin, will be found significant and I believe constant.

6. *Flowers.* The flowers are of less help than might be expected, though the size of the various parts, particularly the petals, is certainly significant. *C. hochstetteri* is striking in having only half the normal number of stamens.

Several species are considered here as polymorphic, sometimes with subsidiary variants or groups distinguished, some of which have distinctive geographical ranges. It is hard to decide whether these are different facets of one species or distinct species linked by hybrid intermediates. Observation in the field may well yield valuable evidence about this problem, which is also certainly well worth investigating by experimental cultivation.

KEY TO EXOTIC SPECIES

Numerous species of *Cassia*, including some native to East Africa, are in cultivation. The native species and those exotics that are sufficiently established are dealt with in the main text of the flora and in the accompanying key. Of the species treated in this way *C. alata*, *C. bicapsularis*, *C. floribunda* and *C. fruticosa* are the most frequently cultivated. However, the following additional species are cultivated in the area or have been reported to be so:

Petioles and rhachides of leaves eglandular:
Petals pink or red (at least at first, sometimes fading to buff, orange or white):

Bracts of inflorescence falling before the flowers open; racemes 10–25 cm. long . . . *C. grandis* L.f. (Native of tropical America)

Bracts of inflorescence persisting while the flowers are open; racemes 3–16 cm. long:

Ovary glabrous; leaflets oblong . . . *C. renigera* Benth.* (Native of Burma)

Ovary pubescent; leaflets various:

Petals up to 1·5 cm. long; anticous stamen-filaments slightly thickened in middle . . . . *C. roxburghii* DC. (*C. marginata* Roxb., *non* Willd.)* (Native of India and Ceylon)

Petals (1·5–)2–3·5 cm. long; anticous stamen-filaments markedly thickened in middle:

Racemes aggregated into corymbose panicles at ends of leafy branches *C. agnes* (De Wit) Brenan (Native of Thailand and Indo-China)

Racemes mostly borne singly (occasionally once or twice branched) at ends of short normally leafless shoots arising from older twigs:

Stipules ± 5 mm. long, narrow and not leafy; leaflets usually acute . . . . *C. nodosa* Roxb.* (Native of tropical Asia)

Stipules ± 12–25 mm. long, leafy; leaflets usually rounded at apex . . . . . *C. javanica* L.* (Native of tropical Asia)

Petals yellow to golden:

Inflorescences usually pendulous racemes borne laterally and either singly or fascicled:

Leaflets with spreading pubescence beneath, especially on midrib . *C. leiandra* Benth.* (Native of South America)

Leaflets with minute appressed puberulence beneath, even on the midrib:

Petals up to ± 1·3 cm. long; anticous stamen-filaments abruptly and markedly thickened in middle; racemes up to ± 30 cm. long . *C. brewsteri* F. v. Muell. (Native of Australia)

Petals 1·6–3 cm. long; anticous stamen-filaments gradually and slightly

* Material of these species not seen. Records from Dale, Introd. Trees Uganda: 18–19 (1953). *C. nodosa* and *C. javanica* require confirmation. Herbarium-material under the former name proves to be *C. agnes*.

thickened in middle; racemes ± 15–60 cm. long . . . *C. fistula* L. (see p. 64)

Inflorescences stiff terminal panicles of ± elongate or corymbose racemes:

Leaflets normally acute, usually ± pubescent beneath; pods almost terete . . . *C. spectabilis* DC. (Native of tropical America)

Leaflets rounded or obtuse and often emarginate at apex, minutely appressed-puberulous beneath; pods flattened:

Stipules subulate, minute . . . *C. siamea* Lam. (Native of tropical Asia)

Stipules broad, falcate . . . *C. timoriensis* DC. (Native of tropical Asia)

Petioles and rhachides of leaves with prominent glands between at least some of the pairs of leaflets:

Leaflets linear or almost so, up to (with East African records) ± 2(–3) mm. wide, in 1–3 pairs . . . . *C. eremophila* Vogel (Native of Australia)

Leaflets lanceolate or oblong to ovate or obovate, 2·5 mm. or more wide:

Pairs of leaflets 2 per leaf; petals 3–4·5 cm. long . . . . *C. splendida* Vogel (Native of South America)

Pairs of leaflets 3–40 pairs; petals up to 3 cm. long:

Leaflets small, up to 9 mm. long; racemes 2-flowered . . . . . *C. polyphylla* Jacq. (Native of the West Indies)

Leaflets 1–6·5 cm. long; racemes 2–20-flowered:

The leaflets tomentose or densely pubescent beneath . . . *C. tomentosa* L. f. (Native of South America)

The leaflets glabrous, subglabrous or puberulous beneath:

Leaflets mostly in 12–40 pairs per leaf . . . . . *C. multijuga* Rich. (Native of South America)

Leaflets in 3–7 pairs per leaf:

Leaflets elliptic or ovate, glaucous beneath . . . . *C. surattensis* Burm. f. (*C. glauca* Lam.) (Native of tropical Asia)

Leaflets obovate, green beneath . *C. coluteoïdes* Collad.* (Native of tropical America)

* *C. bicapsularis* (see p. 71), closely related and also cultivated, differs in having **not** more than 3 pairs of leaflets and short pedicels to 7 mm., not 2 cm. or more.

KEY TO NATIVE AND NATURALIZED SPECIES

A. Petioles and rhachides of leaves eglandular* (occasionally in this group very minute clustered reddish bodies, quite possibly glands, may be present on the rhachis at the insertions of leaflet-pairs, but these are very different from the conspicuous single glands in the other group) [B on p. 53]:

Filaments of 2–3 anticous stamens with a double hairpin- or S-bend above base; pods indehiscent, cylindrical or slightly compressed, elongate, mostly 20–90 cm. long; seeds not markedly apiculate near hilum nor with areoles on the faces:
- Inflorescence-axes 0·5–12 cm. long, not obviously pendulous:
  - Ovaries glabrous; anticous stamen-filaments without obvious swellings; bracts mostly falling before, sometimes some persisting until the flowers open; large or medium rain-forest trees:
    - Leaflets conspicuously emarginate at apex, not at all acuminate, glabrous-margined even when young; petals yellow . . . 1. *C. angolensis*
    - Leaflets not or only slightly emarginate at apex, acute to obtuse or obtusely acuminate, pubescent-margined at least when young; petals white to pink . . . . 2. *C. mannii*
  - Ovaries puberulous to pubescent; anticous stamen-filaments each with a swelling about half-way along; bracts persistent while the flowers are open; shrubs or small trees of woodland and wooded grassland . . 3. *C. abbreviata***
- Inflorescence-axes 15–42 cm. long, pendulous or not:
  - Petals white; racemes pendulous; bracts falling while flower-buds are young; leaflets minutely appressed-puberulous beneath . 4. *C. burttii***
  - Petals yellow:
    - Racemes pendulous, mostly simple, single and lateral; bracts persisting till the flowers are open; leaflets densely appressed-puberulous beneath . . . . 5. *C. sieberana*
    - Racemes aggregated into a large, usually erect, pyramidal, terminal panicle:
      - Bracts persisting until flowers open; leaflets with crisped non-appressed pubescence beneath . . . . 6. *C. thyrsoïdea*
      - Bracts all falling while the flower-buds are young; leaflets (in East Africa) glabrous except at extreme base . . . 7. *C. afrofistula*

Filaments of all stamens straight or nearly so and subequal in length; pods dehiscent, flattened, or sometimes winged or ridged along each valve,

* Four species in groups A and B may flower without leaves being present: 1, *C. angolensis*, with yellow flowers, anticous stamen-filaments strongly arcuate, and glabrous ovaries; 3, *C. abbreviata*, similar but with puberulous to pubescent ovaries; 4, *C. burttii*, with white flowers; 19, *C. singueana*, with yellow flowers and the anticous stamen-filaments straight or slightly curved.

** A hybrid between *C. abbreviata* and *C. burttii*, intermediate in its inflorescences and petal-colour, has been found; see p. 60.

comparatively short, 2·5–19 cm. long; seeds various:

Leaflets in two pairs; stems with glandular-based setae in the indumentum; seeds without areoles, narrowed but not acuminate or apiculate near hilum . . . . 26. *C. absus**

Leaflets in 3–18 pairs; stems eglandular; seeds with a well-marked areole on each face and acuminate or apiculate near hilum (unknown in 14, *C. humifusa*):

Pods 12–19 cm. long with a wing along middle of each valve; bracts subtending flower-buds 2·5–3 cm. long; leaves mostly 30–75 cm. long; leaflets 5–19 cm. long and 2·5–12 cm. wide . . . . 8. *C. alata*

Pods 3–12 cm. long, unwinged, with at most a ridge of crests along middle of valves; bracts subtending flower-buds up to 2·7 cm. long but usually much smaller; leaves 3–40 cm. long; leaflets various:

Stipules broadly ovate-cordate, 8–12 mm. wide 11. *C. didymobotrya*

Stipules linear to ovate-triangular, at most 3 mm. wide:

Pods with a ridge of longitudinal crests along middle of each valve, curved; leaflets ± obovate; stems (in East Africa) with fine appressed puberulence; petals (in East Africa) 8·5–9 mm. long . . . 9. *C. italica*

Pods not longitudinally ridged along middle, usually straight but sometimes curved; leaflet-shape and indumentum of stems various; petals 7–28 mm. long:

Leaflets generally emarginate or retuse at apex, not mucronate, broadly elliptic to suborbicular, 1·8–6·3 cm. wide . 12. *C. ruspolii*

Leaflets rounded to acute or subacuminate at apex, not emarginate or retuse, often mucronate, variable in shape, usually 0·3–1·6 cm. wide, rarely wider:

Stems not prostrate; shrubs or small trees to 7 m. high:

Stipules 3–5 mm. long; pods 1·6–2·6 cm. wide; stems (in East Africa) with appressed puberulence only 10. *C. senna*

Stipules 5–11 mm. long; pods 1–1·8 cm. wide; stems with fine short spreading pubescence . . 13. *C. longiracemosa*

Stems prostrate; herb, sometimes slightly woody; stems pubescent;

* *C. absus* has an organ on the leaf-rhachis between the lower pair of leaflets. Recent experts on *Cassia* have sometimes considered this a gland (De Wit), sometimes as an "acicular appendage (? glandular)" (Steyaert). Steyaert placed *C. absus* among the species with eglandular petioles and rhachides. I agree with De Wit's view, but as there seems to be a genuine difference of opinion, and the organ is small and inconspicuous, I have accounted for *C. absus* in both parts of the key.

leaflets acute to subacuminate at apex . . . . 14. *C. humifusa*

B. Petioles and/or rhachides of leaves with conspicuous glands (sometimes only a single one present, at apex or base of petiole):

Sepals obtuse or rounded at apex; leaflets shortly petiolulate:
- Rhachis of leaves glandular (i.e. with projecting glands between one or more of the pairs of leaflets); petiole eglandular:
  - Indumentum of plant with numerous glandular-based spreading setae; herb; leaflets in 2 pairs per leaf . . . . 26. *C. absus*
  - Indumentum of plant without glandular-based setae; shrubs or trees (except 22, *C. obtusifolia*, which has 3 pairs of leaflets per leaf):
    - Leaflets gradually narrowing or tapering upwards to a usually acute or subacuminate apex:
      - Leaves all with 2 pairs of leaflets; pods dehiscent along one suture only . . 15. *C. fruticosa*
      - Leaves with 3–13 pairs of leaflets (only rarely and sporadically a reduced leaf with 2 pairs); pods indehiscent or not dehiscing as above:
        - Stipules linear, quickly falling off; branchlets glabrous; pods subterete, up to 10 cm. long . . . . . . 16. *C. floribunda*
        - Stipules leafy, semi-cordate to reniform, ± persistent; branchlets subglabrous to pubescent or tomentose; pods ± compressed, 12 cm. long or more . 18. *C. petersiana**
    - Leaflets not gradually narrowing or tapering upwards, rounded or sometimes abruptly narrowed and obtuse or rarely subacute at apex:
      - Stipules very narrow, 0·3–0·75 mm. wide, usually linear, subulate or lanceolate:
        - Pedicels of flowers 0·3–0·5 cm. long; seeds not areolate; shrub with (2–)3 pairs of leaflets per leaf; racemes 3–15-flowered . . . . . . 17. *C. bicapsularis*
        - Pedicels of flowers (0·5–)1·5–5 cm. long; seeds areolate:
          - Leaflets in 3 pairs per leaf; racemes 1–2-flowered; herb or undershrub 22. *C. obtusifolia*
          - Leaflets in 4–12 pairs per leaf; racemes 2–many-flowered; shrubs or trees:
            - Pedicels of flowers 2·2–4(–5) cm. long; pods subcylindric or slightly compressed, valves stiff and rather hard; branchlets glabrous to densely pubescent . . . 19. *C. singueana**

* A possible hybrid between *C. petersiana* and *C. singueana* is known (see p. 73).

Pedicels of flowers 1·5–1·8 cm. long; pods flattened, valves papery; branchlets minutely appressed-puberulous . . . . . 20. *C. baccarinii*
Stipules broadly reniform, 7–22 mm. wide, leafy . . . . 21. *C. auriculata*
Rhachis of leaves eglandular; petiole glandular near base:
Leaflets without hairs except on the ciliolate margins; sepals glabrous to thinly pubescent outside; areoles on flattened faces of seed:
Bracts very acute; gland on petiole hemispherical, ovoid or subglobose; peduncles (in East Africa) 0·3–0·5 (very rarely –0·8) cm. long . . . . 23. *C. occidentalis*
Bracts obtuse to subacute; gland on petiole clavate or cylindrical; peduncles 0·8–2·5 cm. long . . . . 24. *C. sophera*
Leaflets densely pilose on both surfaces; outer sepals densely pilose outside; areoles on narrow marginal part of seed . . . 25. *C. hirsuta*
Sepals * acute or acuminate at apex; leaflets sessile or separated from the rhachis merely by the very short pulvinus:
Petiolar gland ± distinctly stipitate or raised on a short column:
Perennial herbs:
Stipitate glands both on the petiole and rhachis of leaf (cf. also 30, *C. hildebrandtii*):
Leaflets in 3–13 pairs per leaf:
Midrib of leaflet somewhat excentric but not close to one margin; lateral nerves branching mostly towards margin; petals 10–14 mm. long . 28. *C. grantii*
Midrib of leaflet very strongly excentric, running within 1·5 mm. of anticous margin of leaflet; venation reticulate; petals 17 mm. long. (Known only from Mufindi in Tanganyika) . . 29. *C. sp. A*
Leaflets in (15–)20–43 pairs per leaf; at most a single stipitate gland sometimes present on the rhachis . . . . . 35. *C. katangensis*
Stipitate gland on the petiole, but none on the rhachis of the leaf (rarely a few sessile glands on the rhachis in 30, *C. hildebrandtii*):
Stalk or column of gland very short, 0·1–0·75 mm. long:
Leaflets in 4–13(–14) pairs per leaf, rounded or obtuse and mucronate or subacute at apex; stems prostrate or sometimes erect:
Stalk of petiolar gland 0·3–0·7 mm. long; stems usually ± densely

* Much allowance must be made when using the key from this point onwards. See note on p. 48. Diagnostic features of stipules, petiolar gland and rhachis of various species of this group are illustrated in fig. 10 (p. 57).

pubescent, rarely puberulous; well-developed leaves usually more than 2·5 cm. long . . . . . 30. *C. hildebrandtii*

Stalk of petiolar gland 0·1–0·2 mm. long; stems appressed-puberulous; leaves 0·5–2·2 cm. long; stems always prostrate . . . . . . 45. *C. usambarensis*

Leaflets in (15–)20–43 pairs per leaf, acute or acuminate at apex; stems erect . 35. *C. katangensis*

Stalk of gland rather long, 1–2·5 mm.:

Midrib of leaflet somewhat excentric particularly in lower part of leaflet, but not marginal, giving off lateral nerves on both sides; pedicels 1–2·5 cm. long 31. *C. zambesica*

Midrib of leaflet marginal, with lateral nerves on anticous side none or very inconspicuous; pedicels 0·4–0·6 cm. long . . . . 32. *C. fallacina*

Annual herbs:

Pedicels and sepals with rather long spreading hairs, petiolar gland on a stalk up to ± 0·2 mm. long; leaf-rhachis channelled but not crenate-crested along upper side; leaflets 4–8·5 mm. long; pods 4–5 mm. wide; stems apparently erect or ascending 33. *C. fenarolii*

Pedicels and sepals appressed-puberulous; petiolar gland on a stalk 0·25–0·8 mm. long; leaf-rhachis crenate-crested along upper side; leaflets 3–5 mm. long; pods 2·5–3·5 mm. wide; stems prostrate or erect . . . . 34. *C. gracilior*

Petiolar gland sessile:

Rhachis of leaf longitudinally channelled, but not or scarcely toothed or crenate-crested along upper side (to p. 58):

Stamens 4–5; pedicels 3·5–7 mm. long; petals 4–5 mm. long . . . . . . 43. *C. hochstetteri*

Stamens 8–10; pedicels and petals of various lengths:

Pedicels 0·2–0·4(–0·5) cm. long; petals 3·5–4·5 mm. long; petiolar gland 2–4 mm. long; leaflets almost symmetrical, with midrib central or almost so, especially above . . . . . 27. *C. nigricans*

Pedicels 0·6–2·5(–2·9) cm. long; petals 4–19 mm. long; petiolar gland variable in size, in most species less than 2 mm. long; leaflets usually ± asymmetrical, sometimes almost symmetrical:

Petiolar gland large, 1–1·5 mm. wide, sunken in or flush with the surface of the petiole and separating widely the sides of the channel running along the upper side of the petiole . . . . 37. *C. comosa*

Petiolar gland 0·2–1 mm. wide, not or only slightly separating the sides of the

channel running along the upper side of the petiole:
- Gland on petiole very small, 0·2–0·25 mm. in diameter; leaflets in 5–11 (–13) pairs; prostrate perennial . 45. *C. usambarensis*
- Gland on petiole 0·25–2·0(–2·5) mm. long, 0·25–1 mm. wide; leaflets usually (not always) more than 13 pairs; plants usually not prostrate perennials except in 46, *C. ghesquiereana*:
  - Stipules ± falcate and semi-cordate at base; leaflets ending in a prickle-like point bent to one side; midrib of leaflet very excentric . 39. *C. falcinella*
  - Stipules straight or almost so, oblique at base:
    - Plant perennial, prostrate; petals small, 6·5 mm. long . . 46. *C. ghesquiereana*
    - Plants annual, or if perennial then with erect or ascending stems:
      - Petiolar gland (on well-developed leaves) 1–2(–2·5) mm. long:
        - Perennial with ± tufted stems arising annually from a ± thickened woody rootstock; pods 3·5–4 mm. wide . . . . 36. *C. parva*
        - Annual with a single main stem which however may become somewhat woody below; pods 5–6 mm. wide 40. *C. kirkii*
      - Petiolar gland 0·5–1 mm. long:
        - Midrib of leaflet somewhat excentric, with the lateral nerves on the anticous side present and visible:
          - Perennial with ± tufted stems arising annually from a ± thickened woody rootstock. . 36. *C. parva*
          - Annuals; stems sometimes somewhat woody below but rootstock not becoming woody and thickened:
            - Pedicels (0·6–)1–2·3 cm. long; petals 8–19 mm. long:
              - Gland ± elliptic, cushion-shaped, yellowish; petals 8–15 mm. long . 40. *C. kirkii**

* Only in untypical plants of *C. kirkii* is the gland always less than 1 mm. long. These are mainly found in N. Tanganyika and Kenya, and may be caused by growth under unfavourable conditions (see p. 94).

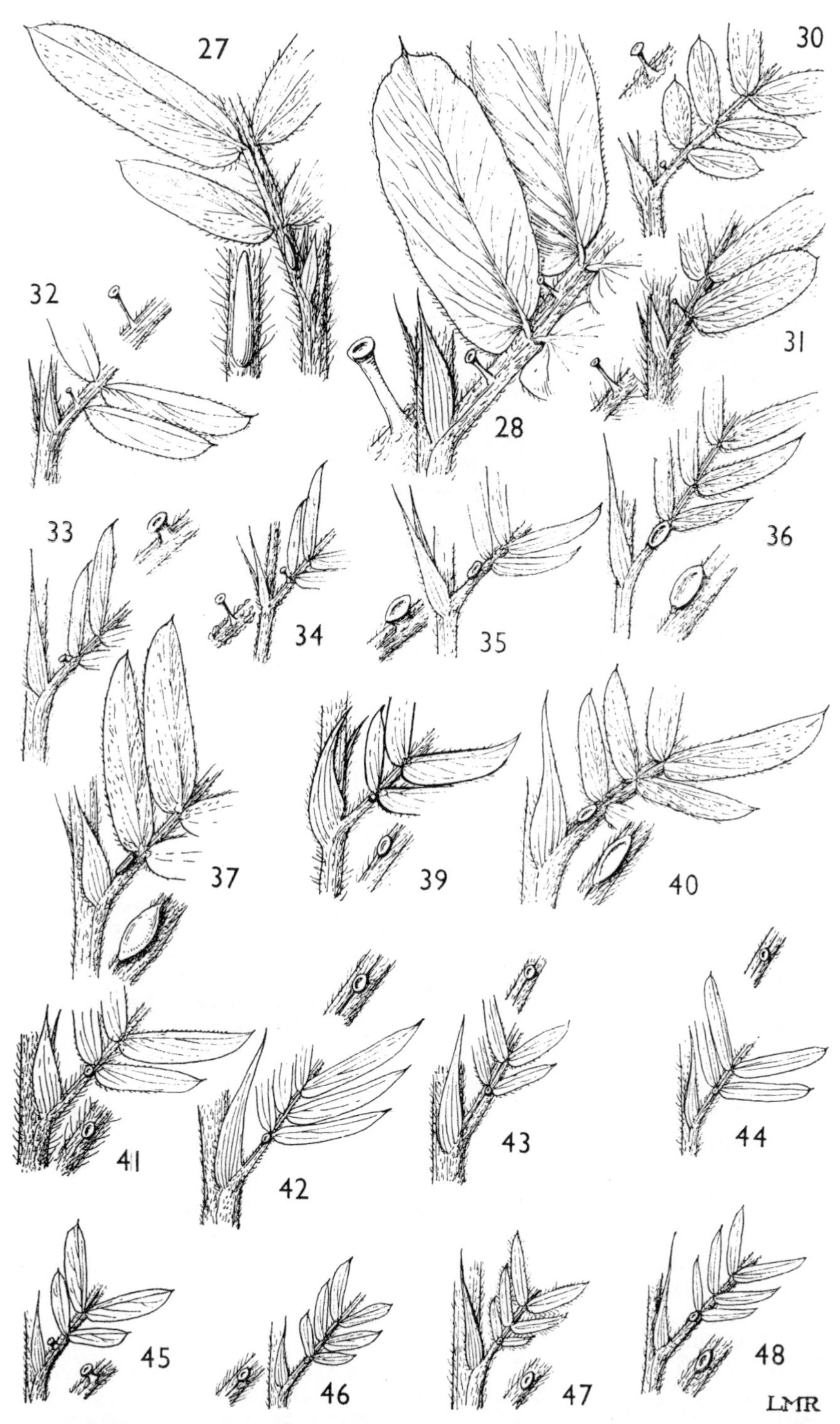

FIG. 10. *CASSIA*—part of stem, stipules, petiole and basal part of leaf of various species, × 2; petiolar gland of each species also shown separately, × 4. Species numbered as in text. **27,** *C. nigricans*; **28,** *C. grantii*; **30,** *C. hildebrandtii*; **31,** *C. zambesica*; **32,** *C. fallacina*; **33,** *C. fenarolii*; **34,** *C. gracilior*; **35,** *C. katangensis*; **36,** *C. parva*; **37,** *C. comosa*; **39,** *C. falcinella* var. *falcinella*; **40,** *C. kirkii* var. *kirkii*; **41,** *C. wittei*; **42,** *C. quarrei*; **43,** *C. hochstetteri*; **44,** *C. exilis*; **45,** *C. usambarensis*; **46,** *C. ghesquiereana*; **47,** *C. polytricha* var. *pulchella*; **48,** *C. mimosoïdes*. 27, from *Tanner* 1383; 28, from *Mgaza* 345; 30, from *Greenway* 9166; 31, from *Tanner* 2882; 32, from *Anderson* 607; 33, from *B. D. Burtt* 2015; 34, from *Milne-Redhead & Taylor* 9285; 35, from *Bullock* 2017; 36, from *Lynes* I.h. 107; 37, from *Milne-Redhead & Taylor* 8257A; 39, from *Haarer* 2269; 40, from *Milne-Redhead & Taylor* 9073; 41, from *Richards* 8814; 42, from *Milne-Redhead & Taylor* 9666; 43, from *Milne-Redhead & Taylor* 10049; 44, from *Tanner* 2013; 45, from *Lugard* 255a; 46, from *Purseglove* 3703; 47, from *Milne-Redhead & Taylor* 9218; 48, from *Milne-Redhead & Taylor* 9667.

Gland circular or nearly so, dark purplish-red (at least when dry); petals 10–19 mm. long . . 41. *C. wittei*
Pedicels mostly 0·6–1 (–1·5) cm. long; petals 7–8·5 mm. long . 42. *C. quarrei*
Midrib of leaflet strongly excentric with no lateral nerves on the anticous side . 47. *C. polytricha*
Rhachis of leaf with a ± distinctly toothed or crenate-crested wing running along upper side:
Leaflets in 5–10 pairs (rarely a few of the lowest leaves on the main stem with up to 20 pairs); gland on petiole 0·25 mm. in diameter; plant apparently annual, with decumbent or ascending stems . . . 44. *C. exilis*
Leaflets in 10–76 pairs:
Plant perennial with apparently annual stems arising from a ± thickened woody rootstock; leaflets mostly in 10–30 pairs 38. *C. sp. B*
Plant usually annual, sometimes with the stems becoming woody above ground-level and allowing the plant to perennate 48. *C. mimosoïdes*

1. **C. angolensis** *Hiern*, Cat. Afr. Pl. Welw. 1: 291 (1896); L.T.A.: 631 (1930); Milne-Redh. in Hook., Ic. Pl., pp. 2–3 sub t. 3368 (1938), in obs.; T.T.C.L.: 98 (1949); Steyaert in F.C.B. 3: 503 (1952); Mendonça & Torre in C.F.A. 2: 176 (1956). Type: Angola, Cuanza Norte, Cazengo, Cacula, *Welwitsch* 1736 (LISU, holo., BM, K, iso. !)

Tree 6–25 m. high, in most ways very similar to *C. mannii*. Young branchlets puberulous to subglabrous. Leaves with petiole and rhachis eglandular, the latter up to 30 cm. long. Leaflets up to 13 pairs, emarginate at apex, glabrous or almost so, without pubescent margins, otherwise as in *C. mannii*. Racemes as in *C. mannii* except for being somewhat more pubescent and with bracts sometimes persistent till the flowers open. Petals golden-yellow, 2–3·0 cm. long, (1·1–, *fide* F.C.B.)1·5–2·0 cm. wide; flowers otherwise as in *C. mannii*. Pods up to 70 cm. long, similar to those of *C. mannii*. Seeds brown, ellipsoid to obovoid, compressed, 8–10 × 5–9 × 6 mm., transversely cracked.

TANGANYIKA. Lushoto District: Amani, 21 Dec. 1932, *Greenway* 3304!; Morogoro District: Nguru Forest, Sept. 1924, *Simmance* 173!; Lindi District: Rondo Plateau, Nov. 1950, *Bryce* in *Eggeling* 6019! & Mchinjiri, 9 Aug. 1951, *Bryce* B11!
DISTR. T3, 6, 8; Congo Republic, Mozambique, Malawi, Zambia and Angola
HAB. Lowland rain-forest; 820–1070 m.

NOTE. Most closely related to *C. mannii*, but distinct by its glabrous-margined leaflets with emarginate apices and by the yellow flowers.

2. **C. mannii** *Oliv.*, F.T.A. 2: 272 (1871); L.T.A.: 631 (1930); Milne-Redh. in Hook., Ic. Pl., t. 3368 (1938); Steyaert in F.C.B. 3: 499 (1952); I.T.U., ed. 2: 59, photo. 7 (1952). Type: Principe, *Mann* 1125 (K, holo. !)

Tree up to 15–25 m. high. Bark dark brown. Young branchlets glabrous to slightly pubescent. Leaves with petiole and rhachis eglandular, the latter

15–34 cm. long. Leaflets in 5–12 pairs, petiolulate, ovate-elliptic to oblong-elliptic, (3·5–)4–7(–8) cm. long, 2–3·2(–4·5) cm. wide, acute to obtuse or obtusely acuminate at apex, not or only slightly emarginate, ± pubescent on midrib, otherwise subglabrous to puberulous, but with pubescent margins at least when young. Racemes 3–12 cm. long. Bracts mostly fallen before the flowers open. Petals pink or white, about 1·5–2·5 cm. long, (0·8–)1–1·4 cm. wide. Stamens 9–10; filaments of 2–3 lower with S-bend near base. Pods cylindrical, 40–90 cm. long, 2–3 cm. in diameter, black, longitudinally and transversely partitioned within. Seeds brown, obovoid, compressed, 9–10 × 6 × 3–4 mm., transversely cracked.

UGANDA. W. Nile District: E. Madi, Zoka Forest, Jan. 1952, *Leggat* 76!; Bunyoro District: Budongo Forest, Waisoke R., Sept. 1933, *Eggeling* 1421 in *F.H.* 1367! & Budongo Forest, Feb. 1935, *Eggeling* 1605 in *F.H.* 1504!

DISTR. U1, 2; Ivory Coast, Nigeria, S. Tomé, Principe, Cameroun Republic, Gabon, Congo and Sudan Republics

HAB. Lowland rain-forest; 850–1370 m.

NOTE. *C. angolensis* is the species most closely related to *C. mannii*; for the distinctions between them see under the former.

Eggeling (in notes attached to his gathering No. 1605) remarks that, with its rosy-pink flowers borne in great profusion, *C. mannii* is one of the most decorative of Uganda's indigenous trees.

Mendonça & Torre in C.F.A. 2: 176–7 (1956) have taken up the name *C. psilocarpa* Welw. in Ann. Conselho Ultram. 1858: 587 (1859), remarking that it may prove to be the earliest name for *C. mannii*. The only descriptive matter given by Welwitsch for *C. psilocarpa* is " Arbor elegans, coma dilatata, (floribus ??) leguminibus cylindricis, 2–3 pedalibus pendulis ". I do not consider this diagnostic or sufficient to validate the name, and therefore maintain *C. mannii*.

3. **C. abbreviata** *Oliv.*, F.T.A. 2: 271 (1871); L.T.A. 632 (1930); T.S.K.: 61 (1936); T.T.C.L.: 98 (1949); Steyaert in F.C.B. 3: 503 (1952); Coates Palgrave, Trees of Central Africa: 93–6 (1956); Brenan in K.B. 13: 231 (1958); F.F.N.R.: 120 (1962). Types: Mozambique, near Lupata, *Kirk* & near Tete, *Kirk* & Malawi, Manganja Hills, *Meller* & Lake Nyasa, Cape Maclear, *Kirk* (all K, syn. !)

Shrub or small tree 3–10 m. high. Bark brownish-grey, rough. Young branchlets glabrous, puberulous or pubescent. Leaves with petiole and rhachis eglandular, the latter 5–25 cm. long. Leaflets in 5–12 pairs, petiolulate, ovate-elliptic to oblong-elliptic, sometimes elliptic-lanceolate, (1·5–)3–6(–7·5) cm. long, (0·8–)1·2–3(–4·5) cm. wide, rounded to obtuse or subacute at apex, usually ± pubescent or puberulous. Racemes 0·5–9 cm. long. Bracts persistent while the flowers are open. Petals yellow, 1·5–3·5 cm. long, 0·7–1·8 cm. wide. Stamens 10; filaments of 3 lower each with an S-bend near base and a swelling about half-way along their length. Pods cylindrical, 30–90 cm. long, 1·5–2·5 cm. in diameter, from velvety to glabrous and blackish, transversely but not longitudinally partitioned within. Seeds embedded in pulp, brown to blackish, 9–12 × 8–9 × 3 mm., not transversely cracked.

KEY TO SUBSPECIES

Leaflets with non-appressed often curled hairs beneath . . . subsp. **abbreviata**
Leaflets with appressed minute straight hairs beneath:
  Inflorescence-axis (in Flora area) with very short spreading pubescence; petals 1·8–3 cm. long . . . subsp. **beareana**
  Inflorescence-axis puberulous with minute appressed hairs; petals 1·5–2 cm. long . . . . . subsp. **kassneri**

subsp. **abbreviata**; Brenan in K.B. 13: 231 (1958)

Branchlets pubescent. Leaflets pubescent beneath with non-appressed often curled hairs. Axis of inflorescence with dense short spreading pubescence. Petals 2–3 × 0·8–1·5 cm. Pods densely tomentellous or velvety.

TANGANYIKA. Mpwapwa, 1 Oct. 1930, *Hornby* 310!; Rungwe District: Bulambia, 25 Oct. 1912, *Stolz* 1625!; Lindi District: Mnacho, 12 Dec. 1942, *Gillman* 1239!
DISTR. **T1**, 4, 5, 7, 8; Mozambique, Zambia, Rhodesia and the Congo Republic
HAB. Woodland and wooded grassland; 220–1520 m.

subsp. **beareana** (*Holmes*) *Brenan* in K.B. 13: 232 (1958); K.T.S.: 100 (1961). Type: East Africa (locality uncertain), *O'Sullivan Beare* (London, Pharmaceutical Society, holo. !, K, fragments !)

Branchlets pubescent or puberulous. Leaflets normally (always in East Africa) puberulous beneath with minute inconspicuous appressed straight hairs often only 0·1 mm. long. Axis of inflorescence normally (always in East Africa) with very short spreading pubescence. Petals 1·8–3 × 0·9–1·5 cm. Pods densely puberulous to sometimes glabrous (as in the type).

KENYA. Kilifi District: Kibarani, 23 Feb. 1945, *Jeffery* 96!; Lamu District: Kiunga, 5 Dec. 1946, *J. Adamson* 280 in *Bally* 5971!
TANGANYIKA. Handeni District: Sindeni Hills, 14 Sept. 1933, *B. D. Burtt* 4830!; Mpanda District: Mlala Hills, 27 Oct. 1959, *Richards* 11574!; Morogoro District: 9 Nov. 1932, *Wallace* 443!
DISTR. **K7**; **T3**, 4, 6; Congo Republic, Somali Republic (S.), Mozambique, Zambia, Rhodesia, Botswana, the Transvaal and South West Africa
HAB. Uncertain but probably similar to that of subsp. *abbreviata*—also recorded from " margin of evergreen forest on rock outcrop " (*Burtt* 4830); 460–1200 m. (probably occurs at lower altitudes also)

SYN. *C. beareana* Holmes in Pharm. Journ. 68 (ser. 4, 14): 42 (1902); L.T.A.: 631 (1930); T.T.C.L.: 98 (1949)

NOTE. *J. Adamson* 280 in *Herb. Bally* 5971 in *C.M.* 18108 (EA !, K !), Kenya, Kiunga, shows a pod (with the EA specimen) which is sparsely puberulous or subglabrous.

subsp. **kassneri** (*Bak. f.*) *Brenan* in K.B. 13: 234 (1958); K.T.S.: 100 (1961). Type: Kenya, Machakos District, Makindu R., *Kassner* 598 (BM, holo. !, K, iso. !)

Branchlets puberulous with minute appressed hairs. Leaflets puberulous beneath as in subsp. *beareana*. Axis of inflorescence clothed like the branchlets. Petals 1·5–2 × 0·7–1 cm. Pods densely and minutely grey-puberulous.

KENYA. Northern Frontier Province: 25 km. NE. of Wajir, 27 May 1952, *Gillett* 13362!; Kitui District: Mutha Plains, Aug. 1938, *Joana* in *Bally* 7429!; Kwale District: between Samburu and Mackinnon Road, 30 Aug. 1953, *Drummond & Hemsley* 4059!
TANGANYIKA. Lushoto District: Mnazi, 22 Dec. 1929, *Greenway* 1971!
DISTR. **K1**, 4, 7; **T3**; not known elsewhere
HAB. Wooded grassland, deciduous bushland and dry scrub with trees; 350–1000 m.

SYN. *C. kassneri* Bak. f. in J.B. 67: 194 (1929); L.T.A.: 632 (1930)

NOTE (species as a whole). The flowers of *C. abbreviata* have a sweet scent, variously described as being like that of lilac, violets and sweet peas. The bark of subsp. *kassneri* is used for tanning (*Gillett* 13362) and a decoction of the roots of subsp. *beareana* was reported to be a remedy for blackwater fever.

From *C. angolensis* and *C. mannii* the presence of a marked swelling in each of the three long anticous stamen-filaments will separate *C. abbreviata*. The three subspecies into which the latter is divided are well-marked. Subsp. *abbreviata*, as far as the area of this Flora is concerned, is restricted to Tanganyika at altitudes above 900 m.; while subsp. *beareana* also has a wide range in Tanganyika, but has only been recorded as growing below 760 m. The latter also extends northwards into Kenya, but only near the coast. Inland in Kenya, and probably in drier habitats than those of the other two subspecies, the subsp. *kassneri* occurs, and extends just into northern Tanganyika.

Heady 1248 (EA !), from Kenya, Kitui District, Athi-Tiva, is puzzling in that its indumentum is that of subsp. *kaessneri*, but its petals are up to about 2 cm. long and 1·4 cm wide—thus in size more like those of subsp. *beareana*.

3 × 4. **C. abbreviata** subsp. **abbreviata** × **burttii**

Graceful shrub up to 4·5 m. high, differing from *C. abbreviata* in the ± pendulous often longer inflorescences 12–35 cm. long, the flowers sulphur-yellow when open (deep yellow in bud), the pods (1 examined) 1–1·6 cm. in diameter in different places and almost completely sterile.

It differs from *C. burttii* in the shortly pubescent young branchlets, the leaflets in 8–11 pairs with (when mature) a rather dense indumentum of raised (not appressed except when the leaves are young) hairs beneath, the inflorescences shorter than usual in *C. burttii* and terminating leafy shoots (not lateral), the axes and pedicels pubescent with spreading hairs, the bracts persistent while the flowers are open, the petals sulphur-yellow when open, deep yellow in bud, up to ± 2·6 cm. long and 1·5 cm. wide, and the pods with some indumentum outside, rather hard-walled and almost completely sterile.

TANGANYIKA. Lindi District: about 74 km. W. of Lindi, 9·5 km. W. of Mtama and 1·5 km. W. of Nyangao R., 14 Dec. 1955, *Milne-Redhead & Taylor* 7654!
DISTR. **T**8; not recorded elsewhere
HAB. *Albizia* woodland, one plant only seen; 160 m.

NOTE. The specific parentage of this, with which I agree, was suggested by the collectors themselves. Colour-photographs of the hybrid and *C. burttii* were made. The raised indumentum on the lower side of the mature leaflets most strongly indicates subsp. *abbreviata* as one parent, although the altitude is lower than usual for this subspecies. It is hard to describe the indumentum on the outside of the pod of the hybrid, as it is old and its surface weatherworn.

4. **C. burttii** *Bak. f.* in J.B. 73: 80 (1935); T.T.C.L.: 96 (1949). Type: Tanganyika, Morogoro District, Wami Road, *B. D. Burtt* 5032 (BM, holo.!, EA, FHO, K, iso.!)

Shrub or small tree 1–6 m. high. Young branchlets minutely and inconspicuously puberulous. Leaves with petiole and rhachis eglandular, the latter 15–24 cm. long. Leaflets in 5–8 pairs, petiolulate, elliptic or ovate-elliptic, 3·5–10(–12) cm. long, 1·8–4·6(–7) cm. wide, minutely and inconspicuously appressed-puberulous beneath. Racemes 15–50 cm. long, pendulous, simple or branched, mostly lateral. Bracts falling while the buds are young. Petals white, 1–2·4 cm. long, 0·6–1·2 cm. wide. Stamens 10; filaments of 3 lower each with an S-bend near base and an abrupt conspicuous swelling above the bend. Pods ± 60 cm. long, 1–1·5 cm. in diameter, similar to those of 7, *C. afrofistula*. Seeds embedded in pulp, chestnut-brown, ovate-orbicular or suborbicular, 8–10 × 7–9 × 3–5 mm., with slight reticulate cracking.

TANGANYIKA. Kilosa District: Kidodi, Oct. 1952, *Semsei* 1000!; Morogoro, Dec. 1953, *Eggeling* 6774!; Lindi District: about 72 km. W. of Lindi and 8 km. W. of Mtana near Nyangao R., 14 Dec. 1955, *Milne-Redhead & Taylor* 7650!
DISTR. **T**6, 8; Mozambique
HAB. Woodland, wooded grassland, thickets; 15–800 m.

NOTE. The long gracefully drooping racemes of white-petalled flowers make this a most elegant and attractive plant. It hybridizes with *C. abbreviata*; see after the latter.

5. **C. sieberana*** *DC.*, Prodr. 2: 489 (1825); L.T.A.: 632 (1930); Steyaert in F.C.B. 3: 500 (1952); I.T.U., ed. 2: 60, fig. 13 (1952). Type: Senegal, *Sieber* 48 (G, holo., K, iso.!)

Shrub or small tree 2–12 m. high (said in F.C.B. to reach 20 m.). Bark grey to brown or almost black. Young branchlets rather densely pubescent. Leaves with petiole and rhachis eglandular, the latter (10–)13–29 cm. long. Leaflets in usually 4–12 pairs (to 14, *fide* I.T.U.), petiolulate, elliptic to ovate-elliptic or sometimes obovate-elliptic, (2·5–)3·5–9·5 cm. long, 2–5 cm. wide, densely appressed-puberulous beneath, ± crisped-puberulous above. Racemes 15–42 cm. long, pendulous, mostly simple, single and lateral. Bracts persist-

* This spelling, in place of the more familiar "*sieberiana*", has to be adopted in accordance with Article 73 of the Intern. Code Bot. Nomencl. (1961).

ing till the flowers are open. Petals yellow, (1–)2–3·5 cm. long, (0·7–)1–2·5 cm. wide. Stamens 10; filaments of 3 lower each with an S-bend near base, without a swelling above. Pods cylindrical, 30–90 cm. long, 1–1·7 cm. in diameter, glabrous, blackish-brown, cylindrical, transversely but not longitudinally partitioned within. Seeds not embedded in pulp, brown, ellipsoid or ovoid-ellipsoid, 8–9 × 4·5–7 × 2·5–4 mm., with slight reticulate cracking.

UGANDA. W. Nile District: between Moyo and Rumogi, Mar. 1935, *Eggeling* 1813 in *F.H.* 1645! & Terego, Mar. 1939, *Hazel* 717!; Acholi District: Abera Forest Reserve (near Gulu), 30 Mar. 1945, *Greenway & Eggeling* 7263!

DISTR. U1; Senegal and Gambia eastwards to Nigeria, the Congo Republic, Sudan Republic and Uganda

HAB. Wooded grasslands; ± 1070 m.

NOTE. The drooping racemes of ornamental yellow flowers have frequently caused this tree to be compared with laburnum; indeed it has been called African Laburnum.

6. **C. thyrsoïdea** *Brenan* in K.B. 13: 234 (1958). Type: Tanganyika, Songea District, *Milne-Redhead & Taylor* 8272 (K, holo.!, BR, iso.!)

Tree 4·8–18 m. high (or, *fide Burtt* 4534, a semi-scandent shrub to 4·5 m.). Young branchlets densely pubescent. Leaves with petiole and rhachis eglandular, the latter 14·2–26 cm. long. Leaflets in 6–8 pairs, petiolulate, elliptic or ovate-elliptic, 4·5–11·5 cm. long, 2·1–6 cm. wide, with crisped but not appressed pubescence beneath especially on midrib and nerves, sometimes tomentose. Inflorescences similar to those of 7, *C. afrofistula*, with racemes to 15–27 cm. long, but with dense olivaceous pubescence, sometimes with a somewhat silvery sheen on the sepals. Bracts persisting till the flowers are open, 8–11 mm. long. Petals yellow, ± 1·5–2·6 cm. long and 1–2·2 cm. wide. Stamens 10; filaments of 3 lower each with an S-bend near base and a slight swelling above the bend. Pods subcylindrical, attenuate at apex, somewhat so at base, ± 32–42 cm. long, 1–1·5 cm. in diameter, sparsely pubescent to subglabrous, blackish-brown, transversely but not longitudinally partitioned within. Seeds embedded in pulp, brown, ± 7 mm. long, 6 mm. wide and 3 mm. thick.

TANGANYIKA. Mpwapwa, 12 Jan. 1930, *Hornby* 156! & Mpwapwa District, on path to Kiboriani, 25 Jan. 1933, *B. D. Burtt* 4534!; Songea District: 32 km. E. of Songea by R. Mkukira, 19 Jan. 1956, *Milne-Redhead & Taylor* 8272!

DISTR. T5, 8; Malawi

HAB. Riverine forest; 900–1370 m.

SYN. [*C. sieberana* sensu T.T.C.L.: 97 (1949), pro majore parte et quoad specim. cit., *non* DC.]

NOTE. The relationship of this species appears to be closest with *C. afrofistula*, from which it differs clearly in the denser indumentum and persistent longer bracts. In addition, *C. afrofistula* is coastal, while *C. thyrsoïdea* is not.

7. **C. afrofistula** *Brenan* in K.B. 13: 236 (1958); K.T.S.: 101 (1961). Type: Tanganyika, Tanga District, Sawa, *Faulkner* 1956 (K, holo.!, BR, iso.!)

Shrub or small tree 1·2–6 m. high (? more). Young branchlets minutely puberulous (in East Africa), soon glabrous. Leaves with petiole and rhachis eglandular, the latter 7–25 cm. long. Leaflets in 4–9 pairs, petiolulate, ovate to ovate-elliptic, 2–10 cm. long, (1–)1·5–5·2 cm. wide, obtuse or subacute at apex, in Flora area glabrous except at extreme base. Inflorescence a usually erect large pyramidal terminal panicle up to 30–40 cm. long or more, composed of terminal and lateral racemes (8–)15–38 cm. long. Bracts all falling while the flower-buds are young, 4–8 mm. long. Petals bright yellow, 1·5–3 cm. long, 1–2 cm. wide. Stamens 10; filaments of 3 lower each with an

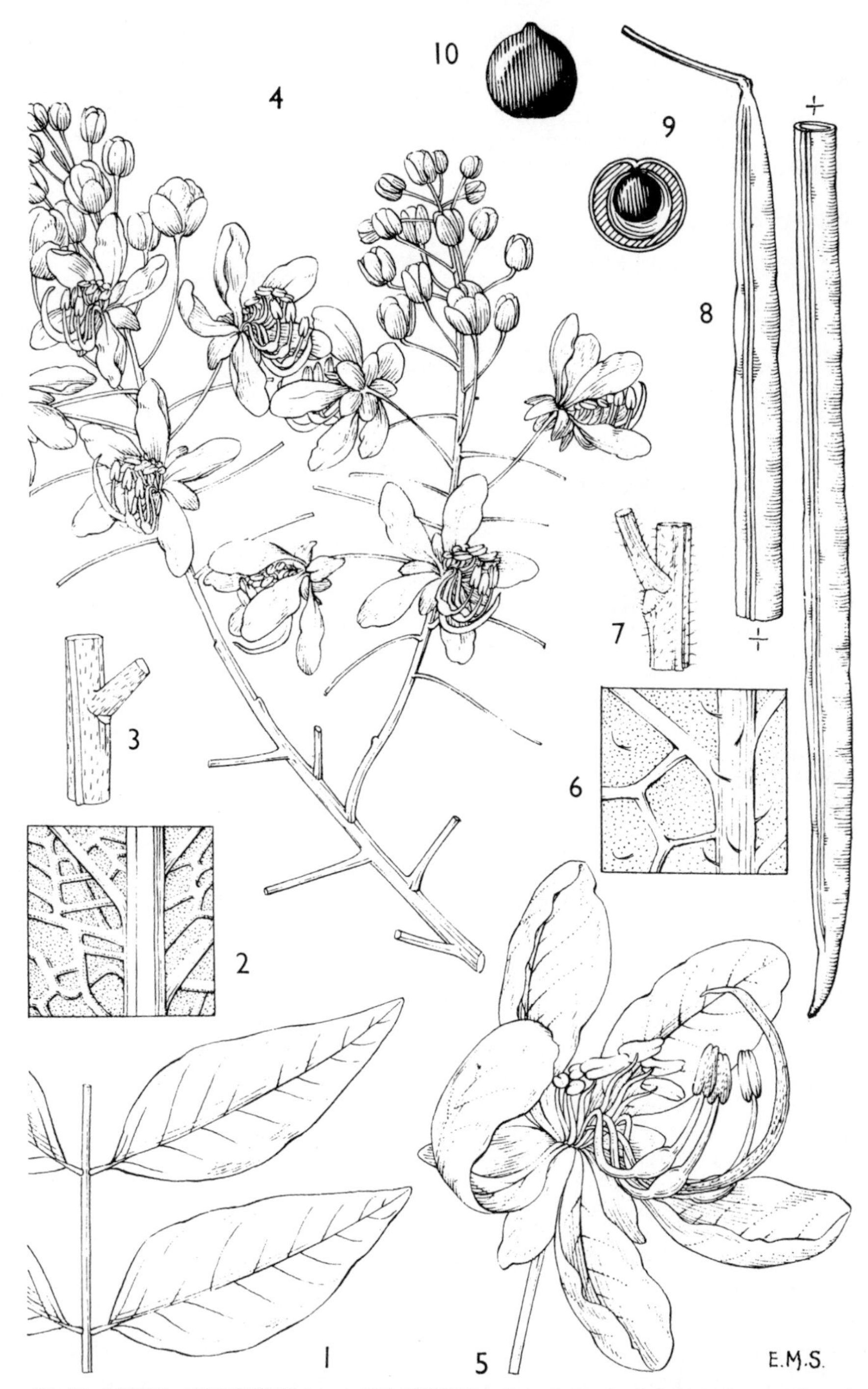

FIG. 11. *CASSIA AFROFISTULA* var. *AFROFISTULA*—**1,** part of leaf, $\times \frac{2}{3}$; **2,** part of lower surface of leaflet, $\times$ 10; **3,** part of inflorescence-rhachis with base of pedicel, $\times$ 6; **4,** part of inflorescence, $\times \frac{2}{3}$; **5,** flower, $\times$ 2. *C. AFROFISTULA* var. *PATENTIPILA*—**6,** part of lower surface of leaflet, $\times$ 10; **7,** part of inflorescence-rhachis with base of pedicel, $\times$ 6; **8,** pod, $\times \frac{1}{2}$; **9,** transverse section of pod, $\times$ 1; **10,** seed, $\times$ 2. 1–5, from *Faulkner* 1956; 6–10, from *Faulkner* 343. Reproduced by permission of the Bentham-Moxon Trustees.

S-bend near base and an abrupt conspicuous swelling a little more than half-way along their length. Pods subcylindrical or very slightly compressed, usually ± attenuate at base and apex, 20–60 cm. long, 1–1·5 cm. in diameter, glabrous, blackish-brown, transversely but not longitudinally partitioned within. Seeds embedded in pulp, brown or blackish-brown, 6–8 mm. in diameter, 4 mm. thick. Fig. 11, p. 63.

KENYA. Kilifi District: coast N. of Malindi, Boula, 5 Nov. 1945, *J. Adamson* 192 in *Bally* 6092! & Malindi, Oct. 1951, *Tweedie* 996!; Tana River District: Mambosasa, *R. M. Graham* Z.303 in *F.H.* 1791!

TANGANYIKA. Tanga District: Tanga, Ras Kasone, May 1942, *Lindeman* 1025!; Mafia I., 3 Apr. 1933, *Wallace* 743!; Lindi District: Ngongo, 25 May 1943, *Gillman* 1483!

ZANZIBAR. Zanzibar I., 1927, *Toms* 123! & Fumba, 28 Jan. 1929, *Greenway* 1233! & Chwaka road, 9 Dec. 1954, *R. O. Williams* 189!

DISTR. **K**1, 7; **T**3, 6, 8; **Z**; **P** (*fide* U.O.P.Z.: 175); Mozambique

HAB. Probably coastal evergreen bushland; common in the dry coral areas of Zanzibar and Pemba (U.O.P.Z.: 175); 0–90 m.

SYN. [*C. sieberana* sensu T.S.K.: 61 (1936), *non* DC.]
[*C. fistula* sensu T.T.C.L.: 97 (1949), pro parte, *non* L.]
[*C. beareana* sensu U.O.P.Z.: 175 (1949), *non* Holmes]

NOTE. The above description relates to var. *afrofistula*. The var. *patentipila* Brenan in K.B. 13: 238 (1958) occurs in Mozambique and should be looked for in SE. Tanganyika. It differs in having the inflorescence ± shortly pubescent with spreading hairs, not appressed-puberulous, and in mostly having some usually sparse crisped pubescence on the lower side of the leaflets (fig. 11/6–10).

C. afrofistula has been frequently confused with true *C. fistula* L., which is probably a native of tropical Asia, but is cultivated as an ornamental tree in East Africa. *C. fistula* has normally pendulous racemes 1–3 together, and leaflets almost equal at base, with more numerous lateral nerves and minutely appressed-puberulous beneath. The pods of *C. fistula* are 1·5–2·5 cm. in diameter with hard woody walls, while those of *C. afrofistula* can easily be cracked between finger and thumb. This greater fragility of the pods also distinguishes *C. afrofistula* from *C. angolensis*, *C. mannii*, *C. abbreviata* and *C. sieberana*.

8. **C. alata** *L.*, Sp. Pl.: 378 (1753); L.T.A.: 637 (1930); T.T.C.L.: 96 (1949); U.O.P.Z.: 175 (1949); Steyaert in F.C.B. 3: 507 (1952); Dimitri & Alberti in Rev. Invest. Agric. 8: 29, fig. 12 (1954); De Wit in Webbia 11: 231 (1955). Type: a cultivated plant in *Herb. Clifford* (BM, syn.!)

Herb or shrub 1–6 m. high. Young stems puberulous with minute spreading hairs, eglandular. Leaves mostly 30–75 cm. long; petiole and rhachis eglandular. Stipules obliquely triangular, 0·6–1·3 cm. long, 0·3–1 cm. wide. Leaflets in 5–7(–13) pairs, oblong-elliptic to obovate-elliptic, mostly 5–19 cm. long and 2·5–12 cm. wide, rounded and usually retuse at apex, usually with a short mucro 0·5–1·5 mm. long, puberulous especially on midrib, lateral nerves and margins. Racemes (including peduncle) 15–70 cm. long. Bracts subtending flower-buds large, orange, 2–2·8 cm. long, 1·2–1·7 cm. wide. Petals bright yellow, 1·7–2·2 cm. long, 0·9–1·2 cm. wide. Stamens normally 10; filaments straight or nearly so; 2 anthers large, 5 medium, 3 small. Pods 12–17(–19) cm. long, 1·3–2 cm. wide (to 3 cm. across wings), dehiscent, transversely septate, with a crenate-margined wing running longitudinally along the middle of each valve. Seeds flattened at right angles to the axis of the pod, deltoid-rhombic, acuminate to the hilum, with a longitudinal ridge along each face, 6–8 mm. long, 4–5 mm. wide, 1·5–2·5 mm. thick.

TANGANYIKA. Tanga District: Potwe, Muheza, 23 June 1957, *Tanner* 3606!; Morogoro, Ruvu, June 1930, *Haarer* 1860!

ZANZIBAR. "Wild in moist situations in Zanzibar and Pemba" (*fide* U.O.P.Z.: 175 (1949)). Without precise locality or date, 1927, *Toms* 207!

DISTR. **T**3, 6; **Z**; **P**; pantropical, originally from South America

HAB. Uncertain, but in the Flora area doubtless always an escape from cultivation; near sea-level to 460 m.

NOTE. *Cassia alata* is a well-known remedy for ringworm and other skin diseases, and has been cultivated at Amani under the name " Ringworm Plant ", and also in Kenya. The large leaflets, large orange bracts and winged pods are conspicuous features of this easily recognized species.

De Wit, in Webbia 11 : 231 (1955), states that the leaflets are in 8–24 pairs. I have found at the most 13 pairs per leaf and I suspect some mistake in De Wit's figures. Equally, I can find no evidence that the pods of *C. alata* are ever as long as 25 cm. as stated in F.C.B. 3 : 507 (1952).

9. **C. italica** (*Mill.*) *F. W. Andr.*, F.P.S. 2 : 117 (1952) ; Mendonça & Torre in C.F.A. 2 : 178 (1956) ; Brenan in K.B. 13 : 239 (1958). Type : whereabouts uncertain ; not found at BM

Perennial herb or small shrub 15–60 cm. high or often prostrate. Young stems with fine ± appressed puberulence (or, but not in Flora area, with spreading pubescence), eglandular. Leaves 3–12 cm. long ; petiole and rhachis eglandular. Stipules small, narrowly triangular to ovate-triangular, 3–9 mm. long, 1–2·5 mm. wide near base. Leaflets in (3–)4–6(–7) pairs, obovate-elliptic to obovate-oblong, 1–4·3 cm. long, 0·4–2(–2·7) cm. wide, rounded to subtruncate or emarginate at apex, occasionally subacute, minutely appressed-puberulous (or, but not in East Africa, pubescent). Racemes (including peduncle) 2–25 cm. long. Bracts subtending flower-buds 3–5 mm. long, 3–3·5 mm. wide. Petals yellowish-white to bright yellow, 0·85–2 cm. long, 0·35–1 cm. wide. Stamens 9–10 ; filaments straight or nearly so ; 2 anthers large, 4–5 medium, 3 small. Pods oblong, flattened, upwardly falcate, 3–6 cm. long, 1·3–2 cm. wide, blackish, transversely venose, with a ridge of raised crests running along the middle of each valve. Seeds ± compressed, in same plane as pod, ovate, apiculate near hilum, 6–7 mm. long, 4 mm. wide, about 1·25 mm. thick, reticulate-rugose, with a small areole 1–1·25 × 0·5 mm. on each face.

DISTR. (of species). Drier regions throughout Africa, extending through south-western Asia to India

SYN. (of species). *Senna italica* Mill., Gard. Dict., ed. 8, No. 2 (1768)
Cassia aschrek Forsk., Fl. Aegypt.-Arab.: cxi, 86 (1775); T.T.C.L.: 96 (1949). Type: Yemen, Môr, *Forsskål* (C, holo. !)
C. obovata Collad., Hist. Cass.: 92, t. 15/A (1816); L.T.A.: 636 (1930), *nom. illegit.* Type: as *C. italica*

subsp. **micrantha** *Brenan* in K.B. 13: 241 (1958). Type: Kenya, Turkana District, *Padwa* 144 (K, holo. !)

Racemes (including peduncle) 2–8 cm. long, shorter than the subtending leaves. Flowers small, pale. Sepals 5–8 mm. long. Petals 8·5–9 mm. long, 3·5–4·5 mm. wide. Large anthers 5·5–6 mm. long, medium ones 2·5–3·5 mm. long, small ones 1·25 mm. long.

UGANDA. Karamoja District: Lalachat, Feb. 1936, *Eggeling* 2805 ! & Kangole, Nov. 1957, *Wilson* 397 !
KENYA. Turkana District: 40 km. SW. of Lodwar, 12 May 1953, *Padwa* 144 !; Baringo District: Kamasia, Maji ya Moto, 16 July 1945, *Bally* 4523 !; Masai District: Ol Lorgosailic plains, 20 July 1947, *Bally* 5137 !
TANGANYIKA. Shinyanga, 25 Jan. 1933, *Bax* 111 ! & New Shinyanga, 10 May 1945, *Greenway* 7440 !; Masai District: Ketumbaine, 7 Jan. 1936, *Greenway* 4287 !
DISTR. (of subsp.). **U**1; **K**1–3, 6; **T**1, 2; Senegal, Mali, the Tibesti Mts. in the Sahara, Ethiopia, Socotra, Somali Republic (N.), Botswana, Angola, South West Africa and ? the Transvaal; also in India
HAB. Grassland; 460–1830 m.

NOTE. Typical *C. italica*, with larger flowers than in subsp. *micrantha* and racemes mostly longer than the leaves, has not so far been found south of Ethiopia and the Somali Republic. *C. italica*, quite possibly typical, was, however, in cultivation at Amani in 1942 (*Hill* in *E.A.H.* 11291 !), but the specimen is in fruit only and its exact identity is thus uncertain.

10. **C. senna** *L.*, Sp. Pl. : 377 (1753), ? pro parte excl. *β Senna italica* . . . ; Brenan in K.B. 13 : 243 (1958). Type : uncertain, not found in Linnaean or

Clifford Herbaria. *Senna alexandrina sive foliis acutis* Morison, Pl. Hist. Univ. Oxon. 2: 201, t. 24/1 (1715), cited by Linnaeus, is clearly *C. senna* as interpreted here.

Shrub 0·3–3 m. high. Young stems with appressed puberulence or short spreading pubescence, eglandular. Leaves (3–)5–15 cm. long; petiole and rhachis eglandular. Stipules subulate, linear or narrowly triangular, 3–5 mm. long, up to 1 mm. wide. Leaflets in (3–)4–8(–12) pairs, lanceolate to narrowly elliptic or elliptic, 1·5–5·2(–6) cm. long, 0·4–0·9(–1·5) cm. wide, narrowly acute to rounded at apex, with or without a slender mucro up to 1 mm. long, appressed-puberulous or pubescent. Racemes (including peduncle) 5–30 cm. long. Bracts subtending flower-buds 0·5–1·1 cm. long, 0·35–0·5 cm. wide. Petals yellow or orange-yellow, 0·7–1·7 cm. long, 0·7–0·9 cm. wide. Stamens 10; filaments straight or nearly so; 2 anthers large, 5 medium, 3 small. Pods (3–)4–7 cm. long, 1·6–2·6 cm. wide, shortly oblong, flattened, slightly and upwardly falcate or sometimes almost straight, dehiscent, transversely septate, not winged or crested, papery. Seeds ± compressed, in same plane as pod, oblong, apiculate near hilum, 6–7 mm. long, 4 mm. wide, 2 mm. thick, reticulate or rugose, with a small areole 1–2 × 0·5–0·7 mm. on each face.

var. **senna**; Brenan in K.B. 13: 243 (1958)

Leaflets mostly narrowed to an acute or subacute apex, with a distinct slender mucro 0·5–1 mm. long. Pedicels glabrous to sparingly pubescent. Pods sparingly puberulous or pubescent.

KENYA. Kilifi, Mfumbini beach, 27 Sept. 1945, *Jeffery* 329!
DISTR. **K7**; in the N. extending from the central Sahara (Silet) eastwards to Arabia and India, southwards in Africa to the Sudan Republic, Eritrea, French Somaliland, Somali Republic (N.) and (? introduced) in Mozambique
HAB. Uncertain; said to grow in bush on sand

SYN. *C. angustifolia* Vahl, Symb. Bot. 1: 29 (1790); Tschirch, Handb. Pharmakognose 2(2): 1408–1412 (1917); L.T.A.: 637 (1930). Type: Yemen, *Forsskål* (C, holo., BM, iso.!)
C. acutifolia Del., Fl. Aegypt. Ill.: 61, t. 27 (1813); L.T.A.: 637 (1930). Type: from Egypt, whereabouts uncertain, ? at P

NOTE. This species in a wide sense, but particularly var. *senna*, is the main source of the well-known medicine of that name.
The status of the plant in East Africa is quite uncertain; it may well be an introduction (it has been cultivated at Amani). The solitary specimen from the Flora area is thinly appressed-puberulous, with narrow very acute leaflets up to about 4–5 cm. long and 6–7 mm. wide, and slightly falcate pods about 3·5–5 cm. long and 1·7–1·9 cm. wide.

var. **obtusata** *Brenan* in K.B. 13: 244 (1958). Type: Somali Republic (N.), 45° 15′ E., 8° 58′ N., *Gillett* 4100 (K, holo.!)

Leaflets mostly rounded to obtuse at apex, mostly without a mucro or with only a short one to ± 0·5 mm. long. Pedicels and pods ± densely pubescent or puberulous.

KENYA. Northern Frontier Province: Lorian, end of Golana Gof, 19 Jan. 1943, *Bally* 2109! & between Banessa and Ramu, 23 May 1952, *Gillett* 13281!
DISTR. **K1**; Somali Republic
HAB. Dry scrub with trees; ± 550–1280 m.

NOTE. This, unlike var. *senna*, is certainly native in the Flora area. The var. *obtusata* is intermediate in its characters between *C. senna* var. *senna* and *C. holosericea* Fres. from the Sudan Republic, Ethiopia, Somali Republic (N.), Arabia, and India. The latter has spreading indumentum, while that of var. *obtusata* is appressed. The pods of *C. holosericea* are also narrower. For further details, see Brenan in K.B. 13: 244–45 (1958).

11. **C. didymobotrya** *Fres.* in Flora 22: 53 (1839); L.T.A.: 638 (1930); T.S.K.: 61 (1936); T.T.C.L.: 97 (1949); Steyaert in F.C.B. 3: 504, t. 36 (1952); I.T.U., ed. 2: 59 (1952); De Wit in Webbia 11: 241 (1955); Mendonça

Fig. 12. *CASSIA DIDYMOBOTRYA*—**1,** part of branch with leaves and inflorescences, × $\frac{2}{3}$; **2,** stipule, × 1; **3,** sepal, × 1; **4,** petal, × 1; **5–7,** three different sorts of stamen, × 1; **8,** ovary and style, × 1; **9,** pod, × $\frac{2}{3}$; **10,** seed, × 2. 1–8, from *Milne-Redhead & Taylor* 10919; 9, 10, from *T. H. E. Jackson* 304.

& Torre in C.F.A. 2: 177 (1956); K.T.S.: 101 (1961); F.F.N.R.: 120 (1962). Type: Ethiopia, *Rueppell* (FR, holo.)

Shrub 0·6–4·5(–9) m. high. Young stems pubescent, sometimes villous, rarely subglabrous, eglandular. Leaves mostly 10–30(–36) cm. long; petiole and rhachis eglandular (or only with tiny reddish clustered bodies at leaflet insertions). Stipules broadly ovate-cordate, finely acuminate, 1–2·5 cm. long, 0·8–1·2 cm. wide. Leaflets in 8–18 pairs, oblong-elliptic, 2–6·5 cm. long, 0·6–2·5 cm. wide, mostly rounded to obtuse at apex, with a distinct slender mucro 1–3 mm. long, ± pubescent on both surfaces. Racemes (including peduncle) 11–35(–40) cm. long. Bracts subtending flower-buds 0·9–2·7 cm. long, 0·5–1·4 cm. wide. Petals bright yellow, 1·8–2·7 cm. long, 1–1·6 cm. wide. Stamens 10; filaments straight or nearly so; 2 anthers large, 5 medium, 3 small. Pods oblong, flattened, 8–12 cm. long, 1·5–2·5 cm. wide, dehiscent, transversely septate, not winged. Seeds ± compressed in same plane as pod, oblong, apiculate near hilum, 8–9 mm. long, 4–5 mm. wide, 2·5 mm. thick, with an oblong areole 4 × 1·5 mm. in centre of each face. Fig. 12, p. 67.

UGANDA. Bunyoro District: Rusangura, June 1937, *Eggeling* 3351!; Kigezi District: Kachwekano Farm, Dec. 1949, *Purseglove* 3136!; Teso District: Serere, Aug. 1932, *Chandler* 847!

KENYA. SE. Elgon, Nov. 1930, *T. H. E. Jackson* 304 in *C.M.* 2940!; Meru, Apr. 1951, *Hancock* 102!; Machakos/Masai Districts: Chyulu Hills, 9 July 1938, *Bally* 1177 in *C.M.* 8219!

TANGANYIKA. Kilimanjaro, N. slopes, 29 Nov. 1932, *C. G. Rogers* 127 in *C.M.* 16635!; Iringa, 18 Sept. 1936, *Emson* 609!; Songea District: Hanga Farm, by R. Hanga, 27 June 1956, *Milne-Redhead & Taylor* 10919!

DISTR. **U**1–4; **K**3–7; **T**1–8; Congo Republic, Sudan Republic, Ethiopia, Mozambique, Malawi, Zambia, Rhodesia and Angola

HAB. By lake-shores, streams, rivers and other damp places in upland rain-forest, grassland, woodland, etc.; sometimes also common in old cultivations, and found in hedges near buildings; 900–2440 m.

SYN. *C. nairobensis* [Aggeler & Musser, Los Angeles, California, seed catalogue: 63 (1930), *nomen subnudum*] L. H. Bailey, Hortus Second: 146 (1941) & Man. Cult. Pl., ed. 2: 586 (1949), *sine descr. lat.* No type cited, but authentic specimens seen—cultivated in California, *L. H. & E. Z. Bailey* 7780 & 7952 (BH!)

NOTE. The indumentum of *C. didymobotrya* varies a good deal in length and density, and the bracts and stipules are variable in size. There seems no reason, however, for recognizing these variants taxonomically.

C. didymobotrya is said to smell strongly of mice (*Milne-Redhead & Taylor* 10919).

12. **C. ruspolii** *Chiov.* in Atti R. Accad. Ital., Mem. Cl. Sc. Fis., etc., 11 (Pl. Nov. Aethiop.): 29 (1940). Type: Ethiopia, Ueb Karenle valley at Ito Danna, about 42°45′ E., 6°20′ N., *Ruspoli & Riva* 1663 (FI, holo.!)

Suffruticose, up to 1·2 m. high, with thickened underground rootstock. Stems with minute appressed puberulence, eglandular. Leaves 10–40 cm. long; petiole and rhachis eglandular or only with tiny clustered bodies at leaflet insertions. Stipules very small, obliquely ovate-triangular, 2–3·5 mm. long, 1·5–2·5 mm. wide near base. Leaflets mostly in 4–10 pairs, broadly elliptic to suborbicular, sometimes oblong-elliptic or the terminal ones somewhat obovate, 2·5–9 cm. long, 1·8–6·3 cm. wide, generally emarginate or retuse at apex and without a mucro, minutely puberulous at least beneath. Racemes (including peduncle) 23–53 cm. long. Floral bracts not seen. Petals yellow, 2–2·5 cm. long, (0·6–)1·3–1·7 cm. wide. Stamens 10; filaments straight or nearly so; 2 anthers large, 5 medium, 3 small. Pods oblong, flattened, straight, 7·5–9 cm. long, 1·5–2·2 cm. wide, slightly venose, not crested. Seeds compressed, in same plane as pod, ± ovate, apiculate near hilum, 7–9 mm. long, 3·5–5 mm. wide, 1·5 mm. thick, reticulate, with an areole 4·5–5 × 0·5 mm. on each face.

KENYA. Northern Frontier Province: between Dandu and Gaddaduma, about 39° 45′ E. 3° 28′ N., 18 May 1952, *Gillett* 13231!
DISTR. **K1**; Ethiopia, Somali Republic
HAB. "Shallow soil over rock outcrop in alluvial flats"; 820 m.

NOTE. *C. ruspolii* is outstanding on account of its long racemes, long leaves and large emarginate or retuse leaflets.

13. **C. longiracemosa** *Vatke* in Oesterr. Bot. Zeitschr. 30: 80 (1880); L.T.A.: 639 (1930); K.T.S.: 102 (1961). Type: Kenya, Ndara and elsewhere in Teita, *Hildebrandt* 2506 (B ? holo.†, BM, K, iso.!)

Shrub to 3 m. high, or sometimes a small tree to 7 m. Stems with short spreading pubescence, eglandular. Leaves (3·5–)9–20 cm. long; petiole 0·9–2·1 cm. long, like the rhachis eglandular or only with tiny clustered bodies at leaflet insertions. Stipules narrowly subulate-triangular, 5–11 mm. long, 1–2·5 mm. wide near base, usually 2–5-nerved. Leaflets in 5–13 pairs, elliptic or slightly obovate, 0·7–2·5(–3·5) cm. long, 0·4–1·4(–2·2) cm. wide, subacute or rounded at apex, usually with a distinct mucro 0·5–2 mm. long, pubescent on both surfaces. Racemes (including peduncle) 12–37 cm. long; flowering axes usually sparsely pubescent. Bracts subtending flower-buds 7–10 mm. long, 2–4·5 mm. wide. Petals bright yellow, 1·4–2·8 cm. long, 0·8–1·5 cm. wide. Stamens 9–10; filaments straight or nearly so; 2 anthers large, 4–5 medium, 3 small. Pods oblong, flattened, straight or slightly curved, 3·5–7·5 cm. long, 1–1·8 cm. wide, purplish to blackish, slightly venose, not crested, at most umbonate over each seed. Seeds ± compressed, in same plane as pod, ± ovate, apiculate near hilum, 6–7·5 mm. long, 4–4·5 mm. wide, 1·5–2·5 mm. thick, reticulate, with a small areole 2–2·5 × 0·5–0·75 mm. on each face.

UGANDA. Karamoja District: Kanamugit, *Eggeling* 2992! & Kangole, Nov. 1941, *Dale* U.175!
KENYA. Northern Frontier Province: about 105 km. NE. of Wajir, 27 May 1952, *Gillett* 13355!; Kitui District: Mumoni Hill, Aug. 1937, *Gardner* in *F.H.* 3685!; Kwale District: between Samburu and Mackinnon Road, 31 Aug. 1953, *Drummond & Hemsley* 4071!
TANGANYIKA. Masai District: Shombole Mt., near Lake Natron, 23 Sept. 1944, *Bally* 3801!; Moshi District: Himo plains near Pangani Bridge, 13 Jan. 1945, *Bally* 4200!; Moshi/Teita (Kenya) Districts: Lake Chala, 21 Jan. 1936, *Greenway* 4451!
DISTR. **U1**; **K1**, 2, 4, 7; **T2**; Ethiopia and Somali Republic (S.)
HAB. Dry scrub with trees, deciduous bushland and semi-desert scrub; 300–1620 (–2130) m.

NOTE. Related to *C. didymobotrya*, *C. senna* and *C. italica*. Easily separated from the first by the stipules, bracts and pods; from the second by the longer stipules, usually broader leaflets, and narrower pods; from the third by the non-crested pods and (in East Africa) long racemes and spreading indumentum.

The leaflets of *C. longiracemosa* vary in shape from elliptic to orbicular. The var. *nummularifolia* Chiov. in Ann. Bot., Roma 13: 391 (1915) (Somali Republic (S.), Baidoa, *Paoli* 957, FI, holo.!) is merely an extreme with small suborbicular leaflets, to which some specimens from the Flora area, e.g. *L. C. Edwards* E44, Kenya, Kitui (EA!), approach closely.

14. **C. humifusa** *Brenan* in K.B. 13: 246 (1958); K.T.S.: 102 (1961). Type: Kenya, Northern Frontier Province, 29 km. NE. of Wajir, *Gillett* 13361 (K, holo.!, BR, EA, iso.!)

Herb, or sometimes slightly woody, with prostrate densely pubescent eglandular stems up to 80 cm. or more long. Leaves 5–14·5 cm. long; petiole (0·3–1·4 cm. long) and rhachis eglandular or only with tiny clustered bodies at leaflet insertions. Stipules narrowly lanceolate-triangular, or the larger ones oblong-triangular, 1-nerved, 6–13 mm. long, 1·5–3 mm. wide near base. Leaflets in 6–12 pairs, lanceolate or elliptic-lanceolate, 1·1–2·6(–3·5) cm.

long, 0·3–1(–1·6) cm. wide, acute or subacuminate and often shortly mucronate at apex, shortly pubescent on both sides. Racemes (including peduncle) 11–32 cm. long; flowering axis densely pubescent. Bracts subtending flower-buds ovate-acuminate, 3·5–5 mm. long, 2·5–3 mm. wide. Petals yellow, 1·6–2 cm. long, 0·9–1·1 cm. wide. Stamens 10; filaments straight or nearly so; 2 anthers large, 5 medium, 3 small. Ripe pods not seen, probably similar to those of *C. longiracemosa*.

KENYA. Northern Frontier Province: Wajir, 24 Dec. 1943, *Bally* 3730 in *C.M.* 16258! & Jan. 1949, *Dale* K.705! & Jan. 1955, *Hemming* 430!
DISTR. **K**1, apparently endemic and so far found only at or near Wajir
HAB. Dry scrub with trees; 210–610 m.

NOTE. Probably related to *C. longiracemosa* but differing in the prostrate stems, acute or subacuminate leaflets, single-nerved stipules, densely pubescent inflorescence-axes and, usually but not always, shorter petioles.

15. **C. fruticosa** *Mill.*, Gard. Dict., ed. 8, No. 10 (1768); T.T.C.L.: 97 (1949); Dale, Introd. Trees Uganda: 18 (1953); De Wit in Webbia 11: 247 (1955). Type: Mexico, Vera Cruz, *Houstoun* (BM, holo.!)

Shrub or small tree 2–7·5 m. high. Young branchlets minutely appressed-puberulous. Leaves: petiole eglandular; rhachis glandular between the lower of the two pairs of leaflets. Stipules usually small, linear-lanceolate to subulate, soon falling. Leaflets asymmetrically ovate (almost symmetrical in some South American forms, but not in East Africa), (2·5–)4–16(–22) cm. long, 1·7–7·2(–11) cm. wide, narrowed or subacuminate above to an acute or subacute sometimes obtuse apex, minutely and inconspicuously appressed-puberulous on both surfaces. Flowers usually pale or dull yellow, in short racemes aggregated near branchlet-ends. Sepals rounded at apex. Petals obovate to obovate-suborbicular, (1–)2–3·3 cm. long. Stamens: 7 fertile (or 6, *fide* De Wit) with short straight filaments and subequal anthers, the three lower anthers curved, the others less so; 3 upper stamens reduced, sterile, or (*fide* De Wit) absent. Pods subterete, brown, beaked, dehiscing along one suture only, 15–27 cm. long, 1–1·3 cm. in diameter, not septate within. Seeds many, embedded in sticky pulp, brown or blackish, subreniform, ± 4–5 × 2–3 mm.; areoles absent.

KENYA. Kilifi, 16 Sept. 1949, *Jeffery* 671!
TANGANYIKA. Lushoto District: Amani, 20 Jan. 1931, *Greenway* 2820!
DISTR. **K**7; **T**3; native apparently in the West Indies and Central and South America; now widely cultivated in the tropics and sometimes, as in the Flora area, escaping
HAB. " Growing amongst bushes and tall grass on a steep slope. Not very common " (*Greenway* 2820); in grass and bush on sandy soil (*Jeffery* 671); from ? near sea-level to 610 m.

SYN. *C. bacillaris* L. f., Suppl. Pl.: 231 (1781); Benth. in Trans. Linn. Soc. 27: 521, t. 62 (1871); L.T.A.: 635 (1930). Type: Surinam, *Dahlberg* (LINN, holo.!)

NOTE. The range of this species as a native is rather uncertain, as the American material is not homogeneous and needs revision. *C. fruticosa* is in cultivation at various places in Uganda, Kenya and Tanganyika.

16. **C. floribunda** *Cav.*, Descr.: 132 (1802); De Wit in Webbia 11: 245 (1955). Type: cultivated in Madrid Botanic Garden, originally from Mexico, Puebla de los Angeles. Whereabouts of type uncertain, not at MA

Herbaceous undershrub, shrub or small tree usually up to 3(–4·5) m. (or ? more) high. Branchlets glabrous, green. Leaves: petiole eglandular; rhachis glandular between all, or all but the topmost pair of leaflets. Stipules linear, very quickly falling off. Leaflets in normally 3–4 pairs (very rarely and sporadically 2 pairs by reduction, and in America as many as 5 pairs), lanceolate to ovate, almost symmetrical, (3–)4–11·3 cm. long, (1·6–)2–4 cm. wide, acutely and gradually tapering or subacuminate to apex, glabrous. Racemes axillary and apparently terminal, often aggregated near branchlet-ends. Sepals rounded at apex. Petals obovate to obovate-suborbicular, bright yellow, 1–1·5(–2) cm. long. Stamens: 2 with large anthers and long filaments; 1 with medium anther and filament; 4 with rather smaller anthers and very short filaments; 3 with usually reduced empty rounded anthers and very short filaments. Pods subterete, brown, very shortly or not beaked, slowly dehiscent, 6·2–10 cm. long, 1–1·5 cm. in diameter, longitudinally and transversely septate within. Seeds many, not embedded in pulp, olive, asymmetrically obovate, 5–6 × 3·5–4 mm.; areoles absent.

UGANDA. Toro District: Kyaka, Ibambalo Local Forest Reserve, 4 July 1950, *Osmaston* 662!; Ankole District: Kyamuhunga, June 1938, *Eggeling* 3660! & Igara, Bushenyi, Dec. 1938, *Cree* 160!

KENYA. Kiambu District: Thika, banks of Chania R., 29 June 1947, *Bogdan* 830!; Meru, May 1951, *Hancock* 107!

TANGANYIKA. Arusha District: E. Meru, Ngongongare, June 1953, *Procter* 232!; Lushoto District: Amani, 18 Dec. 1928, *Greenway* 1043!; Njombe District: Lupembe, 5 Feb. 1946, *van Rensburg* 265!

DISTR. **U**2; **K**3, 4, 6; **T**1–3, 7; widespread in the tropics, but probably native of America only; doubtless an alien in East Africa

HAB. Various: grassland, wooded grassland, river-banks, papyrus-swamps, upland dry evergreen forest; also in hedges, by roadsides, and in disturbed or waste ground; 910–3200 m.

SYN. *C. laevigata* Willd., Enum. Hort. Berol.: 441 (1809); Benth. in Trans. Linn. Soc. 27: 527 (1871); L.T.A.: 634 (1930); T.T.C.L.: 97 (1949); Steyaert in F.C.B. 3: 511 (1952); Dimitri & Alberti in Rev. Invest. Agric. 8: 18, fig. 6 (1954). Type: cultivated in Berlin Botanic Garden (? B-W, holo.)

NOTE. De Wit (in Webbia 11: 246 (1955)) described the seeds of *C. floribunda* as being 8 mm. long, but I suspect a mistake here. According to Corner, Wayside Trees of Malaya, ed. 2, 1: 389 (1952), *C. floribunda* can attain 10 m. in height.

17. **C. bicapsularis** *L.*, Sp. Pl.: 376 (1753); L.T.A.: 635 (1930); T.T.C.L.: 96 (1949); Steyaert in F.C.B. 3: 511 (1952); De Wit in Webbia 11: 235 (1955); F.F.N.R.: 120 (1962). Type: *Herb. Linnaeus* 528.10 (LINN, syn.!)

Shrub, often spreading, scrambling or climbing, 1·5–9 m. high. Stems glabrous (in East Africa—in South America sometimes puberulous). Leaves: petiole eglandular; rhachis with a prominent clavate or subglobose gland between the lowest pair of leaflets. Stipules very small, caducous, up to ± 0·6 mm. wide. Leaflets in (2–)3 pairs, obovate-elliptic to suborbicular, 0·9–3·1(–3·9) cm. long, 0·7–2·1(–2·4) cm. wide, rounded at apex, glabrous (in East Africa). Racemes 3–15-flowered, the peduncle usually well-developed and obvious. Pedicels 0·3–0·5 cm. long in flower, 0·5–0·7 cm. long in fruit. Sepals rounded at apex. Petals yellow or orange, obovate, 1–1·3 cm. long. Stamens: 3 lower with large anthers, central with short, laterals with longer filaments; 4 with medium anthers and very short filaments; 3 upper with reduced anthers, at least the lateral ones obhastate. Pods oblong-linear, straight, rounded at apex, not or tardily dehiscent, cylindrical, brown, (5–)8–15 cm. long, 1–1·5 cm. wide. Seeds many, olive-brown, compressed, elliptic to subreniform, 5–6 × 3·5–4·5 mm.; areoles absent.

UGANDA. Ankole District: Gayaza, *Ford* 611!; Teso District: Serere, Dec. 1931, *Chandler* 274!; Entebbe, Oct. 1931, *Eggeling* 40 in *F.H.* 206!

KENYA. Nakuru District: 19 km. on Nakuru–Eldama Ravine road, 15 Sept. 1948, *Bogdan* 2065!; Meru, June 1951, *Hancock* 115!; Teita District: Voi, 5 Feb. 1953, *Bally* 8744!

TANGANYIKA. Mwanza, Aug. 1932, *Rounce* 201!; Masai District: Mto Wa Mbu, 13 Sept. 1954, *Matalu* 3193!; Morogoro District: N. Uluguru Reserve above Morningside, June 1953, *Semsei* 1256!

DISTR. **U**2–4; **K**3–5, 7; **T**1, 2, 6; native of West Indies and western South America; cultivated in other parts of the tropics and sometimes escaped

HAB. Grassland, bushland, old cultivations, roadsides, etc., originally planted or naturalized; often used as a hedge-plant; near sea-level to 2130 m.

NOTE. Owing to insufficient information being given about the status of some of the specimens cited above, it is possible that the localities given are not all places where the species is truly naturalized. The East African distribution must also be accepted with caution for the same reason. See also footnote on p. 50.

18. **C. petersiana** *Bolle* in Peters, Reise Mossamb. Bot. 1: 13 (1861); L.T.A.: 633 (1930); T.S.K.: 61 (1936); T.T.C.L.: 97 (1949); U.O.P.Z.: 179 (1949); I.T.U., ed. 2: 60 (1952); Steyaert in F.C.B. 3: 508 (1952); K.T.S.: 102 (1961); F.F.N.R.: 119 (1962). Type: "Mozambique, Querimba I. and Mozambique", *Peters* (B, holo.†. A specimen at Kew (lower valley of R. Shire, *Meller*) is annotated as having been compared with the original specimen of *C. petersiana*)

Shrub or tree 0·6–12 m. high. Bark rough, fissured. Branchlets from subglabrous to densely pubescent or sometimes tomentose. Leaves: petiole eglandular; rhachis with a prominent projecting but rather caducous gland between some or all of the 4–10(–13) pairs of leaflets. Stipules conspicuous, leafy, semi-cordate to reniform, often caudate-attenuate at one end, to 2·5 cm. long, ± persistent. Leaflets lanceolate or ovate-lanceolate, sometimes ovate or ovate-elliptic, almost symmetrical, (2–)2·8–10·3 cm. long, (0·7–)0·9–4·2 cm. wide, gradually ± acuminate, rarely acumen obscure and short. Racemes corymbose, aggregated into ± corymbose terminal panicles. Sepals rounded at apex. Petals bright yellow, obovate, 1·5–3·2 cm. long, unequal, the largest somewhat oblique and dentate-margined. Stamens: 3 with large anthers and long filaments, 4 with medium anthers and short filaments, 3 with small rounded anthers and short filaments. Pods ± compressed, linear, 12–25 cm. long, 0·9–1·5 cm. wide, indehiscent, the valves ultimately breaking away from the prominent and transversely cracked sutures. Seeds many, somewhat compressed, ovoid or suborbicular, 5–7 × 4–6 mm., brown, with an olive areole ± 4–5 × 2·5 mm. on each face or marginal.

UGANDA. W. Nile District: Payida, 28 Nov. 1941, *A. S. Thomas* 4077!; Mbale District: Bufumbo, Nov. 1932, *Chandler* 1001!; Mengo District: Bukomero rest camp, Sept. 1932, *Eggeling* 541 in *F.H.* 916!

KENYA. Baringo District: Kabarnet, Nov. 1930, *Dale* 2433!; N. Kavirondo District: Kakamega Forest Station, 17 Sept. 1949, *Maas Geesteranus* 6268!

TANGANYIKA. Bukoba, Kabirizi, Apr. 1950, *Watkins* 405 in *F.H.* 3095!; Rufiji District: Mafia I., Kirongwe, 26 Aug. 1937, *Greenway* 5169!; 2·5 km. W. of Songea, 18 May 1956, *Milne-Redhead & Taylor* 10354!

ZANZIBAR. Zanzibar I., Haitajwa, 7 June 1930, *Vaughan* 1326! & Panga Juu, 10 June 1930, *Vaughan* 1341!

DISTR. **U**1–4; **K**3, 5; **T**1, 4, 6–8; **Z**; eastern Africa from Ethiopia and the Sudan Republic southwards to Mozambique and the Transvaal, extending westwards to Central African Republic (Haut-Ubangi) and Cameroun Republic; also in Madagascar

HAB. In or on edge of rain-forest, riverine forest, deciduous woodland, coastal evergreen bushland and wooded grassland; 12–2130 m.

VARIATION. Occurring over an extensive range of habitat and altitude, this widespread species is very variable. Although some of the variants show a correlation with geography, they do not seem sufficiently clearly defined to justify their being given names. In East Africa there are three principal ones:

1. With subglabrous stems and leaflets. On Zanzibar and in the neighbourhood of the Tanganyika coast usually at low altitudes up to about 300 m. (though *Schlieben* 4076, from the Uluguru Mts. at 1400 m., is this variant). *Greenway* 5169, *Vaughan* 1326 and 1341 all belong here, and also no doubt the type of *C. petersiana*.

2. With the stems ± densely pubescent and the lower surface of the leaflets varying from thinly appressed-hairy to rather densely pubescent. The commonest form in Uganda, Kenya and Tanganyika from about 300 m. upwards. All the Uganda and Kenya specimens cited above belong here, as does *Watkins* 405 in *F.H.* 3095 from Tanganyika.

3. With the stems and lower surface of the leaflets tomentose. Restricted to S. Tanganyika, between about 200 and 1400 m. alt. Examples are: Mbeya District, Mbozi–Abercorn road, near frontier, 30 Mar. 1932, *St. Clair-Thompson* 1117!; Rungwe District, Bulambia, 15 Mar. 1913, *Stolz* 1946!; Kilwa District, Madaba, 20 June 1932, *Schlieben* 2480! This variant is *C. petersiana* var. *tomentosa* Bak. f., L.T.A.: 634(1930).

Particularly in territories to the south of the Flora area perplexing intermediates occur between the above three categories, together with other variations (e.g. forms with small narrow leaflets in the Transvaal and Mozambique) that are absent from East Africa. *Milne-Redhead & Taylor* 10354, cited above, is intermediate between categories 2 and 3.

Further evidence is required about the pods of the three variants described above. Those of (1) are usually narrow, up to about 1·1 cm. wide, while those of (2) are often up to 1·3–1·5 cm. wide, but the material is insufficient and it is not always certain whether the pods available are really mature.

The bracts of *C. petersiana* vary very greatly in width, from lanceolate to reniform, sometimes in the same inflorescence. Pairs of glands, rather similar to those on the leaf-rhachis, occur on the inflorescence-axis, occupying the position of stipules in relation to the bracts.

18 × 19. **C. petersiana** × **singueana**?

Tree 6 m. high, differing from *C. petersiana* in the leaflets being mostly rounded to obtuse at apex, with rather fewer lateral nerves in about 6–8 pairs. It differs from *C. singueana* in the ± foliaceous stipules up to ± 5 mm. wide near base and in the often ovate-elliptic leaflets.

Uganda. Acholi District: Orom, 11 Oct. 1947, *Dawkins* 285!
Kenya. Tana River District: Garsen, S. bank of Tana, 16 Sept. 1958, *Hacker* 167B!
Distr. **U1**; **K7**; also in Zambia
Hab. Woodland; about 1070 m. (*Dawkins* 285)

Note. The identification of these specimens is not beyond doubt. It is just possible that they are most abnormal forms of *C. petersiana*, although unlikely. Further investigation in the field is needed.

19. **C. singueana** *Del.*, Cent. Pl. Afr.: 28 (1826); Del. in Caillaud, Voy. à Méroé 4: 27 (1827); T.T.C.L.: 98 (1949); I.T.U., ed. 2: 60, t. 4 (1952); Steyaert in F.C.B. 3: 509 (1952); Mendonça & Torre in C.F.A. 2: 179 (1956); K.T.S.: 102, t. 8 (1961); F.F.N.R.: 120 (1962). Type: Ethiopia, Singué [Jebel Singe], *Caillaud* (MPU, holo.)

Shrub or small tree 1–15 m. high. Branchlets glabrous to densely pubescent. Leaves: petiole eglandular (i.e. without conspicuous glands—minute reddish ones may be present); rhachis with a conspicuous fusiform or stipitate gland between each pair of leaflets except often the terminal. Stipules subulate, caducous, 0·5–0·75 mm. wide. Leaflets (5–)6–10(–12) pairs, elliptic, oblong-elliptic or obovate-elliptic, (1·5–)2·5–6·3(–7·5) cm. long, (0·7–)1·4–2·7(–3) cm. wide, rounded and often emarginate at apex, rarely subacute, glabrous or nearly so to densely pubescent. Racemes 6- to many-flowered, usually pedunculate and ± corymbose, often aggregated towards branchlet-ends and often produced when the plant is leafless; their axes bear conspicuous glands similar to those on the leaf-rhachides. Flowers yellow, ⚥ or sometimes ♀ (without stamens). Pedicels 2·2–4(–5) cm. long. Sepals rounded at apex. Petals obovate to suborbicular, 1·5–3(–3·5) cm. long. Stamens: 3 lower with large anthers and long filaments; 4 with somewhat shorter anthers and short filaments; 3 upper with reduced anthers. Pods linear, straight or somewhat twisted, torulose, subcylindric or slightly compressed, 5·5–26 cm. long, 0·7–1·0 cm. wide, indehiscent, with stiff and rather hard valves, glabrous to ± pubescent, rounded to abruptly acute and often apiculate at apex. Seeds dull brown, almost circular, flattened, ± 5–6 mm. in diameter, with a small areole 2–2·5 × 1–1·5 mm. on each face. Fig. 13, p. 74.

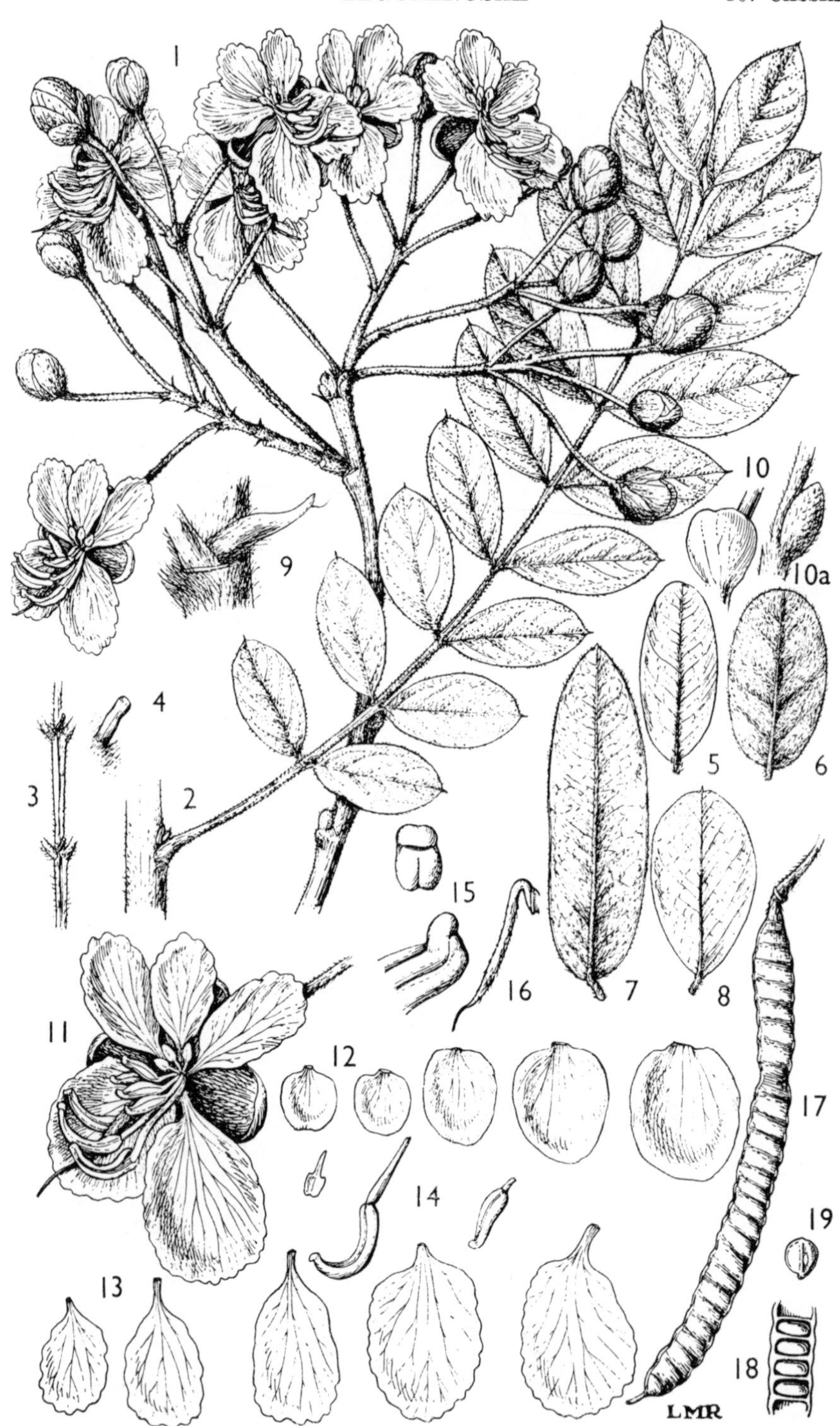

FIG. 13. *CASSIA SINGUEANA*—**1,** inflorescence (produced on leafless branchlet), × $\frac{2}{3}$; **2,** leaf and part of branchlet, × $\frac{2}{3}$; **3,** glands on rhachis of leaf, × 1; **4,** single gland from rhachis of leaf, × 6; **5–8,** undersides of leaflets from four different specimens, to show range of size, shape and indumentum, × $\frac{2}{3}$; **9,** gland on axis of inflorescence, × 8; **10, 10a,** bract and basal part of pedicel, from two different specimens, to show range of shape and indumentum, × 2; **11,** open flower, × 1; **12,** sepals, × 1; **13,** petals, × 1; **14,** stamens, one each of the three sizes, × 1; **15,** top of anther, viewed from above and side, × 6; **16,** ovary, × 1; **17,** pod, × $\frac{2}{3}$; **18,** part of pod, cut longitudinally, × $\frac{2}{3}$; **19,** seed, × $\frac{2}{3}$. 1–4, 9, 11–16, from *Milne-Redhead & Taylor* 11180; 5, from *Brasnett* 312; 6, from *Anderson* 928; 7, from *Tanner* 2906; 8, from *Stolz* 2393; 10, from *Richards* 10138; 10a, from *Willan* 29; 17, from *Bally* 7578; 18, 19, from *Lynes* I.g. 22.

UGANDA. Acholi District: Padibe, Apr. 1943, *Purseglove* 1354!; Karamoja District: Warr, 4 Nov. 1939, *A. S. Thomas* 3159!; Teso District: Serere, Dec. 1931, *Chandler* 347!

KENYA. Northern Frontier Province: Balambala, 23 Jan. 1943, *Bally* 2181!; Kiambu District: Theta road, 5 July 1952, *Kirrika* 199!; Kwale District: Shimba Hills, Mwele Mdogo Forest, 6 Feb. 1953, *Drummond & Hemsley* 1149!

TANGANYIKA. Shinyanga, 11 Oct. 1949, *Windisch-Graetz* in *Bally* 7578!; about 8 km. S. of Dodoma, 19 July 1956, *Milne-Redhead & Taylor* 11180!; Rungwe District: Kyimbila, 19 Dec. 1913, *Stolz* 2393!

DISTR. **U1**, 3; **K1**, ?2, 3–5, 7; **T1**–8; widespread in tropical Africa, except in the rain-forest regions; also on the Comoro Is.

HAB. Woodland, wooded grassland and bushland; frequently noted on termite mounds; near sea-level to 2130 m.

SYN. *C. goratensis* Fres. in Flora 22: 53 (1839); L.T.A.: 634 (1930); T.S.K.: 61 (1936). Type: Ethiopia, *Rueppell* (FR, holo.)

C. zanzibarensis Vatke in Oesterr. Bot. Zeitschr. 30: 77 (1880); L.T.A.: 635 (1930); T.T.C.L.: 99 (1949). Type: Tanganyika, Bagamoyo District, R. Wami and R. Kingoni, *Hildebrandt* 904 (B, holo.†, BM, iso.!)

C. goratensis Fres. var. *glabra* Bak. f., L.T.A.: 634 (1930). Types: no specimens cited

C. goratensis Fres. var. *flavescens* Bak. f., L.T.A.: 634 (1930). Types: Tanganyika, Tanga District, Misoswe (Mizozue), *Holst* 2225 (K, syn.!) & Tanganyika, without locality, *Busse* 169 (BM, syn.!, K, isosyn.!) & *Busse* 392 (BM, syn.!, K, isosyn.!)

C. singueana Del. var. *glabra* (Bak. f.) Brenan in K.B. 4: 77 (1949); T.T.C.L.: 99 (1949)

C. singueana Del. var. *flavescens* (Bak. f.) Brenan in K.B. 4: 77 (1949); T.T.C.L.: 98 (1949)

VARIATION. *C. singueana* shows a wide range of variation which is difficult to analyse clearly. It appears that characters such as presence or lack of indumentum, number, size and shape of leaflets and size of flowers are varying to some extent independently. Specimens from a single region often look similar, but may pass by perplexing intermediates into a different variant occupying an adjacent region. Furthermore the same or similar variants may occur in regions widely separated geographically. It seems therefore preferable not to recognize any named infraspecific taxa, but to attempt a general simplified account of what occurs in East Africa.

In Uganda forms prevail with comparatively short leaflets about 2–4 cm. long and up to or slightly more than twice as long as wide. Their general shape is elliptic, with their greatest breadth about the middle, but less commonly tendencies to an obovate or slightly ovate shape may be shown. The indumentum on the leaflets varies from densely pubescent to almost glabrous. The inflorescence axes are ± grey-pubescent and the sepals ± so outside.

In inland Kenya forms occur similar to those of Uganda, and these extend to the coast. There, however, a rather distinctive variant also occurs, with rather long elliptic leaflets up to 6 cm. in length and usually more than twice as long as broad and the indumentum on the outside of the sepals characteristically dense, conspicuous and yellowish (var. *flavescens*, see above).

This variant also extends along the coast of Tanganyika. A specimen of this variant, *Hughes* 240, from Tanganyika, Tanga District, Same–Gonja road, has valuable notes on the bark, which is there stated to be medium-grey, smooth, marked with characteristic broadly obtriangular lenticels 0·6 cm. long, slight vertical ridges and from time to time marked horizontal ones. It is at present uncertain if these characters are shown also by the other variants of *C. singueana*.

Inland in Tanganyika two other principal forms occur, one with densely pubescent short obovate leaflets which is found in Kenya also, and a glabrous one in south-western Tanganyika (var. *glabra*, see above) which is scarcely separable from similar but apparently infrequent forms in Kenya and Uganda.

20. **C. baccarinii** *Chiov.* in Ann. Bot., Roma 13: 390 (1915); L.T.A.: 638 (1930); K.T.S.: 101 (1961). Type: Somali Republic (S.), between Baghei and Audianle, *Paoli* 946* (FI, lecto.!)

* There are two specimens at Florence annotated as *C. baccarinii*: one, *Paoli* 827 from Bardera, is *C. holosericea* Fres.; the other, *Paoli* 946, is the lectotype of *C. baccarinii* cited above. Owing to some strange oversight, Chiovenda cited a single " type " of *C. baccarinii*, consisting of the locality of 946 combined with the number of the other specimen. The description of *C. baccarinii* seems, however, entirely based on 946.

Shrub 1·2–2·1 m. high. Branchlets minutely appressed-puberulous, as are the petioles, rhachides and pedicels. Leaves: petiole eglandular (i.e. without conspicuous glands—minute reddish ones may be present); rhachis with a conspicuous fusiform or stipitate gland between the lowest pair or several pairs of leaflets. Stipules subulate, caducous, 0·3–0·75 mm. wide. Leaflets normally 4–6 pairs (only occasional reduced leaves with 2–3 pairs), elliptic to obovate-elliptic, 0·9–3(–3·4) cm. long, 0·4–1·6(–1·8) cm. wide, rounded and often ± emarginate at apex, minutely appressed-puberulous on both surfaces. Racemes 2–6(? more)-flowered, clearly pedunculate, corymbose; their axes often with conspicuous glands similar to those on the leaf-rhachides. Pedicels (in flower) 1·5–1·8 cm. long. Sepals rounded at apex. Petals yellow, obovate to suborbicular, 1·3–2 cm. long. Stamens: 3 lower with large anthers and long filaments; 4 with somewhat shorter anthers and short filaments; 3 upper with reduced anthers. Pods oblong-linear, slightly curved, flattened, 5–7·5 cm. long, 1–1·4 cm. wide, indehiscent, with papery valves, scarcely undulate, appressed-pubescent, rounded to abruptly acute at apex. Seeds olive, normal size and shape uncertain, but with a small areole 1·5 × 0·5–0·8 mm. on each face.

KENYA. Northern Frontier Province: Mandera region, near Baloboleh [? Balballa], Jan. 1949, *Dale* K702! & Dandu, 22 Mar. 1952, *Gillett* 12616! & about 5 km. S. of Takabba, 21 May 1952, *Gillett* 13253!

DISTR. **K1**; Somali Republic

HAB. Dry scrub with trees; 610–760 m.

NOTE. Chiovenda described *C. baccarinii* as having 8 anthers—3 longer and 5 shorter, but I find what is at any rate the normal arrangement as in the description above. Pods of this species showing well-developed seeds are wanted.

21. **C. auriculata** *L.*, Sp. Pl.: 379 (1753); L.T.A.: 639 (1930); T.T.C.L.: 98 (1949); De Wit in Webbia 11: 234 (1955). Type: Ceylon, *Hermann* (BM, syn.!)

Shrub or small tree 1–5(–7·5) m. high. Branchlets pubescent. Leaves: petiole eglandular; rhachis with a narrow subulate or fusiform gland between each (except sometimes the terminal pair) of the (4–)6–13 pairs of leaflets. Stipules ± persistent, leafy, broadly reniform, 7–22 mm. wide, one side produced into a subulate point. Leaflets oblong-elliptic to obovate-elliptic, (0·8–)1–2·5(–3·7) cm. long, 0·4–1(–1·2) cm. wide, rounded and mucronate at apex, in the Flora area ± puberulous on margins and midrib beneath (sometimes ± pubescent, especially beneath). Racemes corymbose, 2-8-flowered, aggregated into ± rounded terminal panicles. Sepals rounded at apex. Petals yellow, 1·7–3 cm. long. Stamens: 3 lower with large anthers and long filaments; 4 with medium anthers and short filaments; 3 upper with reduced anthers. Pods oblong-linear, straight, flattened, 6–10(–18) cm. long, 1·2–1·6(–2·3) cm. wide, indehiscent, with papery valves, transversely undulate between seeds, shortly pubescent. Seeds purplish-brown, compressed, ovate-oblong, 7–9 mm. long, 4–5 mm. wide; a distinct areole on each face 3–3·5 × 0·5–0·75 mm.

TANGANYIKA. Morogoro District: 16 km. along Wami road, 11 May 1933, *B. D. Burtt* 4672!; Uzaramo District: 98 km. from Dar es Salaam on main road to Morogoro, 21 June 1955, *Welch* 295!; Ulanga District: between the Mahenge Plateau and the confluence of R. Kilombero and R. Luwegu, 11 June 1932, *Schlieben* 2326!

DISTR. **T6**; India, Ceylon, Burma; cultivated here and there in the tropics; in Africa also in the Sudan Republic, but probably alien there

HAB. Woodland and wooded grassland, in *Acacia clavigera* subsp. *usambarensis*-*Combretum* country (T.T.C.L.: 98); 150–610 m.

SYN. *C. densistipulata* Taub. in P.O.A. C: 200 (1895); Ghesq. in Rev. Bot. Appliq. 14: 244 (1934). Type: Tanganyika, Morogoro District, Kimambira, *Stuhlmann* 104 (B, holo. †)

NOTE. I am unable to follow Ghesquière's contention (Rev. Bot. Appliq. 14: 244 (1934)) that *C. densistipulata* is specifically distinct from *C. auriculata*. If *C. auriculata* is truly native in East Africa then its geographical distribution is remarkable; however, the possibility of its early introduction by man into Tanganyika from the East should be considered. There is a specimen from **T7**, Iringa District, Ngerere (*Busse* 3034 in EA !), but there is no evidence to indicate whether it was from a wild plant or not.

22. **C. obtusifolia** *L.*, Sp. Pl.: 377 (1753); De Wit in Webbia 11: 254 (1955); Brenan in K.B. 13: 248 (1958); F.F.N.R.: 120 (1962). Type: Dillenius, Hortus Eltham.: 71, t. 62 (1732) (lecto. !). A specimen grown from seed collected in Cuba, near Havana, *Herb. Dillenius* (OXF, typo. !)

Annual or perennial herb or undershrub 0·6–2(–2·4) m. high. Stems ± pubescent. Leaves: petiole eglandular; rhachis with a prominent cylindrical gland between each of the 1–2 lower pairs of leaflets. Stipules linear or filiform. Leaflets in 3 pairs, obovate, (1–)1·5–5(–6) cm. long, (0·5–)1–3(–3·9) cm. wide, rounded or abruptly narrowed above to a usually mucronate apex, ± pubescent to subglabrous on surfaces. Racemes 1–2-flowered, the peduncle usually (not always) very short, the flowers thus usually appearing axillary. Pedicels (0·5–)1·5–3·5 cm. long in flower, 1·5–4·5 cm. long in fruit. Sepals rounded at apex. Petals yellow, obovate, 0·95–1·9 cm. long. Stamens: 3 lower largest, with anthers 3–4·5(–7) mm. long and narrowed shortly below their apex like the neck of a bottle; 4 stamens somewhat smaller; 3 very small and reduced. Pods linear, straight or ± curved, tapering at base and apex, (6–)13–23 cm. long, (0·4–)0·5–0·6(–0·7) cm. wide, dehiscent, subterete and ± angled longitudinally. Seeds many, brown, ± rhombic or ovoid, 4·5–6·5 × 2–4 mm.; areoles linear, very narrow, 3–4·5 × 0·3–0·5 mm.

UGANDA. Karamoja District: Kangole, 22 May 1940, *A. S. Thomas* 3464 !; Teso District: Serere, July–Aug. 1932, *Chandler* 852 !; Busoga District: 7 km. W. of Kaliro, Bugonzo, 17 Sept. 1952, *G. H. S. Wood* 401 !

KENYA. Uasin Gishu District: Kipkarren, Mar. 1932, *Brodhurst Hill* 677 in *C.M.* 3701 !; S. Kavirondo District: Kisii, Sept. 1933, *Napier* 5325 !; Mombasa, Feb. 1930, *R. M. Graham* DD 774 in *F.H.* 2285 & in *C.M.* 1362 !

TANGANYIKA. Shinyanga, *Koritschoner* 1698 !; Mpwapwa, 25 Mar. 1929, *Hornby* 88 !; Songea District: Mbamba Bay, 5 Apr. 1956, *Milne-Redhead & Taylor* 9480 !

ZANZIBAR. Zanzibar I., Bweleo, 28 Jan. 1929, *Greenway* 1223 !; Pemba I., Vitongoge, 4 Aug. 1929, *Vaughan* 443 !

DISTR. **U**1–4; **K**2–5, 7; **T**1–8; **Z**; **P**; pantropical, extending northwards into southern U.S.A.

HAB. Grassland, and a weed in cultivated ground, by roadsides and on waste ground near habitations; often recorded as growing near pools, in river-beds and near water; ? near sea-level to 1680 m.

SYN. *C. tora* L. var. *obtusifolia* (L.) Haines, Bot. Bihar & Orissa: 304 (1922)
[*C. tora* sensu auct. mult., e.g. L.T.A.: 636 (1930); Steyaert in F.C.B. 3: 512 (1952); Dimitri & Alberti in Rev. Invest. Agric. 8: 14, fig. 4 (1954); Mendonça & Torre in C.F.A. 2: 180 (1956), *non* L.]

NOTE. *C. obtusifolia* has been very generally confused with *C. tora* L., a species confined to Asia, from India to China and Fiji, with the possible exception of one Congo specimen (*Devred* 3951) and the single East African specimen discussed below. *C. tora* differs from *C. obtusifolia* in the shorter pedicels, about 0·5–1 cm. in flower and not exceeding 1·5 cm. in fruit, in the three large anticous anthers being abruptly rounded at apex and not narrowed into a neck, and in the seeds (usually narrow, to 2·7 mm. wide) having areoles 1·5–2 mm. wide; the areole is thus not at all linear as in *C. obtusifolia* and may indeed cover most of each face of the seed. *C. tora* always has a gland between each of the two lower pairs of leaflets, while in all parts of the world except Africa *C. obtusifolia* has, with rare exceptions, only one gland per leaf; in Africa, however, there may be one or two, even on the same plant.

Now there is a single specimen in the East African Herbarium that may be the true *C. tora*: Tanganyika, Mafia I., May 1907, *Zimmermann* 1452 ! It has two glands per leaf, pedicels of immature fruits 1–1·3 cm. long, and the largest anthers abruptly rounded. There are no ripe seeds; the plant has never been re-collected; and there is no evidence of its status on Mafia. I am therefore unwilling to admit *C. tora* to the flora at present, but it should be searched for again on Mafia I.

23. **C. occidentalis** *L.*, Sp. Pl.: 377 (1753); L.T.A.: 635 (1930); T.T.C.L.: 97 (1949); U.O.P.Z.: 179, fig. (1949); Steyaert in F.C.B. 3: 513 (1952); Dimitri & Alberti in Rev. Invest. Agric. 8: 22, fig. 8 (1954); De Wit in Webbia 11: 256 (1955); Mendonça & Torre in C.F.A. 2: 181 (1956); F.F.N.R.: 119 (1962). Type: a cultivated plant in *Herb. Clifford* (BM, syn.!)

Erect herb, sometimes slightly woody, 0·15–2 m. high. Stems subglabrous. Leaves up to ± 20(–25) cm. long; petioles shortly above base with a large sessile squat hemispherical to ovoid or subglobose gland; rhachis eglandular. Leaflets in (3–)4–5(–6) pairs (in East Africa), ovate to ovate-elliptic or (sometimes) lanceolate, (2·5–)5–12 cm. long, rarely more, (1·5–)2–4 cm. wide, acute or acuminate* at apex, glabrous except for ciliolate margins and inconspicuous scattered glands beneath. Racemes from upper axils, very short, almost umbellate; peduncles 3–5 (very rarely and exceptionally to 8) mm. long (in East Africa). Bracts acute. Sepals normally glabrous outside. Petals yellow, obovate, 0·9–1·5 cm. long. Stamens 10; 2 anticous rather large on long filaments, 4 smaller on shorter filaments, 3 (one of them anticous) much smaller. Pods slightly curved upwards or sometimes nearly straight, linear, (5–)8–12·5(–13) cm. long, 0·5–1·0 cm. wide, not or tardily dehiscent, compressed, brown, subglabrous, septate, many-seeded. Seeds compressed, lying at right angles to long axis of pod, grey-brown, ovate-suborbicular, 4·5–5 × 3·75–4·5 mm.; testa minutely pimpled, with an elliptic areole ± 2·5 × 1·5 mm. on each face. Fig. 14.

UGANDA. W. Nile District: Obongi, June 1952, *Leggat* 93!; Teso District: Serere, June 1932, *Chandler* 787!; Old Entebbe, Jan. 1932, *Eggeling* 177 in *F.H.* 397!

KENYA. Kiambu, Aug. 1932, *Mainwaring* 2192 in *C.M.* 5677!; Teita District: Voi, 7 May 1931, *Napier* 971 in *C.M.* 2253!; Tana R., *Battiscombe* 257 in *C.M.* 13926!

TANGANYIKA. Shinyanga District: Tinde, Mwangilye Hill Forest Reserve, Jan. 1953, *Gane* 57!; Mafia I., Kirongwe, 25 Aug. 1937, *Greenway* 5164!; Songea District: Mbamba Bay, 5 Apr. 1956, *Milne-Redhead & Taylor* 9479!

ZANZIBAR. Zanzibar I., Fumba beach, 9 Feb. 1930, *Vaughan* 1214!; Pemba I., Mkoani, 9 Aug. 1929, *Vaughan* 472!

DISTR. **U**1–4; **K**3–5, 7; **T**1–8; **Z**; **P**; pantropical, possibly originating in tropical America

HAB. Usually a weed of cultivation, roadsides and waste ground near villages and buildings; also recorded from grassland and lake-shores; 0–1740 m.

NOTE. The " plant at times emits a very unpleasant smell " (Greenway in T.T.C.L.: 97). For this reason it has no doubt got its name " Stinking Weed " (U.O.P.Z.: 179).

In East Africa a constant and easily recognized species, only to be confused with *C. sophera*, under which species the distinctions are given.

Fyffe 83 (K!), Entebbe Botanic Garden, has rather narrower leaflets in up to 10 pairs, peduncles about 10–11 mm. long and more turgid pods. The petiolar gland is rather similar to that of *C. occidentalis*, of which it is probably a variety. The specimen is a poor one, and it is likely that it was only in cultivation.

24. **C. sophera** *L.*, Sp. Pl.: 379 (1753); Benth. in Trans. Linn. Soc. 27: 532 (1871); L.T.A.: 636 (1930); Dimitri & Alberti in Rev. Invest. Agric. 8: 26, fig. 10 (1954); De Wit in Webbia 11: 265 (1955). Type: Ceylon, *Hermann* (BM, lecto.!)

Shrub up to 2(–3) m. high. Stems subglabrous. Leaves up to ± 25 cm. long, often considerably shorter; petioles shortly above base with a sessile clavate or cylindrical often somewhat pointed gland; rhachis eglandular. Leaflets in 4–10(–12, *fide* De Wit) pairs, lanceolate to oblong-lanceolate or ovate-lanceolate, 2–7(–9, *fide* De Wit) cm. long, 0·8–2(–2·5) cm. wide, acute or acuminate in the Flora area, elsewhere often obtuse, glabrous except for ciliolate margins and sometimes for a very few inconspicuous scattered glands beneath. Racemes from upper axils, very short, almost umbellate; peduncles 0·8–2·5 cm. long. Bracts obtuse to subacute. Sepals thinly pubescent or

* But may be obtuse or rounded on very juvenile shoots.

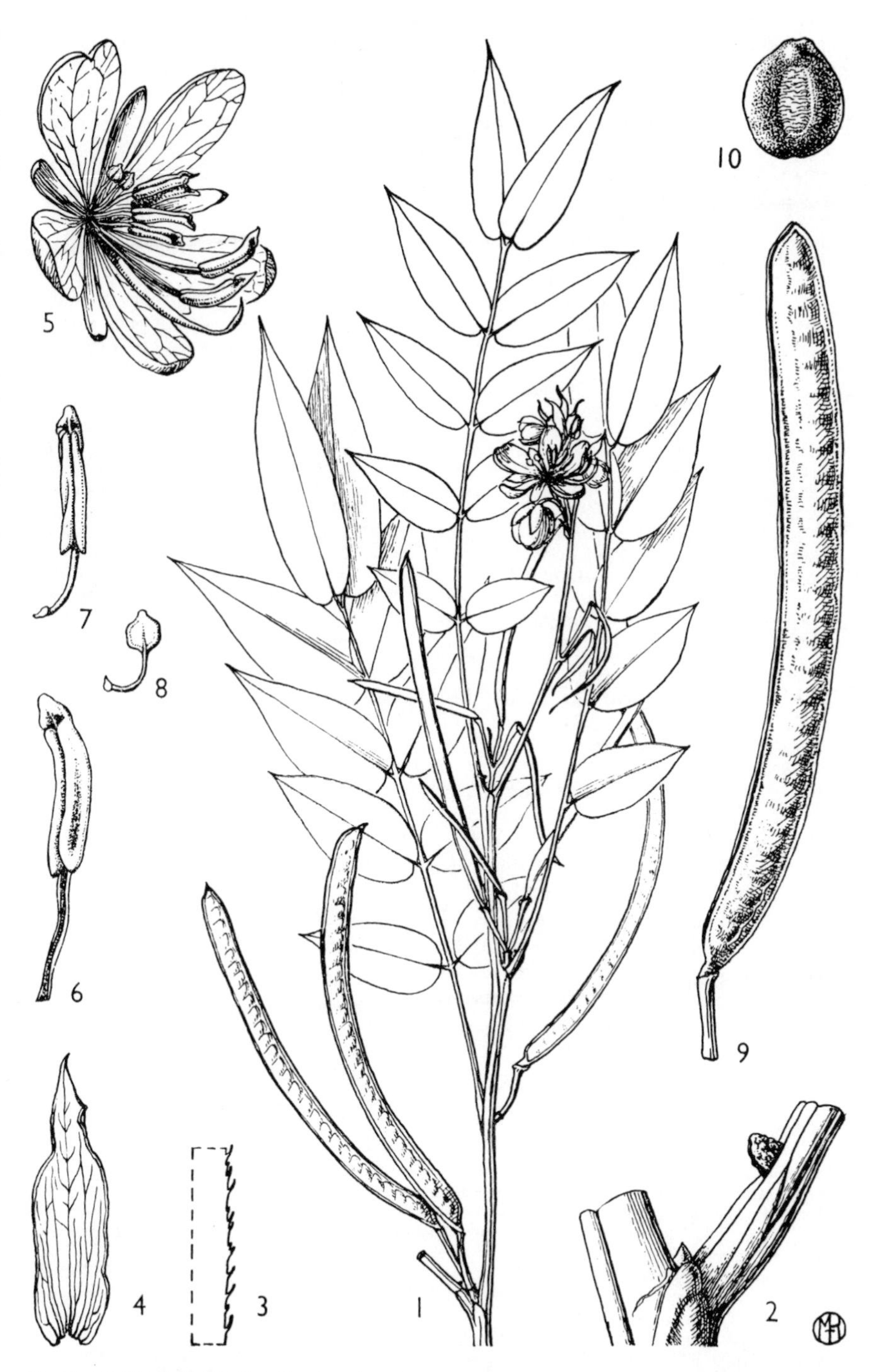

Fig. 14. *CASSIA OCCIDENTALIS*—**1**, part of plant, × ½; **2**, basal part of petiole showing insertion on stem, pulvinus and gland, × 4; **3**, part of leaflet-margin, × 4; **4**, bract, × 2; **5**, flower, × 2; **6**, anticous stamen, × 4; **7**, median stamen, × 4; **8**, posticous stamen, × 4; **9**, pod, × 1; **10**, seed, × 4. 1–3, 5–8, from *Fyffe* 131/24; 5, from *Tweedie* 1033; 9, 10, from *Faulkner* 1648.

puberulous outside (in the Flora area, elsewhere sometimes glabrous). Petals yellow, obovate, 1–1·5 cm. long. Stamens similar to those of *C. occidentalis*. Pod turgid or subcylindrical, straight or very slightly curved upwards, linear, 5–10·4 cm. long, 0·6–1·1 cm. wide, not or tardily dehiscent, brown, glabrous or nearly so, septate, many-seeded. Seeds ± compressed, lying at right angles to the long axis of the pod, grey-brown, ovate to suborbicular, 4–4·5 × 3·5–4·5 mm.; testa almost smooth, with an elliptic areole ± 2–2·5 × 1–1·5 mm. on each face.

KENYA. Mombasa, Feb. 1930, *R. M. Graham* DD 783 in *F.H.* 2282!; Lamu District: Witu, *F. Thomas* 149!

TANGANYIKA. Rufiji District: Mafia I., 31 Mar. 1933, *Wallace* 836!

DISTR. **K7**; **T6**; country of origin doubtful; pantropical, occurring especially in Asia, rarer in America and Africa; in the latter continent frequent in West Africa, rare in East Africa, occurring in Somali Republic as well as the Flora area, but probably always alien.

HAB. Uncertain, presumably occurring at low altitudes; *Graham* in *F.H.* 2282 in "cultivated lands", probably as a weed

NOTE. *C. sophera* is very similar and closely related to *C. occidentalis*. The difficulty in distinguishing the two was discussed long ago by Bentham in Trans. Linn. Soc. 27: 509, 533 (1871). In East Africa *C. sophera* may be easily separated from *C. occidentalis* by the narrower and more raised petiolar glands, the normally more numerous pairs of leaflets, the longer peduncles, the blunter bracts and the more turgid pods. Elsewhere forms of *C. sophera* or *C. occidentalis* occur which break down some of these distinguishing characters. It may be that we are dealing with races of one species, but it would be unwise to assume this without revising this group on a world scale.

C. sophera has been cultivated as a possible cover-crop in Tanganyika at Lyamungu (*Greenway* 4513!)

25. **C. hirsuta** *L.*, Sp. Pl.: 378 (1753); Steyaert in F.C.B. 3: 513 (1952); De Wit in Webbia 11: 250 (1955). Type: a cultivated plant in the Clifford Herbarium (BM, holo.!)

Shrubby herb 0·2–2·4 m. high. Stems pubescent or sometimes pilose, grooved. Leaves up to ± 22 cm. long; petioles shortly above base with a sessile finger-like gland directed towards top of petiole; rhachis eglandular. Leaflets in (2–)3–5(–6) pairs, elliptic to ovate-elliptic, acutely subacuminate or sometimes acute at apex, rather densely pilose on both surfaces. Racemes from middle and upper axils, very short, almost umbellate. Bracts very narrow and acute. Outer sepals densely pilose outside. Petals yellow, obovate, 1–1·7 cm. long. Stamens 10; 2 large, 5 medium (one the central anticous one) and 3 small. Pods somewhat curved downwardly, linear, (8–)12–15(–18) cm. long, 0·3–0·7 cm. wide, dehiscent, with convex valves, brown, densely pilose or pubescent, septate, many-seeded. Seeds somewhat compressed, lying at right angles to the long axis of the pod, dark olive, ovate-suborbicular, ± 3 × 2·5 mm., with a ± elliptic-oblong areole ± 1·75 × 0·75 mm. on each margin.

UGANDA. Mubende District: Kakumiro, Sept. 1936, *Tothill* 2609!

TANGANYIKA. Lushoto District: Amani, 22 May 1929, *Greenway* 1535!; Morogoro District: Kimboza, 30 Mar. 1954, *Padwa* 320! & Matombo, Mar. 1955, *Anatoli* 9!

DISTR. **U4**; **T3**, 6; originally from tropical America, but now established in various parts of the Old World tropics

HAB. A naturalized weed of plantations and cultivated ground in lowland rain-forest areas; said to be very common in the old African cultivations in Kimboza Forest Reserve; 550–1220 m.

NOTE. De Wit (in Webbia 11: 250 (1955)) describes this plant as "stinking". He also (op. cit.: 251) describes the areole as "indistinct, large obovate". I find, however, the areole to be ± elliptic-oblong, rather small and distinct, and placed on the narrow "marginal" part of the seed. In this latter point it differs from all the other East African cassias known to me with compressed areolate seeds; in these the areoles are situated one on each of the flattened faces of the seed.

26. **C. absus** *L.*, Sp. Pl.: 376 (1753); L.T.A.: 639 (1930); T.T.C.L.: 96 (1949); Steyaert in F.C.B. 3: 507 (1952); De Wit in Webbia 11: 279 (1955); Mendonça & Torre in C.F.A. 2: 179 (1956). Type: Hortus Upsaliensis, *Herb. Linnaeus* 528.4 (LINN, syn.!)

Herb, annual or sometimes slightly woody, 0·1–1·2 m. high (to 1·5 m., *fide* F.C.B.), erect or procumbent, usually much branched, sticky because of glandular-based setae in the indumentum. Leaves with eglandular * petiole; rhachis 0·5–1·5 cm. long, between each of the 2 pairs of leaflets with a pale acute gland(?) to ± 1 mm. long. Leaflets obliquely elliptic or obovate, 1–4·5 cm. long, 0·8–3 cm. wide. Racemes 1–13 cm. long, several-flowered. Flowers small, ± 5–7 mm. long, usually yellow, orange, pink or red, said occasionally to be white. Stamens 5, subequal, with straight filaments. Pods (2·5–)4–5·5 cm. long, 0·7–0·8 cm. wide, ± setose-hairy, flattened. Seeds obovate or subrhombic, brown to black, glossy, 4–5·5 × 3·5–4·5 mm. Fig. 15, p. 82.

UGANDA. Karamoja District: Moroto, Sept. 1955, *Wilson* 191!; Busoga District: 3 km. N. of Nkondo gombolola and W. of Nkondo–Kigingi Pier road, 9 July 1953, *G. H. S. Wood* 809!; Mengo District: N. of Bale, Bugerere, 30 June 1956, *Langdale Brown* 2129!

KENYA. Northern Frontier Province: Dandu, 5 May 1952, *Gillett* 13055!; Machakos District: Mtito Andei, Jan. 1950, *Bally* 7714!; Lamu District: W. of Kipini on Witu road, 29 Feb. 1956, *Greenway & Rawlins* 8959!

TANGANYIKA. Moshi, May 1927, *Haarer* 448!; Mpwapwa, 4 Mar. 1929, *Hornby* 74!; Songea District: near R. Luhimba about 28 km. N. of Songea, 6 May 1956, *Milne-Redhead & Taylor* 10102!

DISTR. **U**1–4; **K**1, 4, 5, 7; **T**1–8; widespread in the tropics of the Old World

HAB. Grassland, open places in deciduous bushland and recorded from among the herbaceous vegetation on granite outcrops; also a weed in old cultivations on roadsides, sandy lake-shores and waste ground; 7–1680 m.

NOTE. Beyond differences in amount of branching, habit, size of leaves, etc., such as one might expect in a weedy plant, little variation, except to a certain extent in the density of the indumentum, is shown by this easily recognized species.

27. **C. nigricans** *Vahl*, Symb. Bot. 1: 30 (1790); L.T.A.: 641 (1930); Ghesq. in B.J.B.B. 9: 161 (1932); Steyaert in F.C.B. 3: 518 (1952); Mendonça & Torre in C.F.A. 2: 181 (1956). Type: Yemen, Wadi Surdûd, *Forsskål* (C, holo.)

Herb, apparently annual, erect, simple or branched, 25–45 cm. (–1·8 m.) high. Stems pubescent with short crisped and longer spreading hairs. Leaves ± oblong, 3–10 cm. long, 2–4 cm. wide; gland at top of petiole, sessile, cushion-shaped, 2–4 mm. long, 1–1·25 mm. wide; rhachis eglandular, channelled but not crenate-crested along upper side; leaflets sessile, mostly in 7–15(–18, *fide* L.T.A., but not confirmed) pairs, narrowly oblong, straight, (10–)12–25(–33) mm. long, (2–)2·5–5(–7·5) mm. wide, rounded to obtuse and mucronate, rarely subacute, at apex, shortly and rather densely pubescent on both surfaces; midrib central or almost so, particularly above. Inflorescences supra-axillary (sometimes a second axillary one present), 3–8-flowered; pedicels very short, 2–4(–5) mm. long. Petals small, yellow, 3·5–4·5 mm. long, 1·5–3·5 mm. wide. Pods erect, 1·7–2·4(–4) cm. long, 4–5 mm. wide. Seeds brown, obovate or rhombic, 2·5–4 mm. long, 1·5–2·5 mm. wide, not areolate. Fig. 10/27, p. 57.

UGANDA. W. Nile District: Omogo, 15 Aug. 1953, *Chancellor* 165!; Teso District: Serere, Apr. 1932, *Chandler* 545!; Busoga District: 5 km. N. of Namwiwa, Lubolo, 17 July 1953, *G. H. S. Wood* 831!

KENYA. S. Kavirondo District: Lambwe valley, July 1935, *Napier* 3450 in *C.M.* 6776!

* "Eglandular" here means without the large specialized glands characteristic of the petiole and rhachis of *Cassia*; glandular bristles occur plentifully on the petiole of *C. absus*.

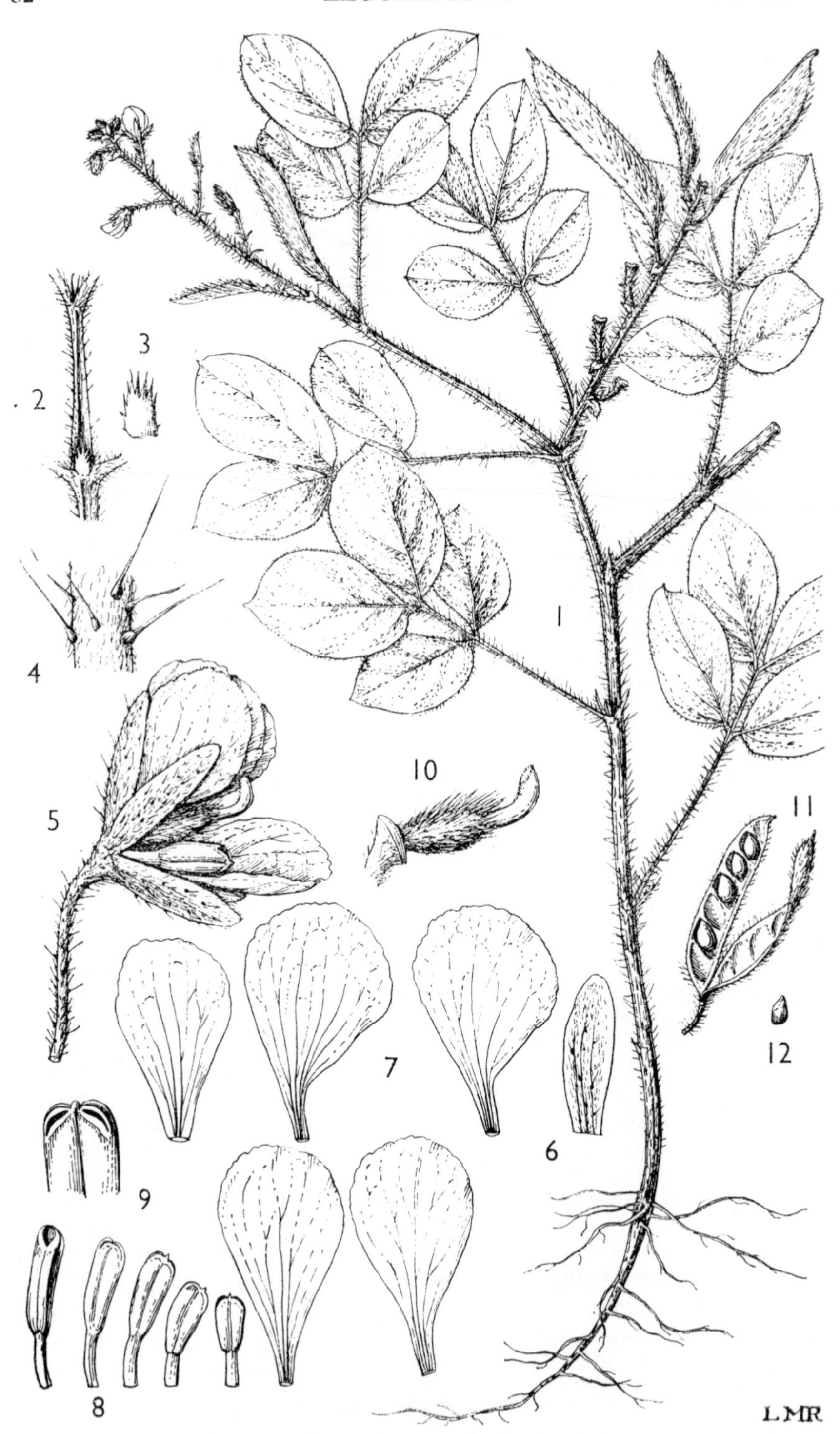

FIG. 15. *CASSIA ABSUS*—**1,** habit, showing root, × $\frac{2}{3}$; **2,** gland on leaf-rhachis, × 4; **3,** single gland, × 8; **4,** hairs on young stem, × 8; **5,** open flower, × 4; **6,** sepal, × 4; **7,** petals, × 4; **8,** stamens, × 4; **9,** top of anther, showing method of dehiscence, × 8; **10,** ovary, × 4; **11,** pod, × $\frac{2}{3}$; **12,** seed, × $\frac{2}{3}$. 1, 11, 12, from *Milne-Redhead & Taylor* 10102; 2–10, from *Polhill & Paulo* 1223.

TANGANYIKA. Mwanza, 1 June 1931, *B. D. Burtt* 2467! & 1933, *Rounce* 277!; Ufipa District: Kasanga, 14 June 1957, *Richards* 10100!

DISTR. **U**1–3; **K**5; **T**1, 4; West Africa from Senegal and Gambia to Nigeria, Central African Republic, Congo Republic, Sudan Republic, Eritrea and Angola; also in Arabia and India

HAB. Grassland, cultivated and fallow ground and roadsides; said (*G. H. S. Wood* 831) to be frequent in N. Busoga District on bare dry sites where there is little competition; 900–1220 m.

SYN. *Chamaecrista nigricans* (Vahl) Standl. in Smithson. Misc. Coll. 68, No. 5: 5 (1917)

NOTE. A very distinct and constant species, outstanding on account of its erect habit, large sessile petiolar gland, almost symmetrical leaflets and its small flowers on short pedicels. The leaves are distichously arranged on the stem, in two planes across the plane of branching (*Chancellor* 165), giving the plant a strange flattened appearance (*B. D. Burtt* 2467).

28. **C. grantii** *Oliv.*, F.T.A. 2: 279 (1871); L.T.A.: 639 (1930); Ghesq. in B.J.B.B. 9: 143 (1932); Mendonça & Torre in C.F.A. 2: 186 (1956). Types: Tanganyika, Morogoro District, Mbuiga, *Grant* & Mozambique, "Maravi country," *Kirk* (both K, syn.!)

Perennial herb with thickened, woody rootstock (? sometimes rhizomatous). Stems radiating from rootstock, prostrate, 3·5–40 cm. long, shortly but rarely densely pubescent to subglabrous. Leaves ± elliptic-oblong or oblong, mostly 1–6(–8·5) cm. long and 1·5–3(–4) cm. wide; gland at top of or in upper half of petiole, broadly cup-shaped, raised on a distinct slender stalk 0·25–1 mm. long; similar glands on the rhachis below the insertion of some (particularly the lower) or all pairs of leaflets; rhachis channelled but not crenate-crested along upper side; leaflets sessile, in (3–)6–11 (very rarely to 13) pairs, oblong or elliptic-oblong, straight or very slightly falcate, 6–20(–25) mm. long, 3–9(–11) mm. wide, rounded to obtuse and mucronate at apex, glabrous to ciliolate or sparingly pubescent, rarely more densely so beneath; midrib somewhat excentric, rarely strongly so, nearer anticous margin, but with numerous lateral nerves on both sides, prominent on both surfaces and branching mostly towards the margins. Inflorescences 1–3-flowered; pedicels 1·5–4·2(–5) cm. long. Sepals acute. Petals yellow, 10–14 mm. long, 6–9 mm. wide. Pods 3–5 cm. long, 4·5–5·5 mm. wide. Ripe seeds not seen. Fig. 10/28, p. 57.

UGANDA. Karamoja District: Mt. Moroto, Apr. 1960, *Wilson* 944!

KENYA. Machakos District: N. Chyulu Hills, eastern slope, 7 May 1938, *Bally* 543 in *C.M.* 8222!; Masai District: Lolgorien, Feb. 1934, *H. Parry* in *C.M.* 6043!

TANGANYIKA. Musoma District: Ikizu, Chamhila [? Chamliho] Hill, 2 Nov. 1953, *Tanner* 1681!; Morogoro District: Mgeta R., between Mgeta and Mlali, 21 Mar. 1953, *Drummond & Hemsley* 1744!; Songea District: Mbamba Bay, *Hay* in *Milne-Redhead & Taylor* 10536!

DISTR. **U**1; **K**4, 6; **T**1, 2, 5, 6, 8; Mozambique, Malawi and Angola

HAB. Hard, bare ground, particularly by roads and paths, in woodland, wooded grassland and grassland; 300–2440 m.

SYN. *C. grantii* Oliv. var. *pilosula* Oliv., F.T.A. 2: 280 (1871); L.T.A.: 640 (1930). Type: Angola, Golungo Alto, *Welwitsch* 1722 (LISU, holo., BM, K, iso.!)
C. kituiensis Vatke in Oesterr. Bot. Zeitschr. 30: 81 (1880); L.T.A.: 640 (1930). Type: Kenya, Kitui, *Hildebrandt* 2811 (B, holo. †, BM, K, iso.!)
C. kituiensis Vatke var. *minor* Taub. in P.O.A. C: 201 (1895), synon. prob. Types: Kenya or Tanganyika, Masai Steppe, *Fischer* 228 & Tanganyika, Shinyanga District, Unyamwezi, *Stuhlmann* 512 (both B, syn. †)
Chamaecrista grantii (Oliv.) Standl. in Smithson. Misc. Coll. 68, No. 5: 5 (1917)

NOTE. *C. grantii* and 29, *C. sp. A* are the only herbaceous cassias in East Africa which normally have stipitate glands both on the petiole and on the rhachis. *C. grantii* has been confused with *C. zambesica* Oliv. and *C. hildebrandtii* Vatke, which seem however to be distinct.

C. grantii is a rather constant species, but two variants found in East Africa are noteworthy. *Hornby* 745 from Tanganyika, Mpwapwa, has the midrib of the leaflet considerably more excentric than usual, while *B. D. Burtt* 2200 from Tanganyika,

Kondoa District, Bereku Ridge, and *Peter* 44277 from the same district, between Beréu (? = Bereku) and Gele (? = Gala), represent a form with the stems with dense spreading pubescence, not short rather sparse curled pubescence. A rather similar plant has been collected in Kenya, Northern Frontier Province, Mt. Kulal, Mar. 1959, *T. Adamson* K3. The possibility that these may have been derived through hybridity with *C. hildebrandtii* is to be considered.

Faulkner 2222, from Tanganyika, Tanga District, Ngomeni, has glands on the rhachis as in *C. grantii*, but in every other way is more like *C. zambesica*, of which it has the long spreading indumentum on the stems. It differs from *B. D. Burtt* 2200 and *Peter* 44277, mentioned above, in the relatively narrower leaflets about 3–4 times as long as wide. Whether this specimen is to be looked on as an abnormal *C. zambesica*, or the product of crossing between that and *C. grantii*, I do not know.

29. **C. sp. A**

Perennial herb, with a woody rootstock and apparently ascending tufted pubescent stems up to ± 20 cm. long. Leaves ± ovate-oblong or elliptic-oblong, ± 1–4 cm. long and 1·1–2·7 cm. wide; gland near top of petiole, flattened or broadly cup-shaped, raised on a distinct slender stalk 0·7–1·5 mm. long; similar glands on the rhachis of some of the leaves below the insertion of some of the lower pairs of leaflets; rhachis channelled but not crenate-crested along upper side; leaflets sessile, in 3–10 pairs, obliquely elliptic or ovate-elliptic, mostly 5–17 mm. long, 4–7(–8) mm. wide, rounded on the posticous, almost straight to narrowly rounded on the anticous side at the mucronate apex, hairy on margins and midrib, otherwise glabrous to sparingly pubescent; midrib very strongly excentric, running close to (within 1·5 mm.) of the anticous margin of the leaflet for its entire length; venation prominent and reticulate on both sides of midrib and both surfaces of leaflet. Inflorescences 1–2-flowered; pedicels 3–9 cm. long. Sepals acute. Petals yellow, 17 mm. long, 9–14 mm. wide. Pods up to at least 4·5 cm. long and 5 mm. wide. Ripe seeds not seen.

TANGANYIKA. Iringa District: Mufindi, Oct. 1931, *Staples* 232! & Mdabulo, Ikanga, 26 Aug. 1952, *Carmichael* 110!
DISTR. T7; not known elsewhere
HAB. Grassland; 1830 m.

NOTE. This is without question closely related to *C. grantii*, although its ultimate status can hardly be decided at present without more material. It differs from *C. grantii* in the very strongly excentric midrib of the leaflets, and in their much more reticulate venation. The flowers are larger and the pedicels usually longer than in *C. grantii*.

30. **C. hildebrandtii** *Vatke* in Oesterr. Bot. Zeitschr. 30: 80 (1880); L.T.A.: 640 (1930). Type: Kenya, Teita District, Ndara Hill, *Hildebrandt* 2464 (B, holo.†, K, iso.!)

Perennial herb, with thick woody rootstock. Stems prostrate or sometimes erect, ± 3–60 cm. long, usually ± densely pubescent, hairs either long and spreading, up to 1–2 mm. long, or shorter and crisped; stems rarely sparsely pubescent or (not so far in Flora area) puberulous. Leaves oblong to ovate-oblong, (1–)1·5–6·5 cm. long, mostly 1–2·5(–3) cm. wide; gland at or near top of petiole, flattened or shallowly cup-shaped, raised on a distinct stalk 0·3–0·7 mm. long; rhachis without glands or rarely with sessile or subsessile ones between the paired leaflets, channelled but not crenate-crested along upper side; leaflets sessile, usually in 4–13(–14) pairs, oblong or lanceolate-oblong, uppermost often somewhat obovate, straight or slightly falcate, mostly (4–)5–15(–18) mm. long, (1·25–)1·5–5·5(–7) mm. wide, rounded or obtuse and mucronate at apex, occasionally subacute, usually ± densely pubescent on both surfaces, rarely sparingly pubescent to glabrous; midrib usually distinctly excentric (sometimes slightly so) and nearer anticous margin, but giving off on both sides ± numerous prominulous or prominent lateral nerves which are unbranched or branch normally only near margins.

Inflorescences 1–4-flowered; pedicels (0·7–)1–3 cm. long. Sepals acute. Petals yellow, 4–12 mm. long, 3·5–10 mm. wide. Pods 2·5–5·2 cm. long, (3·5–)4–5 mm. wide. Seeds brown to blackish, ± 2·2–3 mm. long and 2·5–2·75 mm. wide, not areolate. Fig. 10/30, p. 57.

UGANDA. Ankole District: Ruizi R., 30 Oct. 1950, *Jarrett* 120!; Masaka District: Kabula, Sept. 1945, *Purseglove* 1797!; Mengo District: near Luzira, Apr. 1931, *Snowden* 2016!

KENYA. Northern Frontier Province: W. side of Uaraguess, 3 Dec. 1958, *Newbould* 3107!; Kiambu District: Muguga North, 15 Oct. 1953, *Verdcourt* 1022!; Central Kavirondo District: Sakwa, 30 Sept. 1956, *Padwa* 466!

TANGANYIKA. Bukoba District: Karagwe, Sept.–Oct. 1935, *Gillman* 590!; Masai/Mbulu Districts: Oldeani, 21 June 1935, *R. M. Davies* 1078!; Moshi, Aug. 1927, *Haarer* 597!

DISTR. **U**2, 4; **K**1, 3–7; **T**1–3; Congo Republic, Eritrea, Ethiopia and Somali Republic

HAB. Wooded grassland and grassland; 760–2130 m.

SYN. [*C. grantii* var. *pilosula* sensu Steyaert in F.C.B. 3: 528 (1952), *non* Oliv.]

NOTE. *C. hildebrandtii* is more variable than its immediate relatives. It is not clear whether this variation is intrinsic or whether, perhaps, *C. hildebrandtii* may be able to cross with related species such as *C. grantii* so that the herbarium may give a misleading picture of the taxonomic limits of *C. hildebrandtii.*

Some idea of this variation can be given. The stems are usually prostrate, but may be erect (e.g. *Verdcourt* 1799, from Kenya, Masai District, Ngong Hills); the indumentum is usually dense, but may be comparatively sparse especially on the leaflets (e.g. *Bally* 1759, from Kenya, Kitui, and *Peter* 49734, from Tanganyika, Lushoto District, between Shume and Manolo, in which the indumentum is altogether puberulous and ± appressed); the leaflets are often rather large, but are at times quite small and up to only 7 × 2·5 mm. (e.g. *Sanders* 13, from Tanganyika, Moshi District, Engare Nairobi); the flowers also vary greatly in size, the petals being 4·5–5 mm. long and 2·5–3·5 mm. wide in *Sanders* 13 (already mentioned), but 9–10 mm. long and 6–10 mm. wide in *Verdcourt* 1022, from Kenya, Kiambu District, Muguga North; this may again either be intrinsic or just seasonal. More evidence is needed.

Dummer 5293, from Kenya, Naivasha District, Longonot, and *Seldon* in *E.A.H.* 11839, from Naivasha, Hell's Gates Gorge, are an extreme with glabrous stems, leaves, pedicels and calyces. They also show glands on some of the leaf-rhachides. More material of this is desired.

31. **C. zambesica** *Oliv.*, F.T.A. 2: 280 (1871); L.T.A.: 640 (1930). Type: Mozambique, Shamwara, *Kirk* (K, holo.!)

Perennial herb. Stems prostrate or semi-prostrate, up to ± 40 cm. long, ± densely clothed with straight spreading hairs ± 1–2 mm. long, and normally with a band of shorter pubescence running along one side of the stem. Leaves ovate-oblong to oblong, 1·5–6·5 cm. long, 1·5–2·5(–3) cm. wide; gland at or near top of petiole, peltate and flattened or shallowly cup-shaped, raised on a distinct slender stalk about 1–1·25 mm. long; rhachis without glands, channelled but not crenate-crested along upper side; leaflets sessile, usually in (4–)7–18 pairs, oblong, straight or very slightly falcate, 4–13(–16) mm. long, 1·5–4(–6) mm. wide, rounded to obtuse and mucronate at apex, ciliate on margins, glabrous to pubescent on surfaces; midrib somewhat excentric (nearer anticous margin), particularly in lower part of leaflet but with numerous prominulous lateral nerves on both sides which are unbranched or branch normally only near margins. Inflorescences 1–3-flowered; pedicels 1–2·5 cm. long. Sepals acute. Petals yellow, 7–10 mm. long, 4–10 mm. wide. Pods 1·5–4·5 cm. long, 4–5 mm. wide. Seeds brown, ± 2–3 mm. long and 2·25–2·5 mm. wide, not areolate. Fig. 10/31, p. 57.

KENYA. Kilifi District: Sokoke Forest, 18 June 1945, *Jeffery* 232! & Kibarani, 13 Aug. 1945, *Jeffery* 288!; District uncertain: N. of Mombasa, to Lamu and Witu, 1902, *Whyte*!

TANGANYIKA. Lushoto District: Mashewa, July 1893, *Holst* 8788!; Pangani District: Madanga, Mvumoni, Kiyama, 22 May 1956, *Tanner* 2882!; Lindi District: Nachingwea, 22 May 1955, *Anderson* 1058!

DISTR. **K**7; **T**3, 6, 8; Mozambique and Rhodesia

HAB. Grassland and cultivated land; 80–460 m.

SYN. *C. stuhlmannii* Taub. in P.O.A. C: 201 (1895); L.T.A.: 640 (1930), synon. prob. Type: Tanganyika, probably Morogoro District, between Gerengera [? = Ngerengere] and Mko, *Stuhlmann* 11 (B, holo. †)

NOTE. *C. zambesica* has been sunk under *C. grantii* Oliv. var. *pilosula* Oliv. by Ghesquière in B.J.B.B. 9: 143 (1932). The two are indeed closely related, but *C. zambesica* is distinct in lacking glands on the leaf-rhachis, and in having smaller flowers and usually narrower leaflets. *C. zambesica* is in fact more akin to *C. hildebrandtii*, differing in the considerably longer stalk to the petiolar gland. It is noteworthy that in Kenya and Tanganyika *C. zambesica* is apparently restricted to the eastern districts on or not far from the coast and to comparatively low altitudes.

C. stuhlmannii, of which I have not seen any authentic material, is, as suggested by E. G. Baker (L.T.A.: 640), a probable synonym. Taubert described the leaflets of *C. stuhlmannii* as linear and 1–1·5 mm. long, which is certainly incorrect, but perhaps mm. was a misprint for cm.

Barbosa 1559, from Mozambique, and *Chase* 6360, from Rhodesia, have the shorter pubescence on the stem extending all round it. Similar plants have not so far been collected in East Africa.

32. **C. fallacina** *Chiov.* in Result. Sci. Miss. Stefan.-Paoli Somal. Ital. 1: 68 (1916); L.T.A.: 641 (1930); Ghesq. in B.J.B.B. 9: 144 (1932), pro parte, excl. vars.; Brenan in K.B. 14: 178 (1960). Type: Somali Republic (S.), Giumbo, *Paoli* 286 (FI, holo.!)

Perennial herb or subshrub, with a woody rootstock and spreading or prostrate stems up to 35 cm. long which are shortly crisped-pubescent or puberulous without spreading hairs. Leaves oblong-lanceolate, mostly 1–5 cm. long and 0·7–1·7 cm. wide; gland above middle of petiole, peltate and flattened, raised on a distinct slender yellowish stalk 1–2·25 mm. long (sometimes shorter in lowest leaves); rhachis without glands, channelled but not crenate-crested along upper side; leaflets sessile, usually in 9–27 pairs, oblong, asymmetric, straight or slightly curved, mostly 3–10 mm. long, 1–2 mm. wide, rounded on the posticous, straight or nearly so on the anticous side at the mucronate apex, glabrous except for short hairs on margins and sometimes midrib, sometimes (but typically) with some appressed pubescence on lower surface; midrib very strongly excentric, running along or very close to anticous margin of leaflet for its whole length; lateral nerves on anticous side of midrib none or very inconspicuous, on other side prominulous especially beneath and becoming subparallel. Inflorescences 1–2-flowered; pedicels 4–6 mm. long. Sepals acute. Petals yellow, 5–8 mm. long, 3–6·5 mm. wide. Pods 2–4 cm. long, 3–4 mm. wide. Ripe seeds ± 2 × 1·5 mm. Fig. 10/32, p. 57.

KENYA. Machakos District: Makueni, near Kathonsweni, Jan. 1956, *D. B. Thomas* 324!; Machakos/Teita Districts: Tsavo, ? Feb. 1957, *Shantz & Turner* 4248!; Masai District: Lerujat, about 3 km. from Etepsi [? Oltepesi] on the Magadi road, 16 Feb. 1962, *Glover & Samuel* 2830!; Kwale District: about 1·6 km. from Tanganyika border on Tanga–Mombasa road, 14 Aug. 1953, *Drummond & Hemsley* 3766!

TANGANYIKA. Masai District: Tarangire R., 23 Feb. 1958, *Lamprey* 332!; Mpwapwa District: Kongwa, 2 Feb. 1950, *Anderson* 607!

DISTR. **K**4, 7; **T**2, 5; Somali Republic (S.)

HAB. Grassland; the Kenya specimen was found in open grassland areas in scattered tree grassland; in Tanganyika it was said to be common in grassland on wet grey clay-loam; 100–1160 m.

33. **C. fenarolii** *Mendonça & Torre* in Bol. Soc. Brot., sér. 2, 29: 34, t. 1/B (1955) & in C.F.A. 2: 186, t. 38/B (1956). Type: Angola, Bié, near Chinguar, *Fenaroli* 1108a (Herb. Fenaroli, holo.!)

Herb, apparently annual (described in C.F.A. as perennial). Stems erect or ascending, ± 10–45 cm. high, shortly crisped-pubescent. Leaves ± oblong, up to ± 2·5–7·5 cm. long, 6–11 mm. wide; gland at top of petiole, funnel-shaped, raised on a very short stalk up to ± 0·2 mm. long; rhachis channelled but not crenate-crested along upper side; leaflets sessile, mostly in 11–37 pairs, oblong, slightly falcate, 4–8·5 mm. long, 0·75–1·75 mm. wide, acuminate or shortly aristate at apex, ± ciliate, otherwise glabrous or sparsely hairy beneath; midrib excentric, nearer anticous margin, but with some lateral nerves between; venation ± prominent beneath. Inflorescences 1–2(–3)-flowered; pedicels 1·2–2 cm. long, puberulous, but also with ± numerous straight hairs spreading about at right angles and several times as long as the diameter of the pedicel. Petals yellow, 7–7·5 mm. long, 4–7 mm. wide. Pods 2·9–4·3 cm. long, 4–5 mm. wide. Seeds brownish, ± 3 mm. long and 2 mm. wide, not areolate. Fig. 10/33, p. 57.

TANGANYIKA. Kondoa District: Sambala, 2 May 1929, *B. D. Burtt* 2015!
DISTR. T5; Rhodesia and Angola
HAB. Insufficiently known; "growing in ecotone conditions" (*B. D. Burtt* 2015); 1490 m.

NOTE. *C. fenarolii* is close to *C. gracilior*, but differs in the shortly stipitate glands, wider pods and especially in the long hairs on the pedicels; the latter have a marked tendency to diverge almost horizontally and to curve close to their apices so as to bring the pods into a position about 90°–130° from the pedicel itself. It is hard to assess without fuller materials the relationship of *C. fenarolii* with *C. katangensis*, but the former appears annual, while the latter is certainly perennial.

The identity of *B. D. Burtt* 2015 is by no means certain, as there is very little material of *C. fenarolii* for comparison. *Burtt* 2015 is more robust than the specimens from Rhodesia and Angola, and the indumentum on the outside of the calyx is longer.

Mention should be made here of a series of specimens collected by Peter in **T4**, Kigoma District, and preserved at Berlin. They are *Peter* 36478, from W. of Lugufu, 1060 m., 8 Feb. 1926; 36480, 36562, same locality, altitude, and date; and 37167, from Ujiji, Magaso [Machaso]-Mkuti R., 840–960 m., 19 Feb. 1926. All are clearly conspecific and key down to *C. fenarolii*. They are much branched annuals with prostrate radiating stems and purplish-brown foliage (obvious even in the dried specimens). Otherwise they are similar to *B. D. Burtt* 2015 which, however, is erect and appears to have green leaves. Whether the Peter specimens represent a distinct species or merely a variant is uncertain, until more material can be collected from Kondoa and Kigoma Districts. None of them bears pods.

34. **C. gracilior** (*Ghesq.*) *Steyaert* in B.J.B.B. 20: 248 (1950) & in F.C.B. 3: 517 (1952); Mendonça & Torre in C.F.A. 2: 186 (1956). Type: Congo Republic, Haut-Katanga, Kiambi, *de Witte* 263 (BR, lecto.!, see Steyaert, l.c. (1950))

Annual herb, prostrate or sometimes erect. Stems radiating from rootstock or occasionally simple, 5–30(–45) cm. long. Indumentum an appressed or very short crisped puberulence. Leaves oblong, up to ± 2–6(–8) cm. long, 6–7 mm. wide; gland at top of petiole, peltate or funnel-shaped, raised on a distinct stalk ± 0·25–0·8 mm. long; rhachis crenate-crested along upper side; leaflets sessile, mostly in 17–50 pairs, oblong, 3–5 mm. long, 0·5–0·75 mm. wide, acutely acuminate, glabrous or nearly so or shortly ciliate; midrib running very close to anticous margin; venation not or slightly raised beneath. Inflorescences 1–2-flowered; pedicels 1–2 cm. long, puberulous. Sepals acute. Petals yellow, 3–5·5 mm. long, 1·5–4 mm. wide. Pods (1·7–)2–3·5 cm. long, 2·5–3·5 mm. wide. Seeds grey-brown, ± quadrate, 2 mm. long, 1·2–1·5 mm. wide, not areolate. Fig. 10/34, p. 57.

TANGANYIKA. Ufipa District: near Kalambo Falls, 29 Mar. 1955, *Richards* 5184!; Songea District: below Litenga Hill, 18 Apr. 1956, *Milne-Redhead & Taylor* 9677! & about 3 km. W. of Gumbiro, 9 May 1956, *Milne-Redhead & Taylor* 10026!; Tunduru, *Allnutt* 44!
DISTR. **T4**, 7, 8; Congo Republic, Malawi, Zambia, Rhodesia and Angola (*fide* C.F.A.)
HAB. Grassland and deciduous woodland; in shallow soil near rocks, now also a weed in old cultivated land and by roadsides; 800–1400 m.

SYN. *C. fallacina* Chiov. var. *gracilior* Ghesq. in B.J.B.B. 9: 144 (1932)

VARIATION. Variable in habit, usually prostrate or ascending and branched. *Milne-Redhead & Taylor* 9282, from SE. Tanganyika, about 6 km. E. of Songea, Unangwa

Hill, 22 Mar. 1956, has simple erect stems varying in height from about 9–22 cm. This may well be a form due only to habitat.

35. **C. katangensis** (*Ghesq.*) *Steyaert* in B.J.B.B. 20: 258 (1950) & in F.C.B. 3: 529 (1952). Type: Congo Republic, Welgelegen, *Corbisier* 577* (BR, syn. !)

Perennial herb, probably rhizomatous, with erect stems up to 65 cm. or more high (to 1·25 m. *fide* F.C.B.). Stems simple or nearly so, arising from a ± thickened woody rootstock, pubescent to ± densely spreading-hairy. Leaves linear-oblong to linear, mostly 4–10 cm. long and 1–2(–3) cm. wide; gland near top of petiole, 0·7–1·5 mm. long, raised on a rather indistinct column or stalk ± 0·25–0·5(–0·75) mm. long, the gland often tilted on top; sometimes a second small gland on lower part of petiole; rhachis eglandular, or sometimes with a single gland near base, channelled but not crenate-crested along upper side; leaflets sessile, mostly in (15–)20–43 pairs, obliquely oblong, ± falcate, 4–10(–15) mm. long, 1–2(–3) mm. wide, asymmetrically acute or acuminate, varying from glabrous except for ciliolate margins to ± puberulous or pubescent on both surfaces; midrib strongly excentric, sometimes marginal, but usually with visible lateral nerves between it and the anticous margin. Inflorescences 1–3-flowered [3–5-flowered, *fide* F.C.B.]; pedicels 1·4–3 cm. long, shortly crisped-puberulous and normally also with longer spreading hairs ± 0·5–1 mm. long especially towards base, sometimes densely spreading-hairy. Sepals acute. Petals yellow, 8–10 mm. long, 4–9 mm. wide. Pods 2·5–5 cm. long, 4·5–5·5 cm. wide. Seeds brown, ± 3·5 mm. long, 2 mm. wide, not areolate. Fig. 10/35, p. 57.

TANGANYIKA. Ufipa District: Chapota, 4 Dec. 1949, *Bullock* 2017!; Chunya District: Lupa, 17 Mar. 1962, *Boaler* 521!; Iringa District: 11 km. N. of Iringa, 6 Feb. 1962, *Polhill & Paulo* 1366!

DISTR. **T**4, 7; Congo Republic, Malawi and Zambia

HAB. Deciduous woodland (*Julbernardia-Brachystegia*); 1520–1710 m.

SYN. *C. fallacina* Chiov. var. *katangensis* Ghesq. in B.J.B.B. 9: 145 (1932)

NOTE. Whether the identification of the above specimens with *C. katangensis* will be maintained in the light of future additional evidence is far from certain. At present there are a number of gatherings from Zambia, and some, including the Tanganyika specimens cited above, from other territories, that are extremely close to *C. katangensis* and at present seem best included under that species interpreted in a wide sense. The main diagnostic features are: perennial habit; stems erect, simple or nearly so; petiolar gland shortly and indistinctly stalked; leaflets numerous. The group is, however, heterogeneous and the differences shown by gatherings from different areas may be geographical. More material is much needed.

Bullock 2017 has rather small leaflets 4–8 mm. long and 1–1·25 mm. wide and ± puberulous especially on upper surface, two petiolar glands and pedicels with some spreading hairs. Another specimen, small and inadequate, but perhaps also to be placed in this affinity, is *St. Clair-Thompson* 460, from Tanganyika, Iringa, Signal Hill, all of whose parts above the ground bar the flowers are densely clothed with spreading hairs 0·5–1·25 mm. long.

St. Clair-Thompson 459 (in part), also from Signal Hill, is very close to *C. katangensis*, but the pedicels are shorter (0·6–1·3 cm. long). The stems are inconspicuously puberulous and the leaves subglabrous.

36. **C. parva** *Steyaert* in B.J.B.B. 20: 266 (1950) & in F.C.B. 3: 524 (1952). Type: Congo Republic, Parc National de l'Upemba, Lusinga, *de Witte* 2476 (BR, holo. !)

Perennial herb ± 8–50 cm. high. Stems erect, simple or subsimple or somewhat branched especially near base, pubescent with short curved hairs;

* Steyaert (1950) indicated *Homblé* 206 (BR !) as the holotype of *C. katangensis*, but this cannot be since Ghesquière, on whose variety Steyaert based his species, cited *Homblé* 206 under *C. lechenaultiana* DC. var. *wallichiana* (DC.) Ghesq., deliberately excluding it from *C. fallacina* Chiov. var. *katangensis* Ghesq. *Corbisier* 577 is the only specimen cited by both Ghesquière and Steyaert under *katangensis*.

longer curved or spreading hairs usually also present, sometimes plentiful. Leaves: linear-oblong, (2·5–)3–7(–9) cm. long, 0·7–2 cm. wide; gland at or near top of petiole, sessile, circular to elliptic, flattened or depressed on upper side, often somewhat curved, medium to large, 0·5–1·75(–2, rarely to 2·5) mm. long, 0·7–1 mm. wide, usually somewhat raised above the surface of the petiole or lying on it, brown to dark purplish-red; rhachis eglandular, channelled but not crenate-crested along upper side; leaflets sessile, usually in 7–32 pairs, oblong, narrowly oblong or lanceolate, 3–11 mm. long, 1·25–3 mm. wide, acuminate to abruptly mucronate, appressed-pubescent or sometimes spreading-pubescent on one or on both surfaces or glabrous or nearly so except for ciliate margins; midrib excentric (towards anticous margin) but giving off ± prominulous lateral nerves towards both margins. Inflorescences usually supra-axillary, 1–2(–3)-flowered; pedicels 1·2–2·5 cm. long, pubescent and with longer spreading hairs. Petals yellow, 7–12 mm. long, 4–9·5 mm. wide. Pods ± 3·5–4·5 cm. long, 3·5–4 mm. wide. Seeds brown, subrhombic, ± 2·5–3 mm. long and 2 mm. wide, not areolate. Fig. 10/36, p. 57.

KENYA. Machakos/Masai Districts: Chyulu Hills, 22 Apr. 1938, *Bally* 288 in *C.M.* 8223!
TANGANYIKA. Mbulu District: Kampi ya Faru–Udehei, 4 Aug. 1926, *Peter* 43621!; Ufipa District: Ufipa, Malonje Farm, 13 Mar. 1957, *Richards* 8691!; Iringa District: Mt. Tarik, 17 Feb. 1932, *Lynes* I.h.107! & Iringa, Signal Hill, 20 Feb. 1932, *St. Clair-Thompson* 459 (in part)!; Njombe District: Lupembe region, Feb. 1931, *Schlieben* 249 & 249 bis!
DISTR. **K** ? 4 or 6; **T**2, 4, 7; Congo Republic, Zambia and Rhodesia
HAB. Grassland, deciduous woodland; 1580–2100 m.

SYN. *C. falcinella* Oliv. var. *longifolia* Ghesq. in B.J.B.B. 9: 163 (1932) saltem quoad spec. *Schlieben* 249, 249 bis supra cit. (BR, syn.!, 249 at BM, isosyn.!)

NOTE. *C. parva* is one of the least satisfactorily defined species of the group. The material is not homogeneous and seems to comprise a series of forms intermediate between and to some extent linking up *C. comosa* on one hand and *C. katangensis* on the other. It is possible (there is no evidence at the moment) that *C parva* may be the product of hybridization between *C. comosa* and other species. *C. parva* is separable from *C. katangensis* by the petiolar gland being sessile and usually larger, and from *C. comosa* by the usually smaller less sunken petiolar gland. In *C. comosa* the glands are mostly 2 mm. or more long and sunken in or flush with the surface of the petiole, while in *C. parva* only occasional glands are as much as 2 mm. or more long and they are somewhat raised above or sitting on the surface of the petiole; also the leaflets of *C. parva* are narrower than is usual in *C. comosa*. The distinctive features of *C. parva* are not always clear, and the future may well show that it cannot be maintained as a valid species.

37. **C. comosa** (*E. Mey.*) *Vogel*, Syn. Gen. Cassiae: 65 (1837); Ghesq. in B.J.B.B. 9: 153 (1932); Steyaert in B.J.B.B. 20: 251 (1950). Type: South Africa, E. Cape Province, between Umzimvubu R. [Omsamwubo] and Umsikaba R. [Omsamcaba], *Drège* (? B, holo.†)

Perennial herb with erect, simple or not much branched stems 15–100 cm. high arising from a woody rootstock. Stems from glabrous to pubescent or densely spreading-hairy. Leaves oblong to linear-oblong, (2·5–)4–13 cm. long, 1–3·5 cm. wide; gland at or below top of petiole, sessile, elliptic, cushion-like, depressed in middle, 1·5–4 mm. long, 1–1·5 mm. wide, usually sunken in or flush with the surface of the petiole; rhachis eglandular, channelled but not crenate-crested along upper side; leaflets sessile, in 10–41 pairs, asymmetrically oblong or oblong-elliptic, somewhat narrowing towards apex, mostly 6–22 mm. long, 1·25–7 mm. wide, acute to almost rounded and mucronate or apiculate, glabrous or nearly so, often ± ciliolate; midrib somewhat excentric (towards anticous margin) but giving off numerous rather prominent lateral nerves towards both margins. Inflorescences mostly strongly supra-axillary, 1–3(–5)-flowered; pedicels 0·8–2 cm. long, subglabrous to densely spreading-hairy. Petals yellow, 8–18 mm. long, 4–13 mm.

wide. Pods 4–6·5 cm. long, 5–6 mm. wide, Seeds brown, elliptic, 3·5–4 mm. long, 2·5–2·75 mm. wide, not areolate. Fig. 10/37, p. 57.

DISTR. (of species as a whole). Congo Republic, Tanganyika, Mozambique, Malawi and South Africa

SYN. *Chamaecrista comosa* E. Mey., Comm. Pl. Afr. Austr.: 160 (1836)

var. **capricornia** *Steyaert* in B.J.B.B. 20: 252 (1950) & in F.C.B. 3: 525 (1952). Type: Congo Republic, Katanga, Elisabethville, *F. A. Rogers* 10184 (BR, holo.!)

Stems ± pubescent. Inflorescences 1–3-flowered; pedicels pubescent.

TANGANYIKA. Mbulu District: Pienaar's Heights, 31 Dec. 1930, *B. D. Burtt* 2726! & Ufiome Mt. near Babati, 14 Jan. 1955, *F. G. Smith* 1360!; about 15 km. E. of Songea, 17 Jan. 1956, *Milne-Redhead & Taylor* 8257! & about 13·5 km. E. of Songea, 1 Feb. 1956, *Milne-Redhead & Taylor* 8257A!
DISTR. T2, 4, 8; Congo Republic and South Africa
HAB. Upland bushland and deciduous woodland (*Brachystegia*); 1430–1800 m.

NOTE. The unusually large petiolar gland in *C. comosa*, separating widely the sides of the channel along the upper side of the petiole and often giving a flattened appearance to the petiole where the gland is situated, is most characteristic.

38. **C. sp. B**

Perennial herb, 10–15 cm. high. Stems simple or subsimple, arising from a ± thickened woody rootstock, puberulous with curved appressed hairs; longer spreading hairs absent. Leaves mostly linear-oblong, ± 2·5–6·5 cm. long and 1·0–1·5 cm. wide; gland at top of petiole, (0·25–)0·75–1·25 mm. long, sessile, subcircular to elliptic, blackish-purple when dried; rhachis eglandular, with a low bluntly toothed wing running along upper side; a tooth just below the insertion of each pair of leaflets; leaflets sessile, mostly in 10–30 pairs, obliquely oblong-elliptic, 3–8 mm. long, 1·5–3 mm. wide, abruptly mucronate at apex, glabrous or nearly so except for ciliolate margins; midrib excentric (toward anticous margin), but giving off prominulous lateral nerves towards both margins. Inflorescences supra-axillary, 1–3-flowered; pedicels 1–2 cm. long, puberulous. Petals yellow, ± 10–12 mm. long, 6–7 mm. wide. Pod (unripe) 3 cm. long, 5 mm. wide. Seeds not seen.

KENYA. Machakos District: Ol Doinyo Sapuk, 6 Jan. 1934, *Napier* 3125 in *C.M.* 6044A!
HAB. Rocky soil in hillside grassland; 2130 m.
DISTR. K4; not known elsewhere

NOTE. This plant has a distinct resemblance to *C. comosa* and *C. parva*, but differs from both by the toothed wing along the leaf-rhachis. In addition it differs from *C. comosa* by the smaller petiolar gland and from *C. parva* by the absence of spreading hairs on the pedicels.
The solitary gathering is a scanty one and the measurements given above, especially stature and petal-size, are subject to modification in the future.

39. **C. falcinella** *Oliv.*, F.T.A. 2: 281 (1871); L.T.A.: 641 (1930); Ghesq. in B.J.B.B. 9: 162 (1932), excl. var. *longifolia*; Steyaert in F.C.B. 3: 520 (1952); Brenan in K.B. 14: 178 (1960). Type: Tanganyika, Bukoba District, Karagwe, *Grant* 445 (K, holo.!)

Annual or perennial herb, sometimes with lower parts of stems or the rootstock woody, erect or prostrate, 10–60 cm. high. Stems simple or branched, rather densely hairy with short crisped hairs and longer spreading ones. Leaves ± elliptic or oblong, 0·7–6·5 cm. long, 0·7–2 cm. wide; stipules ± falcate, semi-cordate; gland at top of petiole, sessile, circular to elliptic, flattened or depressed in middle, 0·3–1·25 mm. long, 0·3–0·6 mm. wide; rhachis eglandular, channelled but not crenate-crested along upper side; leaflets sessile, in 5–17 pairs, oblong, straight except at the asymmetric apex, 4–17 mm. long, 0·75–3 mm. wide, acute at apex which ends in a prickle-like point bent to one side, glabrous except on margins or varying to rather densely pubescent especially beneath; midrib very excentric, close to anticous margin, along which runs an often prominent nerve; nerves between midrib and marginal nerve usually 0 or very inconspicuous. Inflorescences 1–3-flowered; pedicels 1–2·5 cm. long, ± densely clothed with rather long spreading hairs. Petals yellow, 4–10 mm. long, 2–10 mm. wide. Pods 2–4·3 cm. long, 4–5·5 mm. wide. Seeds blackish-brown to black, ± rhombic, 3–3·5 mm. long, 2–3 mm. wide, not areolate.

KEY TO INTRASPECIFIC VARIANTS

Petals rather large, 7–10 mm. long, 4·5–10 mm. wide:
  Surface of stipules thinly pubescent to glabrous; leaflets in up to 10 pairs . . . . var. **falcinella**
  Surface of stipules densely though often shortly pubescent; leaflets in 8–14 pairs . . . . . . var. **intermedia**
Petals smaller, 4–6 mm. long, 2–5 mm. wide . . . var. **parviflora**

var. **falcinella**; Brenan in K.B. 14: 179 (1960)

Perennial, with woody rootstock and erect stems. Leaves 0·7–2 cm. long; surface of stipules thinly pubescent to glabrous; gland small, 0·5–0·7 mm. long, 0·4 mm. wide; leaflets in 5–10 pairs. Petals 8–10 mm. long, 5–10 mm. wide. Fig. 10/39, p. 57.

UGANDA. Ankole District: near Mulema, 31 Mar. 1904, *Bagshawe* 190!; Kigezi District: Kamwezi, Feb. 1948, *Purseglove* 2587!
KENYA. Machakos/Masai Districts: Chyulu Hills, 23 Apr. 1938, *Bally* 313 in *C.M.* 8224!
TANGANYIKA. Bukoba District: Karagwe, near Nyakahanga, Oct. 1931, *Haarer* 2269!
DISTR. **U**2; **K**4/6; **T**1; Congo Republic
HAB. Grassland; 1520–1580 m.

NOTE. *C. falcinella* is said in L.T.A. to occur in Eritrea. I cannot confirm this, and it does not seem likely. There are no specimens from there in the British Museum Herbarium.
C. falcinella shows a wider range of variation than most of the related species. The true status of all the variants recognized here must depend on further study.

var. **intermedia** *Brenan* in K.B. 14: 179 (1960). Type: Tanganyika, Ufipa District, Lake Kwela, 19 km. from Mpui, *McCallum Webster* T1 (K, holo.!)

Perennial with woody rootstock and erect stems up to 1 m. high. Leaves 1·5–4 cm. long; surface of stipules densely though often rather shortly pubescent; gland 0·5–1·25 mm. long, 0·4–0·7 mm. wide; leaflets in 8–14 pairs. Petals 7–8·5 mm. long, 4·5–7 mm. wide.

TANGANYIKA. Ufipa District: Mpui, 15 Mar. 1957, *Richards* 8746! & Lake Kwela, 16 Mar. 1957, *Richards* 8760!; Mbeya District: Mbozi, 7 Apr. 1932, *R. M. Davies* 523!
DISTR. **T**4, 7; Zambia
HAB. Grassland; 1060–1800 m.

var. **parviflora** *Steyaert* in B.J.B.B. 20: 251 (1950) & in F.C.B. 3: 521 (1952); Brenan in K.B. 14: 179 (1960). Type: Rwanda Republic, Gabiro, *Becquet* 613 (BR, holo.!)

Annual or with the lower part of the stems above ground becoming woody. Leaves 1–6 cm. long; surface of stipules shortly pubescent or subglabrous; gland 0·3–0·5 mm. long, 0·2–0·5 mm. wide, round or nearly so; leaflets in 6–17(–18) pairs. Petals 4–6 mm. long, 2–5 mm. wide.

UGANDA. Karamoja District: Napak Mt., June 1957, *Wilson* 358!
KENYA. Trans-Nzoia District: Hoey's Bridge, Aug. 1931, *Jex-Blake* in *C.M.* 2581!; Uasin Gishu District: Kipkarren, *Brodhurst Hill* 50!
TANGANYIKA. Mwanza District: Lubili, Mbarika, 24 May 1953, *Tanner* 1540!; Masai District: Ardai Plains, 29 June 1945, *Greenway* 7504!; about 8 km. S. of Dodoma, 19 July 1956, *Milne-Redhead & Taylor* 11181!; Mpwapwa, 1930, *Hornby* 249!
DISTR. **U**1; **K**3; **T**1, 2, 5; Congo Republic, Zambia, Rhodesia, Botswana and South West Africa
HAB. Grassland, cultivated ground, and also recorded from bank of dry seasonal river in sandy desert bushland; 1070–1800 m.

40. **C. kirkii** *Oliv.*, F.T.A. 2 : 281 (1871), pro parte, excl. specim. *Welwitsch* ; L.T.A. : 642 (1930), pro parte, excl. saltem pl. angol. ; Ghesq. in B.J.B.B. 9 : 150 (1932), pro parte, excl. var. *quarrei* ; T.T.C.L. : 96 (1949) ; Steyaert in B.J.B.B. 20 : 253 (1950), pro parte, excl. specim. *Welwitsch* 1721 & in F.C.B. 3 : 525 (1952) ; Mendonça & Torre in C.F.A. 2 : 183 (1956). Types : Malawi, Manganja Hills, *Kirk* (K, syn. !) & *Meller* (K, syn. !)

Erect annual herb 30–120(–240) cm. high, with the single main stem often becoming rather woody in its lower part and thus perhaps enabling the plant to perennate for a short time. Stems simple or ± branched, glabrous to pubescent or ± clothed with spreading hairs. Leaves oblong to linear-oblong, mostly 2·5–13(–17) cm. long, 1·5–3(–4) cm. wide ; gland near top of petiole, sessile, ± elliptic, cushion-like, depressed in middle, (0·75–)1·2–1·8(–2) mm. long, 0·5–1·0 mm. wide, yellow ; sometimes a smaller second gland present below the upper one ; rhachis eglandular or with some sessile glands towards apex, channelled but not crenate-crested along upper side ; leaflets sessile, in (12–)18–43(–51) pairs, narrowly oblong or linear-oblong, (4–)5·5–17(–20) mm. long, (1–)1·25–3·2(–5) mm. wide, acute or subacute and mucronate, glabrous to pubescent on both surfaces ; midrib somewhat excentric (towards anticous margin) but giving off ± prominent lateral nerves towards both margins. Inflorescences mostly supra-axillary, (1–)2–5-flowered ; pedicels (0·6–)1–2·3 cm. long, very shortly pubescent to densely spreading-hairy. Petals yellow, 8–15 mm. long, 5–11 mm. wide. Pods (3–)5–9 cm. long, 5–6 mm. wide. Seeds brown, ± rhombic, 3·3–5 mm. long, 2–2·5 mm. wide, not areolate. Figs. 10/40, p. 57, & 16.

KEY TO INTRASPECIFIC VARIANTS

Leaflets ± densely pubescent on both surfaces . . var. **kirkii**
Leaflets glabrous or nearly so, except on margins :
  Stems with dense spreading hairs . . . . var. **kirkii**
  Stems puberulous or shortly pubescent with small incurved hairs . . . . var. **guineënsis**
  Stems and other parts of the plant glabrous. . . var. **glabra**

var. **kirkii**

Stems usually ± densely clothed with spreading hairs, especially above ; the spreading hairs rarely sparse or absent, and then the stems densely pubescent with short curved hairs. Leaflets usually ± densely pubescent on both surfaces, sometimes glabrous or nearly so except for rather long cilia, but then stem with dense spreading hairs. Figs. 10/40 & 16.

UGANDA. Toro District : Fort Portal, 23 Nov. 1931, *Hazel* 18 ! ; Mengo District : Entebbe, Sept. 1931, *Eggeling* 3 in *F.H.* 146 ! & Kampala, King's Lake, 4 Sept. 1935, *Chandler & Hancock* 26 !

KENYA. Uasin Gishu District : Kipkarren, *Brodhurst Hill* 466 in *C.M.* 3665 ! ; Kitui District : Ukamba, *Scott Elliot* 6480 ! ; N. Kavirondo District : Kakamega Forest, 9 Dec. 1956, *Verdcourt* 1659 !

TANGANYIKA. Buha District : Kasulu, 29 Mar. 1931, *Rounce* 27 ! ; Morogoro District : Mgeta R. valley above Bunduki, 12 Mar. 1953, *Drummond & Hemsley* 1492 ! ; Songea District : R. Luhira just E. of Songea, 3 Apr. 1956, *Milne-Redhead & Taylor* 9451 !

DISTR. **U**2, 4 ; **K**3–5 ; **T**1–4, 6–8 ; Nigeria, Congo Republic, Malawi, Zambia and Rhodesia

HAB. Grassland, particularly in damp or marshy places, also on forest-edges and in clearings ; 910–2300 m.

SYN. *C. kirkii* Oliv. var. *velutina* Ghesq. in B.J.B.B. 9 : 152 (1932). Types : Uganda, Mengo District, Entebbe, *Eggeling* 3 in *F.H.* 146 (BR, syn., EA, K, isosyn. !) & Tanganyika, Njombe District, Lupembe area, *Schlieben* 335 (BR, syn.!, BM, K, isosyn.!); five other syntypes from Congo Republic also cited (BR)

C. wildemaniana Ghesq. in B.J.B.B. 9 : 154 (1932) ; Steyaert in F.C.B. 3 : 523 (1952), pro parte. Types : Congo Republic, Tshitirunge, *Bequaert* 6012 (BR, lecto. !)

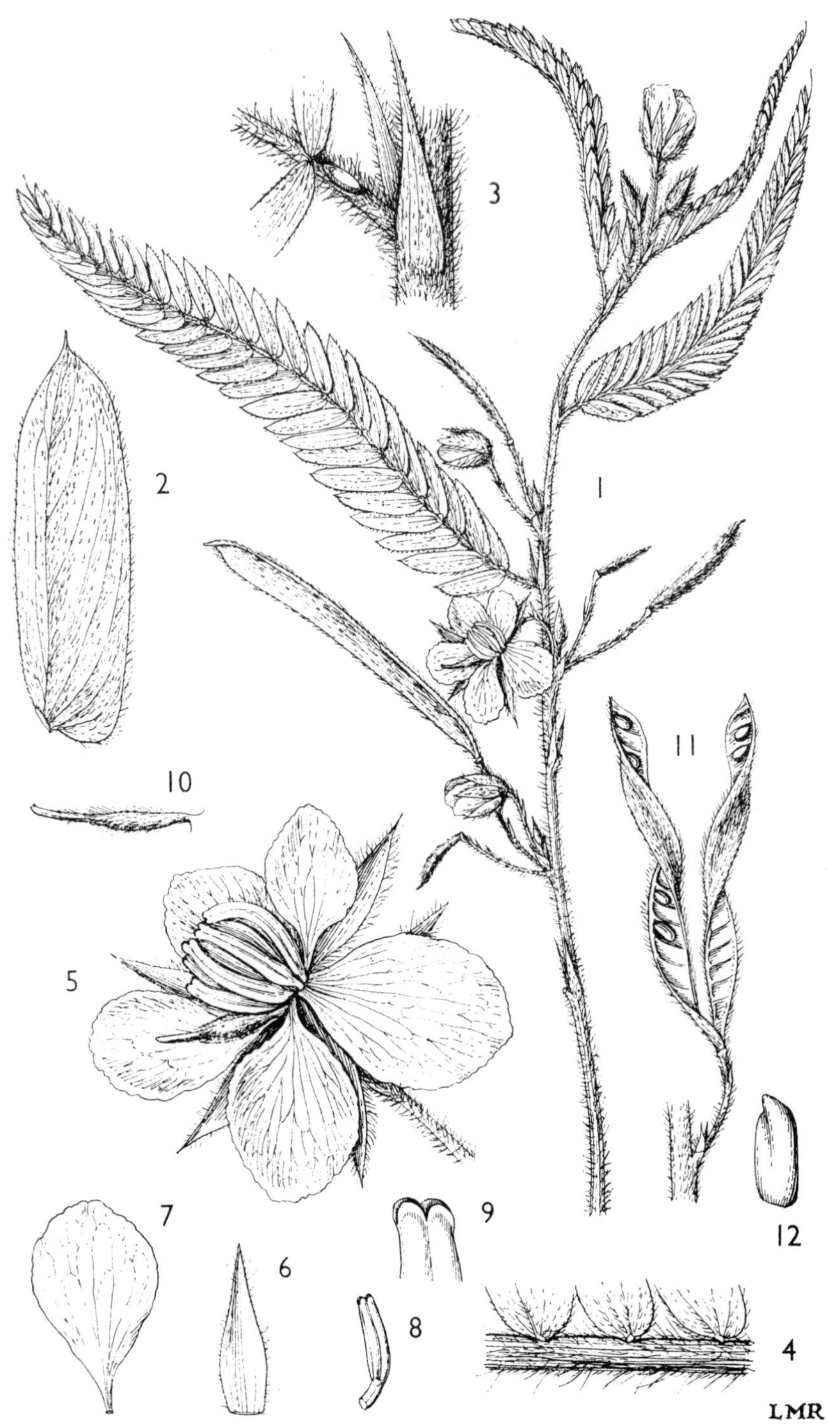

FIG. 16. *CASSIA KIRKII* var. *KIRKII*—**1,** part of stem, showing leaves and flowers, × $\frac{2}{3}$; **2,** leaflet, × 4; **3,** gland on petiole of leaf, × 3; **4,** part of leaf-rhachis, viewed from side, × 4; **5,** open flower, × 2; **6,** sepal, × 2; **7,** petal, × 2; **8,** stamen, × 2; **9,** apical part of anther, showing method of dehiscence, × 8; **10,** ovary, × 2; **11,** pod, × $\frac{2}{3}$; **12,** seed, × 4. 1–4, from *Milne-Redhead & Taylor* 9637; 5–10, from *Rounce* 27; 11, 12, from *Drummond & Hemsley* 1492.

NOTE. The spreading hairs on the stem are plentiful in all the Tanganyika material studied and in most of that from Uganda. They are absent in the following specimens from the latter territory: *Hazel* 18, *Eggeling* 3 in *F.H.* 146 (see above), *Harker* 175, from Ankole District, Sanga, and *Norman* 41, from Masaka District, Sango Bay, Namalala Forest edge. Kenya material usually lacks the hairs, but they are densely present in *Scott Elliot* 6480 (see above), *McDonald* in *A.D.* 1339 from Kericho District, Sotik, and in *Webster* in *Herb. Amani* 9616 from Trans-Nzoia District, Endebess. All specimens seen from the Uluguru and Usambara Mts. in Tanganyika have the indumentum on the stem typically and densely developed, but the leaflets are glabrous or nearly so except for rather long cilia. Elsewhere in Tanganyika var. *kirkii* seems more frequently to have typically pubescent leaflets.

The typically rather large petiolar gland of *C. kirkii* is smaller on various specimens particularly from northern Tanganyika, Kenya and Uganda (*Purseglove* 2886, Kigezi District, Kachwekano). Such plants are the basis of *C. wildemaniana* Ghesq. I can, however, find no other clear distinction from normal *C. kirkii*, and suspect that the smallness of the glands may be due often to exposure causing stunted growth and reduction in size of organs. Similar plants of var. *guineënsis* occur in Kenya (e.g. *Bogdan* 274 and *Mettam* 292 in *C.M.* 730, both from Kiambu District, Kabete) and in Tanganyika (*Johnston* 58, see below). These aberrant forms of *C. kirkii* may be separated from *C. quarrei* by the longer pedicels and larger flowers, and from *C. wittei* by the smaller flowers, the petiolar gland of different shape and probably colour and by the less bushy habit.

It may well be that these difficulties in *C. kirkii* may be explained if hybridization can take place between *C. kirkii* and *C. mimosoïdes*. If this occurs, then plants of the sort described above are precisely what might be expected. The difficulty in defining certain variants of *C. mimosoïdes* itself, particularly Group B, might also be explained in this way. At present this is just speculation, but it would be interesting to see if any field-observations or experimental evidence can be found to support the suggestion.

var. **guineënsis** *Steyaert* in B.J.B.B. 20: 256, t. 9 (1950) & in F.C.B. 3: 526, t. 38 (1952). Type: Congo Republic, Yahila, *Louis* 11215 (BR, holo. !, K, iso. !)

Stems puberulous or shortly pubescent with small incurved hairs; longer spreading hairs absent. Leaflets glabrous or almost so except for margins which are shortly ciliate especially towards base.

UGANDA. W. Nile District: Payida, 25 Aug. 1953, *Chancellor* 190 !; Bunyoro District: near Budongo Forest, 30 Nov. 1938, *Loveridge* 146 !; Mengo District: Kampala, King's Lake, Feb. 1936, *Chandler & Hancock* 155 !
KENYA. Trans-Nzoia District: Kitale, 1931, *Maher* 2041 ! & near Kitale, Maboonde, Aug. 1955, *Tweedie* 1345 ! & Elgon, Oct.–Nov. 1930, *Lugard* 23 !
TANGANYIKA. Kilimanjaro, *Johnston* 58 !
DISTR. **U**1, 2, 4; **K**3, 4, ?5; **T**?1, 2; also from Sierra Leone eastwards to the Cameroun Republic, Gabon, the Congo Republic and Zambia
HAB. Similar to that of var. *kirkii*

NOTE. Steyaert drew attention also to the characteristic red to purplish-brown colour shown by the stems of var. *guineënsis*, at least when dried. There is also a tendency for the leaflets to be not more than 2·5 mm. wide and acute at apex.

var. **glabra** *Steyaert* in B.J.B.B. 20: 257 (1950) & in F.C.B. 3: 528 (1952). Type: Congo Republic, Equateur, Likete on R. Lomela, *Ghesquière* 2711 (BR, holo. !, K, iso. !)

Similar to var. *guineënsis*, but glabrous throughout.

UGANDA. Mengo District: Kyadondo, Magize, Nov. 1914, *Dummer* 1277 ! & Bulumezi, Bombo region, Sept. 1926, *Maitland* 1364 *pro parte* !
DISTR. **U**4; Congo Republic
HAB. Similar to that of var. *guineënsis*

41. **C. wittei** *Ghesq.* in B.J.B.B. 9: 154, t. 2/11–15 (1932)*; T.T.C.L.: 96 (1949); Steyaert in F.C.B. 3: 521 (1952). Type: Congo Republic, Katanga, Kasiki, *de Witte* 443 (BR, lecto. !, BM, iso. !)

Normally an annual herb with erect or arcuate-spreading stems 0·3–1·5 m. high; rarely the stems may become weakly woody above ground-level and thus perhaps enable the plant to perennate for a short time. Stems varying from being densely clothed with rather long spreading hairs to puberulous with short curved hairs, the longer spreading ones being then absent;

* According to Steyaert (l.c.), to be interpreted *pro parte*.

indumentum, at least when dry, often yellowish. Leaves oblong-linear to oblong, 2–9 cm. long, 0·5–3 cm. wide; gland at or near top of petiole, sessile, circular or very broadly elliptic, flattened or somewhat concave on upper side, 0·5–1 mm. long, 0·5–0·75 mm. wide, dark purplish-red, at least when dry; sometimes 2 similar glands together; rhachis eglandular, channelled but not crenate-crested along upper side; leaflets sessile, usually in 13–40 pairs, narrowly oblong to linear-oblong, (4–)6–12(–15, or very rarely to 19) mm. long, (0·75–)1·25–2(–3·5) mm. wide, subacute and mucronate, glabrous except for the minutely ciliolate margins; midrib somewhat excentric (towards anticous margin) but giving off ± prominulous lateral nerves towards both margins. Inflorescences axillary or supra-axillary, 1–3(–4)-flowered; pedicels 1–2·2 cm. long, puberulous or with ± dense spreading hairs. Petals yellow, 10–19 mm. long, 6–14 mm. wide. Pods 5–7·5 cm. long, 5–6 mm. wide. Seeds brown, ± rhombic, ± 3·5–4 mm. long, 2–2·5 mm. wide, not areolate. Fig. 10/41, p. 57.

UGANDA. Kigezi District: Kinaba gap, Dec. 1938, *Chandler & Hancock* 2523! & Kachwekano Farm, Feb. 1950, *Purseglove* 3320!
? UGANDA or KENYA. Elgon, Sept. 1932, *Humphreys* 1128!
TANGANYIKA. Mbeya District: Poroto Mts., 18 May 1957, *Richards* 9803!; Iringa District: Mufindi, *R. M. Davies* L3!; Songea District: Matengo Hills, Lupembe Hill, 29 Feb. 1956, *Milne-Redhead & Taylor* 8763!
DISTR. **U**2; ? **K**; **T**4, 7, 8; Cameroun Republic (Bamenda), Congo Republic, Ethiopia, Mozambique, Malawi and Rhodesia
HAB. Upland grassland and upland evergreen bushland; 1740–2590 m.

SYN. [*C. wildemaniana* sensu F.P.N.A. 1: 376 (1948), saltem pro parte et quoad *de Witte* 1929; verisim. etiam Steyaert in F.C.B. 3: 523 (1952), *non* Ghesq. vel ? quoad *Boutakoff* 70 tantum (BR, syn.!)]

NOTE. Uganda specimens have long spreading hairs on the stem. In **T**6 and 7 they are usually though not always absent. The sole specimen from **T**4, *Richards* 8814, from Ufipa District, Ilemba, has plentiful spreading hairs, is unusually large in its leaves and flowers and, except for not having paired inflorescences, is rather similar to *C. duboisii* Steyaert in B.J.B.B. 20: 262 (1950). The latter, however, may itself be no more than an unusual extreme of *C. wittei*.

Blower in *Eggeling* 6030, from Tanganyika, N. Kilwa District, is very close to *C. wittei*, of which it may be a variant or a related species, and *Busse* 72, without locality, but probably from Zanzibar or Dar es Salaam, seems the same. The petiolar gland of these specimens is circular or almost so, about 0·75–1 mm. long and wide, with a marked tendency for its sides to be curved downwards, so that the gland sits on the petiole like a saddle. The plants are more branched than usual in *C. wittei*, and the indumentum lacks long spreading hairs. The altitudes at which these specimens were collected must be much lower than those at which *C. wittei* is otherwise known to grow. Further material and observation are required.

42. **C. quarrei** (*Ghesq.*) *Steyaert* in B.J.B.B. 20: 264, fig. 26 (1950); Steyaert in F.C.B. 3: 522 (1952). Type: Congo Republic, Katanga, Étoile, *Quarré* 380 (BR, lecto.!)

Annual herb with prostrate and ascending stems forming a low bushy plant about 1–1·5 m. in diameter and under 30 cm. high; sometimes stems erect, branched or subsimple, up to 90 cm. high. Stems varying from puberulous with short curved hairs to subglabrous, without longer spreading hairs. Leaves narrowly or linear-oblong, mostly (3·5–)4·5–9 cm. long, 1–2·3 cm. wide; gland near top of petiole, sessile, ± circular or very broadly elliptic, concave or depressed on upper side, 0·5–0·8, rarely to 1·0 mm. long, 0·5–0·8 mm. wide, dark purplish-red; rhachis eglandular, channelled but not crenate-crested along upper side; leaflets sessile, usually in (17–)25–35 pairs, linear-oblong, 4·5–13 mm. long, 1·25–2·5 mm. wide, acuminate, glabrous on surfaces (except in basal leaflets of leaves); midrib somewhat excentric (towards anticous margin) but giving off ± prominulous lateral nerves towards both margins. Inflorescences mostly supra-axillary, 2–4-

flowered; pedicels 0·6–1(–1·2, rarely in fruit elongating to 1·5) cm. long, puberulous and with longer spreading hairs, sometimes subglabrous. Petals yellow, 7–8·5 mm. long, 3–7 mm. wide. Pods 3–6 cm. long, 3·5–5·5 mm. wide. Seeds brown, ± 3–4 mm. long, 1·5–3 mm. wide, not areolate. Fig. 10/42, p. 57.

KENYA. District uncertain: between Kikuyu and Eldama, 1898, *Whyte* !; S. Elgon, Oct. 1937, *Tweedie* 413 !
TANGANYIKA. Songea District: Kitai, 17 Apr. 1956, *Milne-Redhead & Taylor* 9666 !
DISTR. **K3** or ? 5; **T8**; Congo Republic, Malawi, Zambia, Rhodesia and South Africa (Transvaal, Orange Free State, Natal and Tembuland)
HAB. " Roadsides in open bush country " (*Tweedie* 413); a weed of cultivated ground in *Brachystegia* woodland (*Milne-Redhead & Taylor* 9666); 880–2040 m.

SYN. *Chamaecrista stricta* E. Mey., Comm. Pl. Afr. Austr.: 159 (1836). Type: South Africa, E. Cape Province, R. Bashee, *Drège* (K, iso. !)
Cassia stricta (E. Mey.) Steud., Nom., ed. 2, 1: 308 (1840), *non* Schrank (1819)
C. mimosoïdes L. var. *stricta* (E. Mey.) Harv. in Harv. & Sond., Fl. Cap. 2: 273 (1862)
C. kirkii Oliv. var. *quarrei* Ghesq. in B.J.B.B. 9: 153 (1932), pro parte fide Steyaert
C. capensis Thunb. var. *humifusa* Ghesq. in B.J.B.B. 9: 164 (1932), pro minore parte, quoad *Robyns* 2448 (BR, syn. !)
C. sparsa Steyaert in B.J.B.B. 21: 359, fig. 100 (1951), pro parte quoad holotypum tantum; Steyaert in F.C.B. 3: 523, fig. 45 (1952), excl. cit. " Indes; Chine." Type: Congo Republic, Haut-Katanga, Keyberg, *Schmitz* 3485 (BR, holo. !)

NOTE. Steyaert described *C. quarrei* as up to 2–3 m. high, but I cannot confirm this. I am unable to separate *C. sparsa* from *C. quarrei*. *C. quarrei* resembles very closely indeed the Asiatic and African *C. hochstetteri* Ghesq. (see below). In fact *Roxburgh* s.n. from India, cited by Steyaert under *C. sparsa*, is clearly *C. hochstetteri*, and the Chinese specimen also cited there by Steyaert is from its locality probably the same. *C. quarrei* has 8–10 stamens while *C. hochstetteri* has only 4–5.

43. **C. hochstetteri** *Ghesq.* in B.J.B.B. 9: 155 (1932). Type: Ethiopia, Tigre, Mt. Scholoda, 1837, *Schimper* 66 (BR, lecto., K, isolecto. !)

Annual herb, prostrate or erect. Stems 10–45 cm. long, simple or branched especially towards base, puberulous with short curved hairs which are often yellowish at least when dry; longer spreading hairs absent from stems. Leaves oblong to linear-oblong, 1·5–6 cm. long, 0·7–1·7 cm. wide; gland near top of petiole, sessile, circular or broadly elliptic, 0·2–0·7 mm. in diameter, purplish to blackish at least when dry; rhachis eglandular, channelled but not crenate-crested along upper side; leaflets sessile, mostly in 9–31 pairs, linear-oblong or elliptic-oblong, 4–7·5(–10) mm. long, 1–1·5(–2) mm. wide, rounded to acute and mucronate at apex, glabrous or with a few short cilia on margins particularly near apex; midrib only slightly excentric at least above, the apex of the leaflet being thus subcentral; lateral nerves prominulous beneath. Inflorescences mostly supra-axillary, 1–3-flowered; pedicels 3·5–7(–10) mm. long, puberulous (in Asia sometimes with longer spreading hairs). Petals yellow, 4–5 mm. long, 1·5–5 mm. wide. Stamens only 4–5(8–10 in other related East African species). Pods (1·5–)2·2–4·3 cm. long, 3·5–6 mm. wide. Seeds brown, without areoles. Fig. 10/43, p. 57.

TANGANYIKA. Ngara District: Bugufi, Murugwanza, 20 Jan. 1961, *Tanner* 5624 !; Mbeya Airfield, 11 May 1956, *Milne-Redhead & Taylor* 10049 !; Njombe District: Msima Stock Farm, 1932, *Emson* 313A !
DISTR. **T1**, 7; Ethiopia, Madagascar, India, China, Japan
HAB. Grassland and old cultivations; about 1650–1680 m.

SYN. *Senna dimidiata* Roxb., Fl. Indica 2: 352 (1832). Type: India, Calcutta Botanic Garden, *Roxburgh* (whereabouts of type uncertain—authentic painting no. 1839, K!). *Non Cassia dimidiata* D. Don, Prodr. Fl. Nepal.: 247 (1825) (see note)
Cassia sparsa Steyaert in B.J.B.B. 21: 359 (1951), pro parte, saltem quoad spec. *Roxburgh*, sed excl. holotypum

NOTE. This species is outstanding in having only half the usual number of stamens. The small flowers on short pedicels and the subcentral nerve at the leaflet-apex are also striking features.

The African material of *C. hochstetteri* has pods 3·5–4 mm. wide, while the pods are 4–6 mm. wide in Asia. I can see no other constant difference between specimens from the two continents.

There is a tendency for the inflorescences in the African specimens to be more markedly supra-axillary than they are in Asia, and I have seen no African specimens with spreading hairs on the pedicels, though in Asiatic material they are not infrequent. *C. hochstetteri* has been collected very rarely in Africa, and because of this I prefer not to separate the Asiatic plants at present.

The rejection of the name *Cassia dimidiata* for this species requires some comment. The name was due to Buchanan-Hamilton, who did not, however, publish it; he sent seeds to Roxburgh at Calcutta, and his herbarium-specimen came to D. Don. Don published *Cassia dimidiata* in 1825 from the herbarium-specimen, making no mention of the stamen-number. Roxburgh's *Senna dimidiata* appeared later, in 1832, and the stamens were said to be 4–5. Whatever Buchanan-Hamilton intended by *C. dimidiata*, Don's type at the British Museum (Natural History), which I have examined, shows 8 stamens and is not *C. hochstetteri*, but apparently *C. mimosoïdes* or a close relative, and is also evidently not the same as Roxburgh's *Senna dimidiata*.

44. **C. exilis** *Vatke* in Oesterr. Bot. Zeitschr. 30: 81 (1880); L.T.A.: 641 (1930). Type: Zanzibar I., *Hildebrandt* 963 (? B, holo. †, BM, iso. !)

Apparently annual herb, branching at base into numerous decumbent or ascending stems ± 5–30 cm. long. Stems pubescent with short crisped or longer more spreading hairs. Leaves elliptic to oblong-elliptic, 0·5–2·5(–3) cm. long, 0·8–1·5 cm. wide; gland at top of petiole, sessile, circular, 0·25 mm. in diameter, not infrequently absent from some leaves; rhachis eglandular, crenate-crested along upper side; leaflets sessile, in 5–10 pairs (rarely a few of the lowest leaves on the main stem with up to 20 pairs), narrowly oblong, 3–9 mm. long, 0·9–2 mm. wide, rounded to subacute and shortly mucronate at apex, glabrous or with a few minute inconspicuous hairs on margins only (rarely with a few longer weak hairs on surface); midrib somewhat excentric (towards anticous margin), but giving off lateral nerves towards both margins. Inflorescences supra-axillary, 1-flowered; pedicels 2–4 cm. long, pubescent with short crisped hairs. Petals yellow, ± 6–7 mm. long and 4–5 mm. wide (estimated from damaged material). Pods 3–5 cm. long, 3–4·5 mm. wide. Seeds brown, with minute darker dots in lines, ± rhombic or quadrate, 2·5–3 mm. long, 1·5–1·75 mm. wide, not areolate. Fig. 10/44, p. 57.

TANGANYIKA. Tanga District: Yihirini [Yilihini]–Malamba, 24 Sept. 1918, *Peter* 48190!; Pangani District: Mwera–Kipumbwe, 26 Nov. 1915, *Peter* 48980! & Mkwaja, Mkaramo, 27 July 1955, *Tanner* 2013!; Rufiji District: Mafia I., Nchoroko Kangaga, 23 Mar. 1912, *Braun* in *Herb. Amani* 3650!

ZANZIBAR. Zanzibar I., Oct. 1873, *Hildebrandt* 963!

DISTR. **T3, 6; Z**; not known outside the Flora area

HAB. Grassland on sandy soil (*Tanner* 2013); probably from near sea-level to about 150 m.

SYN. [*C. brevifolia* sensu Ghesq. in B.J.B.B. 9: 162 (1932), pro parte, quoad syn. in obs. *C. exilis*, *non* Lam.]

NOTE. Vatke described *C. exilis* as " lignosa ", and the stems of *Hildebrandt* 963, *Tanner* 2013 and some of the Peter specimens are hardened and weakly woody below, although the plants seem annual. The measurements given by Vatke (leaflets up to 5 mm. long and about 1·5 mm. wide, peduncles 1·5 cm. long) are inexact. The sheet of *Hildebrandt* 963 at the British Museum shows leaflets 2–7 mm. long and peduncles 1·5–2·5 cm. long. Vatke's description of the petiole as eglandular was due to his overlooking the presence of the gland (which is small though clearly visible with sufficient magnification on some of petioles of the sheet of *Hildebrandt* 963 mentioned above). No *Cassia* of the section *Chamaecrista* from continental Africa so far examined by me has constantly eglandular petioles. Special effort should be made to rediscover *C. exilis* in Zanzibar.

45. **C. usambarensis** *Taub.* in P.O.A. C: 201 (1895); L.T.A.: 640 (1930); Brenan in K.B. 14: 181, 183 (1960). Type: Tanganyika, Usambara Mts., *Holst* 140 (B, holo. †)

Perennial herb with branches up to 40 cm. long radiating from a thickened rootstock, forming a carpet. Stems appressed-puberulous. Leaves elliptic, 0·5–2·2 cm. long, 0·7–1·5 cm. wide; gland at top of petiole, usually very shortly (0·1–0·2 mm.) columnar-stipitate, circular, very small (± 0·2–0·25 mm. in diameter), not infrequently absent from some leaves; rhachis eglandular, channelled, but not crenate-crested along upper side; leaflets sessile, in 5–11(–13) pairs, oblong or somewhat wider towards apex, 3–8 mm. long, (0·8–)1–2·5(–3) mm. wide, mostly subacute and mucronate sometimes almost rounded at apex, glabrous except for ciliolate margins; midrib excentric, towards anticous margin but not marginal; some lateral nerves usually visible between midrib and margin. Inflorescences axillary, 1(–2)-flowered; pedicels 1·5–2·3(–2·9) cm. long, appressed-puberulous. Petals yellow, orange or reddish towards base, 7–8·5 mm. long, 3–9 mm. wide. Pods (1·2–)1·5–2·7 cm. long, 3·5–4·5 mm. wide. Seeds brown, ± obovate, 2·5–3 mm. long, 2 mm. wide, not areolate. Fig. 10/45, p. 57.

KENYA. Uasin Gishu District: Eldoret, 11 June 1951, *G. Williams* 236!; Naivasha District: Kinangop, Apr. 1938, *Chandler* 2330!; Kiambu District: Limuru, Dec. 1927, *Lyne Watt* 1187!
TANGANYIKA. Mbulu, Jan. 1935, *Moreau* 23!; Lushoto District: Mtai–Malindi road, near Kidologwai, 19 May 1953, *Drummond & Hemsley* 2653!
DISTR. **K1**, 3, 4, 6; **T2**, 3; not known outside East Africa
HAB. Grassland, especially near rocks and on roadsides; 1760–2590 m.

SYN. *Chamaecrista usambarensis* (Taub.) Standl. in Smithson. Misc. Coll. 68, No. 5: 5 (1917)
[*Cassia grantii* sensu Ghesq. in B.J.B.B. 9: 143 (1932), pro parte, quoad syn. *Cassia usambarensis* et *Chamaecrista usambarensis*, *non* Oliv.]
[*C. brevifolia* sensu Steyaert in B.J.B.B. 20: 266 (1950), pro parte, excl. cit. bibl. et specim. madagasc., *non* Lam.]

46. **C. ghesquiereana** *Brenan* in K.B. 14: 184 (1960). Type: Uganda, Kigezi District, Mt. Mgahinga, *Purseglove* 3703 (K, holo. !, EA, iso. !)

Prostrate perennial herb. Stems radiating from the thickened rootstock, themselves becoming woody below, up to 12–40 cm. long, puberulous or shortly pubescent with short curved hairs, and also with ± numerous longer spreading ones. Leaves narrowly oblong to linear-oblong, 1·2–3·5(–4) cm. long, ± 0·5–0·8(–1·3) cm. wide; gland at or near top of petiole, sessile, circular or elliptic, flattened or convex on upper side, 0·25–1 mm. long, 0·25–0·6 mm. wide, brown to blackish-purple when dry, said to be yellowish when living; rhachis eglandular, channelled but not crenate-crested along upper side; leaflets sessile, in 6–28 pairs, obliquely oblong, somewhat falcate above, 2·5–5(–8) mm. long, 0·5–1(–1·5) mm. wide, acute and mucronate, glabrous on upper surface, glabrous or with a few scattered hairs on lower, ciliate on margins; midrib somewhat excentric (towards anticous margin); lateral nerves slightly raised beneath, those on anticous side of midrib few and obscure or absent. Inflorescences axillary or supra-axillary, 1–2-flowered; pedicels 0·9–1·5 cm. long, clothed like the stem. Petals pale yellow, 6·5 mm. long, 3·5–4 mm. wide. Pods 2–3·8 cm. long, 3·5–4 mm. wide, pubescent with curved but not appressed hairs. Seeds brown, ± rhombic, 2–2·5 mm. long, 1–1·5 mm. wide. Fig. 10/46, p. 57.

UGANDA. Kigezi District: Kachwekano Farm, Apr. 1949, *Purseglove* 2748! & Mt. Mgahinga, June 1951, *Purseglove* 3703!
TANGANYIKA. Mbeya Airfield, 11 May 1956, *Milne-Redhead & Taylor* 10050!; Njombe, Oct. 1931, *Staples* 170!
DISTR. **U2**; **T7**; Congo Republic
HAB. Upland grassland; 1650–2590 m.

SYN. *C. capensis* Thunb. var. *humifusa* Ghesq. in B.J.B.B. 9: 164 (1932), pro minore parte, quoad spec. *Scaetta* 601 tantum (BR, syn. !)

Note. The perennial habit, prostrate stems, small leaflets and flowers, and a general look of *C. mimosoïdes*, though without the toothing or crenation of the leaf-rhachis characteristic of that species, enable *C. ghesquiereana* to be recognized.

The Tanganyika material is still insufficient, the Mbeya specimen being in fruit only and the Njombe one poor. It is possible that the Tanganyika specimens are not identical with those from Uganda and the Congo Republic. The latter have the petiolar gland circular and small, 0·2–0·5 mm. in diameter, while the former have a rather larger and more elliptic gland 0·5–1 mm. long and 0·5–0·6 mm. wide. The seeds of the Tanganyika plant are also slightly larger (2·5 × 1·5 mm., as against 2 × 1 mm.), and sometimes the leaflets also.

47. **C. polytricha** *Brenan* in K.B. 14: 185 (1960). Type: Tanganyika, Songea District, *Milne-Redhead & Taylor* 9402 (K, holo.!)

A variable, annual or perennial herb with simple or somewhat branched, erect or ascending stems up to 50(–100) cm. high. Stems densely clothed with straight spreading hairs 0·25–1·5 mm. long. Leaves oblong-lanceolate, gradually narrowed towards apex, 2–6 cm. long, 0·6–1·7 cm. wide; gland at top of petiole, sessile, orbicular or nearly so, flattened above, 0·5–0·75 mm. long and wide; rhachis eglandular, not or scarcely crenate-crested along upper side; leaflets sessile, in 27–59 pairs (but as few as 10–18 pairs or less in lowest cauline leaves), linear-oblong, (2·5–)3·5–10 mm. long, 0·5–1·5 mm. wide, acute at apex with the point turned towards leaf-apex (or sometimes subacute and mucronate in var. *pauciflora*), conspicuously ciliate on the margins, otherwise glabrous or nearly so; midrib strongly excentric (within 0·25 mm. of anticous margin); lateral nerves prominulous beneath, those on anticous side of midrib absent. Inflorescences axillary or shortly supra-axillary, 1–4-flowered; pedicels 1–2·5 cm. long, densely pubescent with ascending or spreading but not appressed hairs. Petals yellow, obovate, 5–13 mm. long, 3·5–9 mm. wide. Pods linear, 3·2–6 cm. long, 4·5–5 mm. wide, ± spreading-pubescent. Seeds brown, ± rhombic, 2·5–3 mm. long, 1·5–1·75 mm. wide, without areoles.

Key to intraspecific variants

Perennial; petals 8–13 mm. long . . . . . . var. **polytricha**
Annual; petals 5–9 mm. long:
  Inflorescences mostly 2(–3)-flowered . . . . . var. **pulchella**
  Inflorescences mostly 1-flowered . . . . . . var. **pauciflora**

var. **polytricha**; Brenan in K.B. 14: 186 (1960)

Perennial with thickened woody rootstock from which stems arise annually. Inflorescences mostly 2- or more-flowered. Petals 8–13 mm. long.

Tanganyika. Songea District: about 29 km. ENE. of Songea by road to Hanga Farm, 26 Mar. 1956, *Milne-Redhead & Taylor* 9402!; Tunduru District: Nampungu Baraza, Mpangaduka, 16 Apr. 1950, *Tanner* 125!; without precise locality, but probably either Mafia I. or Kilwa, [Nov.–Dec. 1900], *Busse* 447!
Distr. T8; Mozambique and Malawi
Hab. Deciduous woodland (*Brachystegia-Uapaca*); 460–1050 m.

var. **pulchella** *Brenan* in K.B. 14: 187 (1960). Type: Tanganyika, Songea, *Milne-Redhead & Taylor* 9218 (K, holo.!)

Annual. Inflorescences mostly 2(–3)-flowered. Petals 5–7(–8) mm. long. Fig. 10/47, p. 57.

Tanganyika. Songea, near District Office, 16 Mar. 1956, *Milne-Redhead & Taylor* 9218! & near Kitai, 18 May 1956, *Milne-Redhead & Taylor*!
Distr. T8; not known elsewhere, except doubtfully in Rhodesia
Hab. Induced grassland; 1080 m.

var. **pauciflora** *Brenan* in K.B. 14: 187 (1960). Type: Tanganyika, Songea District, *Milne-Redhead & Taylor* 8488 (K, holo.!)

Annual. Inflorescences mostly 1-flowered. Petals 6–9 mm. long.

TANGANYIKA. Rungwe District: Masoko, 25 Sept. 1932, *Geilinger* 2632 !; Songea District: by Kimarampaka stream about 12 km. W. of Songea, 5 Feb. 1956, *Milne-Redhead & Taylor* 8488 !
DISTR. **T7**, 8; Malawi, Zambia and Rhodesia
HAB. Shallow boggy soil overlying " laterite " by streamside; 960 m.

NOTE. *C. polytricha*, the value and status of whose variants can do with further testing, is very close to *C. mimosoïdes* in general facies, but distinguished by the leaf-rhachis not or scarcely crenate-crested along its upper side, and from most forms of *C. mimosoïdes* by the dense spreading indumentum of the stems. The perennial habit of *C. polytricha* var. *polytricha* is also distinctive.

48. **C. mimosoïdes** *L.*, Sp. Pl.: 379 (1753); L.T.A.: 642 (1930); Ghesq. in B.J.B.B. 9: 158, t. 1/24–27 (1932); T.T.C.L.: 96 (1949); U.O.P.Z.: 177, fig. (1949); Steyaert in B.J.B.B. 20: 236, 240, 247, t. 8 (1950) & in F.C.B. 3: 514, t. 37 (1952); De Wit in Webbia, 11: 283 (1955); Mendonça & Torre in C.F.A. 2: 181 (1956). Types: Ceylon, *Herb. Hermann*, vol. 2, pp. 13, 78 (BM, syn. !)

An exceedingly variable prostrate to erect herb up to about 1·5 m. high, usually annual, sometimes with the stems becoming woody above ground-level and enabling the plant to perennate. Stems variable in indumentum, usually puberulous with short curved hairs, sometimes ± densely clothed with longer spreading hairs. Leaves linear to linear-oblong, ± parallel-sided, 0·6–10 cm. long, 0·4–1·5 cm. wide; gland usually at, or near top of petiole, sessile, normally orbicular or nearly so, dish-shaped when dry, 0·4–1·0 mm. in diameter; rhachis eglandular, serrate- or crenate-crested along upper side; leaflets sessile, in 16–76 pairs, obliquely oblong to oblong-elliptic or linear-oblong, (2–)2·5–8(–9) mm. long, 0·5–1·25(–1·9) mm. wide, acute or subacute and shortly mucronate, glabrous or nearly so except for minute inconspicuous ciliolation or sometimes longer ciliation on margins; midrib somewhat excentric (towards anticous margin); lateral nerves obscure to prominulous beneath, those on anticous side of midrib absent or very few. Inflorescences supra-axillary or sometimes axillary, 1–3-flowered; pedicels 0·3–2·5(–3·0) cm. long, usually shortly puberulous, sometimes spreading-hairy. Petals yellow, obovate, 4–13 mm. long, 2–9 mm. wide. Pods linear to linear-oblong, (1·5–)3·5–8 cm. long, 3–5 mm. wide, usually appressed-hairy. Seeds brown, ± rhombic, 2–3 mm. long, 1–2 mm. wide, without areoles. Fig. 10/48, p. 57.

DISTR. (of aggregate species). Widespread in the tropics of the Old World; recorded for America, but needs confirmation; not allowed for America by Britton & Killip in Ann. N. Y. Acad. Sci. 35: 187 (1936)
HAB. (of aggregate species). Clearings in forest, forest margins, wooded grassland, grassland, cultivated and waste ground, sandy river-beds, lake- and sea-shores; 0–2740 m.

VARIATION. The range of variation shown by this species is huge, and to suggest that certain of the extremes belong here together under one species seems at first sight an absurdity. Nevertheless, as Steyaert has pointed out, they are linked by the most baffling intermediates. It is clear that this variation is at any rate partly linked with geography and habitat, but at present I can find no clear way of analysing it. It may be that species formerly distinct have met to produce fertile hybrid swarms, and that *C. mimosoïdes* is such a complex. There is no evidence, however. At present it seems best to divide the range of variation of *C. mimosoïdes* into unnamed groups, although with the realization that many plants will be found which do not conform.

KEY TO THE GROUPS OF CASSIA MIMOSOÏDES

Pedicels short, in fruit usually 5–7(–9) mm., rarely longer in Group C:
  Hairs on pedicels and pods appressed . . . Group E
  Hairs on pedicels and pods spreading:
    Inflorescences mostly single-flowered . . . Group D
    Inflorescences mostly 2- or more-flowered . . Group C

Pedicels longer, in fruit mostly 1–2·5 cm. long (compare also Group C):
  Plant prostrate or ascending at ends . . . Group G
  Plant erect or suberect:
    Inflorescences some or many leafy; or short leafy shoots occupying positions of many inflorescences:
      Flowers large: sepals ± 7–10 mm. long; petals mostly 10 mm. or more long . . . Group A
      Flowers smaller: sepals ± 5–7 mm. long; petals mostly ± 6–8(–10) mm. long . . . Group B
    Inflorescences not or scarcely leafy:
      Flowers small: petals ± 5–7 mm. long . . Group F
      Flowers larger: petals ± 9–11 mm. long:
        Hairs on pedicels appressed . . . . Group A (abnormal forms)
        Hairs on pedicels spreading . . . . Group C

Group A

UGANDA. Bunyoro District: Kabanda, May 1942, *Purseglove* 1231! (doubtful)
KENYA. Teita Hills, May 1931, *Napier* 1073 in *C.M.* 2242!; Kilifi District: Malindi, Oct. 1951, *Tweedie* 992! & Kilifi, 26 Aug. 1937, *Moggridge* 481!
TANGANYIKA. Moshi, May 1927, *Haarer* 392!; Tanga District: 8 km. SE. of Ngomeni, 30 July 1953, *Drummond & Hemsley* 3551!; Morogoro District: Turiani, Apr. 1953, *Semsei* 1141!
ZANZIBAR. Zanzibar I., Tungau [? Tunguu] road, 24 Apr. 1950, *Oxtoby* 1 & without precise locality, *G. de C. Taylor* 2!
DISTR. ? **U2**; **K4**, 7; **T2**, 3, 6; **Z**; Mauritius and Seychelles, probably introduced; closely related plants occur in the Sudan Republic and the eastern Congo Republic
ALT. 0–1370 m.

SYN. *C. mimosoïdes* L. var. *telfairiana* Hook. f. in Bot. Mag. 26, t. 5874 (1870). Type: a cultivated plant at Kew, apparently not preserved

NOTE. The wing along the upper side of the leaf rhachis in Group A has rather low rounded teeth, giving in outline a crenate appearance. The indumentum on the pedicels and calyx varies from appressed to spreading, but this seems without much significance.
Usually the individual inflorescences, or some of them, are characteristically accompanied by leaves arising from their axes; or a short leafy shoot may occupy the position of an inflorescence. This gives a "fascicled" appearance to the foliage, which is a marked feature of this Group. Occasionally the inflorescences are not accompanied by leaves, e.g. in *Jeffery* 612, from Kenya, Kilifi, *E. M. Bruce* 311, from Tanganyika, Uluguru Mts., Bahati, *Greenway* 5170 from Tanganyika, Mafia I., Kirongwe, and in *Oxtoby*, from Zanzibar I., Kiwengwa. At present, I regard these as merely unusual forms of Group A, but this view needs testing.

Group B

UGANDA. W. Nile District: Terego, July 1938, *Hazel* 628!; Teso District: Serere, Dec. 1931, *Chandler* 322!; Mengo District: Kampala, King's Lake, Feb. 1936, *Chandler & Hancock* 154!
KENYA. Northern Frontier Province: Bari Forest, 2 Oct. 1947, *J. Adamson* 426 in *Bally* 6127 in *C.M.* 17472!; Uasin Gishu District: Kipkarren, *Brodhurst Hill* 60 in *C.M.* 2481!; Naivasha District: Longonot, crater-lip, Mar. 1922, *Dummer* 5171!
TANGANYIKA. Musoma, 1933, *Emson* 352!; Rungwe District: Kyimbila, 14 Sept. 1910, *Stolz* 273 & without precise locality, 16 Mar. 1932, *R. M. Davies* 39!
DISTR. **U1–4**; **K1**, 3, ? 5; **T1**, 2, 4, 7; Congo Republic, Sudan Republic, Mozambique, Malawi, Zambia, Angola and South Africa
ALT. 910–1520(–2740) m.

NOTE. There is no clear distinction between Groups A and B. Certain specimens, e.g. *Tanner* 776, from Tanganyika, Mwanza District, Mabale, Mbarika, and *Polhill* 128, from Kenya, Naivasha District, Longonot, might as well be placed in either group. These specimens have sepals 7–9 mm. long and petals about 10–11 mm. long. Nevertheless, Group B shows a significantly more western distribution in our area than A: it is well represented in Uganda, where Group A is only doubtfully present, and conversely it seems to be absent from the maritime coastal areas of Kenya and Tanganyika where Group A is so common.

Group C

KENYA. Mombasa, June 1929, *R. M. Graham* DD453 in *F.H.* 1992 ! & Likoni, July 1932, *Jex Blake* in *Napier* 2267 in *C.M.* 4980 !

TANGANYIKA. Musoma District: Majita, Nyambono, 5 June 1959, *Tanner* 4324 !; Tanga, Jan. 1893, *Volkens* 12 !; Pangani, 9 Feb. 1956, *Tanner* 2610 ! & 8 May 1956, *Tanner* 2828 !

ZANZIBAR. Zanzibar I., without precise locality, 1909, *Last* ! & Kinyasini, 29 Jan. 1929, *Greenway* 1251 !

DISTR. **K**7; **T**1, 3, ? 8; **Z**; not known elsewhere

ALT. 0–450 m. (1370 in **T**1)

NOTE. Group C links Groups A, B, F and G, which have long-pedicellate flowers, with those other groups whose pedicels are shorter. Their length, in Group C, varies between 5 and 13 mm., but on each specimen most are 10 mm. or less. The plants are erect, often freely branched above, and with petals 7–12 mm. long.

Group D

KENYA. Northern Frontier Province: ? Marsabit or Kulal Mts., July 1934, *W. H. R. Martin* 203!; Nairobi, 1 Apr. 1931, *Napier* 839A in *C.M.* 1766! & 3 Feb. 1947, *Bogdan* 306 !

DISTR. **K**1, 4, 6; not known elsewhere

ALT. 1680–1740 m.

NOTE. Although this group has been collected repeatedly at or near Nairobi, really good material is still needed.

Harker 165, from Uganda, Ankole District, Kirahura Junction, is intermediate between Groups D and E. It is erect, with small flowers in 1–2-flowered inflorescences, and pedicels whose hairs are appressed below but spreading though short above the bracteoles. Group D has not been collected in Uganda.

Group E

UGANDA. Karamoja District: Moruatukan, 24 May 1940, *A. S. Thomas* 3550 !

KENYA. Northern Frontier Province: Moyale, 7 July 1952, *Gillett* 13518 !; Machakos District: Kiambere, 24 Nov. 1951, *Kirrika* 142 ! & between Makindu and Kibwezi, 10 May 1955, *Ossent* 94 !

TANGANYIKA. Shinyanga District: Samuye Hills [Samui], Mar. 1936, *B. D. Burtt* 5643 !; Mbulu District: Mbulumbul, 23 June 1944, *Greenway* 6918 !; Kondoa District: Sambala, 28 Mar. 1929, *B. D. Burtt* 2120 !

ZANZIBAR. Zanzibar I., 7 Nov. 1873, *Hildebrandt* 964 !

DISTR. **U**1; **K**1, 4, 7; **T**1–5, 7, 8; **Z**; Sudan Republic, Congo Republic, Mozambique, Rhodesia and the Transvaal; closely related plants occur also in Nigeria, the Ivory Coast, Mali and Madagascar

ALT. 470–1550 m.

NOTE. In East Africa Group E is well-marked and usually constant, having small petals 5–6 mm. long and 2–3·5 mm. wide and narrow pods usually 3·5–4 mm. wide, although as already noted (above) a link with Group D has been found. Elsewhere related plants occur, having somewhat wider pods or larger petals.

Group E comprises plants which are usually erect or ascending. Sometimes, however, they are decumbent or prostrate, e.g. *Richards* 9883, from Tanganyika, Rungwe District, Makate, and *Milne-Redhead & Taylor*, from Tanganyika, Songea District, Lake Nyasa, Mbamba Bay. Such plants much resemble Group G, except for having shorter pedicels.

Group F

UGANDA. Toro District: Ruwenzori, Namwamba valley, Kilembe, 24 Dec. 1934, *G. Taylor* 2614 !; Masaka District: Sesse Is., Bugala I., 8 Oct. 1958, *Symes* 479 !

KENYA. Machakos District: Kitui, 18 Jan. 1942, *Bally* 1541 ! (doubtful, specimen poor)

TANGANYIKA. Tabora, *Lindeman* 595 !; Morogoro District: 42 km. E. of Morogoro, on Dar es Salaam road, 1 Mar. 1955, *Welch* 282 !; Songea District: Kitai, 17 Apr. 1956, *Milne-Redhead & Taylor* 9667 !

DISTR. **U**2, 4; **K**3, ? 4; **T** ? 3, 4, 6–8; widespread in tropical Africa from Gambia to Nigeria and the Sudan Republic and southwards to Angola and Natal; also in Asia from India to Australia

ALT. From probably near sea-level to 1370 m.

SYN. *C. mimosoïdes* L. var. *glabriuscula* Ghesq. in B.J.B.B. 9: 160 (1932), saltem pro max. parte. Type: none cited, but *Goossens* 4879 (from Congo Republic, Gemena) at Kew has been determined by Ghesquière as var. *glabriuscula*

NOTE. Mainly in western and southern Tanganyika, as far as the Flora area is concerned.

Although especially in other territories Group F is well-marked and plentiful, there are in East Africa numerous intermediates linking it with Group B which are hard to place in either group with any certainty.

Group G

UGANDA. W. Nile District: Maracha rest camp, 23 July 1953, *Chancellor* 18!; Ankole District: Ruizi R., 15 Nov. 1950, *Jarrett* 170!; Mengo District: Kome I., 7 July 1951, *Norman* 14!

KENYA. Uasin Gishu District: Kipkarren, *Brodhurst Hill* 532 in *C.M.* 3642!; N. Kavirondo District: Nandi to Mumias, 1898, *Whyte*! (doubtful)

TANGANYIKA. Mwanza, shores of Speke Gulf, 1 June 1931, *B. D. Burtt* 2468!; Ufipa District: near Irwin Farm between Mpui and the Zambia frontier, 22 Mar. 1959, *McCallum Webster* T4!; Lindi District: Nachingwea, 17 June 1953, *Anderson* 921! (doubtful, specimen untypical)

ZANZIBAR. Zanzibar I., Kiwengwa, 29 Jan. 1929, *Greenway* 1241! & Chwaka, 28 Dec. 1961, *Faulkner* 2960!

DISTR. **U**1, 2, 4; **K** ? 1, 3, ? 5; **T**1, 4, ? 8; **Z**; Sierra Leone, Liberia, Congo Republic, Eritrea, ? Sudan Republic, Mozambique, Zambia and Angola; also in India

ALT. 0–2110 m.

SYN. *C. capensis* Thunb. var. *humifusa* Ghesq. in B.J.B.B. 9: 164 (1932), pro parte, quoad *Elskens* 139, *Boutakoff* 45, *Charpois* s.n. (all from Congo Republic, BR, syn.!)

NOTE. A very possibly heterogeneous group of prostrate or decumbent plants with small flowers whose petals are only about 4–7 mm. long, short pods mostly about 1·5–4·5 cm. long, and a tendency for the stem-base to become ± thickened and shortly perennial. In East Africa at least, the indumentum is always short and appressed.

Groups G and F intergrade freely and some specimens are very difficult to place.

C. auricoma Grah. var. *glabra* Ghesq. in B.J.B.B. 9: 158 (1932) is likely to be another synonym of this group.

11. DIALIUM

L., Mant. Pl.: 3 (1767)

Trees or (rarely) large shrubs, not climbing, unarmed. Leaves simply imparipinnate; stipules very quickly falling; conspicuous glands absent from petiole and rhachis; leaflets 3–21, opposite to alternate. Inflorescences of terminal and lateral many-flowered panicles; bracts and bracteoles small, quickly falling. Flowers ⚥, irregular, very rarely (not in East Africa) regular. Sepals 5 (or ? 6, but very rarely and not in Flora area), imbricate. Stamens 2 (in East African species), elsewhere sometimes up to 10; anthers basifixed, dehiscing by lateral slits. Disc (in East African species) well-developed, much wider than the ovary and ± puberulous or pubescent. Ovary small, sessile or shortly stipitate, with 2 ovules. Pod (in East African species) ± ellipsoid to subglobose, not compressed (in species outside the Flora area sometimes ± flattened), indehiscent; exocarp hard, brittle, smooth except for indumentum; mesocarp pulpy, mealy and brown-orange or red when dry. Seeds 1–2, embedded in the mesocarp; testa smooth except for small ± irregular cracks; areoles absent; endosperm present.

A very distinct genus of about 35 species predominantly in the Old World tropics and mostly African. The East African species are readily recognized by the tendency for the leaflets to be alternate, the ample panicles of small, usually greenish to cream flowers with dense short indumentum outside and containing no petals and but 2 stamens, and by the peculiar pods. The appearance of the testa of the seed recalls the "crackling" often seen in the glaze on old pottery; a similar effect may also occur in some species of *Cassia*. The veins of the leaflets of *Dialium* are characteristically marked, at least in East African species, by lines of minute pustules which can be readily seen under a hand lens.

Leaflets ± acuminate or subacuminate at apex, 5–13; anthers 1·5–1·75 mm. long; trees 12–45 m. high:
  Pedicels 1·5–5 mm. long, rusty-brown-puberulous, as are the sepals outside; stamen-filaments 2·5–3 mm. long . . . 1. *D. excelsum*

Pedicels 0·75–1 mm. long, grey-puberulous, as are the sepals outside; stamen-filaments 1·5–2·3 mm. long. 2. *D. holtzii*
Leaflets rounded to obtuse, not acuminate at apex, (5–)7–9; anthers 1·25–1·5 mm. long; tree or shrub 6–15 m. high 3. *D. orientale*

1. **D. excelsum** *Steyaert* in Bull. Soc. Roy. Bot. Belg. 84: 38 (1951). Type: Congo Republic, Yangambi, *Louis* 2881 (BR, holo., BM, K, iso. !)

Tree 18–45 m. high. Bark smooth, pale. Young branchlets glabrous to inconspicuously puberulous. Leaves: petiole and rhachis together (4–)5–8·5 cm. long; leaflets 5–9(–11), lanceolate, with the lowest ones of the leaf tending to be ovate and the terminal one to be elliptic, 2–7(–10) cm. long, 1·2–2·6(–2·9) cm. wide, papery, obtusely subacuminate or acuminate, cuneate or the lowest ± rounded at base, glabrous or almost so; venation ± raised and reticulate on both surfaces. Flowers brown outside, greenish-white inside, in panicles up to 20 × 15 cm.; pedicels 1·5–5 mm. long, rusty-brown-puberulous. Sepals rusty-brown-puberulous outside, 2·5–3 mm. long. Petals 0. Stamens 2; filaments 2·5–3 mm. long. Pods subglobose, 1·1–1·5 cm. long, 1–1·2 cm. wide, puberulous outside. Seeds suborbicular, compressed, 7·5–8 mm. long, 6·5–7·5 mm. wide, dark brown.

UGANDA. Bunyoro District: Budongo Forest, Aug. 1935, *Eggeling* 2136 in *F.H.* 1746! & June 1940, *Eggeling* 3968!; Toro District: Bwamba, Kidongo, Aug. 1937, *Eggeling* 3376!
DISTR. **U2**; Congo Republic
HAB. Lowland rain-forest; ± 760–900 m.

SYN. *D. sp.* very near *D. bipindense* Harms sensu Burtt Davy & Bolton, Check-lists For. Trees & Shrubs Brit. Emp. 1 (Uganda Prot.): 31 (1935); I.T.U.: 34 (1940)
D. sp. sensu I.T.U.: 34 (1940)
[*D. bipindense* sensu I.T.U., ed. 2: 65 (1952), *non* Harms]

NOTE. The distinction made by Steyaert (Bull. Soc. Roy. Bot. Belg. 84: 34 (1951)) between the stipitate ovary of *D. excelsum* and the sessile ovary in *D. holtzii* and *D. orientale* appears to me imaginary. I find the ovary in all three species to be either sessile or very shortly and obscurely stipitate.
In I.T.U., ed. 2: 66 (1952), *Eggeling* 5484 is said to be possibly conspecific with "*D. bipindense*". In fact, however, it is a species of *Pancovia* (*Sapindaceae*).

2. **D. holtzii** *Harms* in E.J. 49: 427 (1913); L.T.A.: 647 (1930); T.T.C.L.: 102 (1949); Steyaert in Bull. Soc. Roy. Bot. Belg. 84: 37 (1951). Type: Tanganyika, Livule R. near Mhondo [? Morogoro District, Liwale R. near Mhonda], *Holtz* 1182 (B, lecto. †) & Bagamoyo District, Chakenge [Chakengi], *Holtz* 1117 (B, syn. †)

Tree up to 12–25 m. high. Bark smooth, grey or creamy-grey-brown. Young branchlets puberulous to glabrous. Leaves: petiole and rhachis together (5–)7·5–18 cm. long; leaflets (9–)11–15, oblong-lanceolate, with the lowest ones of the leaf ± ovate and the terminal one ± elliptic, (1·3–)2·3–7(–8·7) cm. long, (1–)1·5–2·8(–3·5) cm. wide, papery to subcoriaceous, obtusely acuminate or subacuminate, rounded or broadly cuneate at base, sparsely and inconspicuously pubescent to glabrous; venation ± raised and reticulate on both surfaces. Flowers white, cream or yellow, in panicles up to 30 × 15 cm. or more; pedicels 0·75–1 mm. long, grey-puberulous. Sepals grey-puberulous outside, 2·5 mm. long. Petals 0. Stamens 2; filaments 1·5–2·3 mm. long. Pods brown, similar to those of *D. orientale*. Seeds suborbicular, 7·5–8 mm. long, 7 mm. wide, brown.

TANGANYIKA. Lushoto District: Longuza, 29 Sept. 1936, *Greenway* 4628!; Morogoro District: Turiani, Liwale R., Mar. 1956, *Semsei* 2412!; Lindi District: Rondo Plateau, Mchinjiri, Mar. 1952, *Semsei* 694!
DISTR. **T3**, 6–8; Mozambique

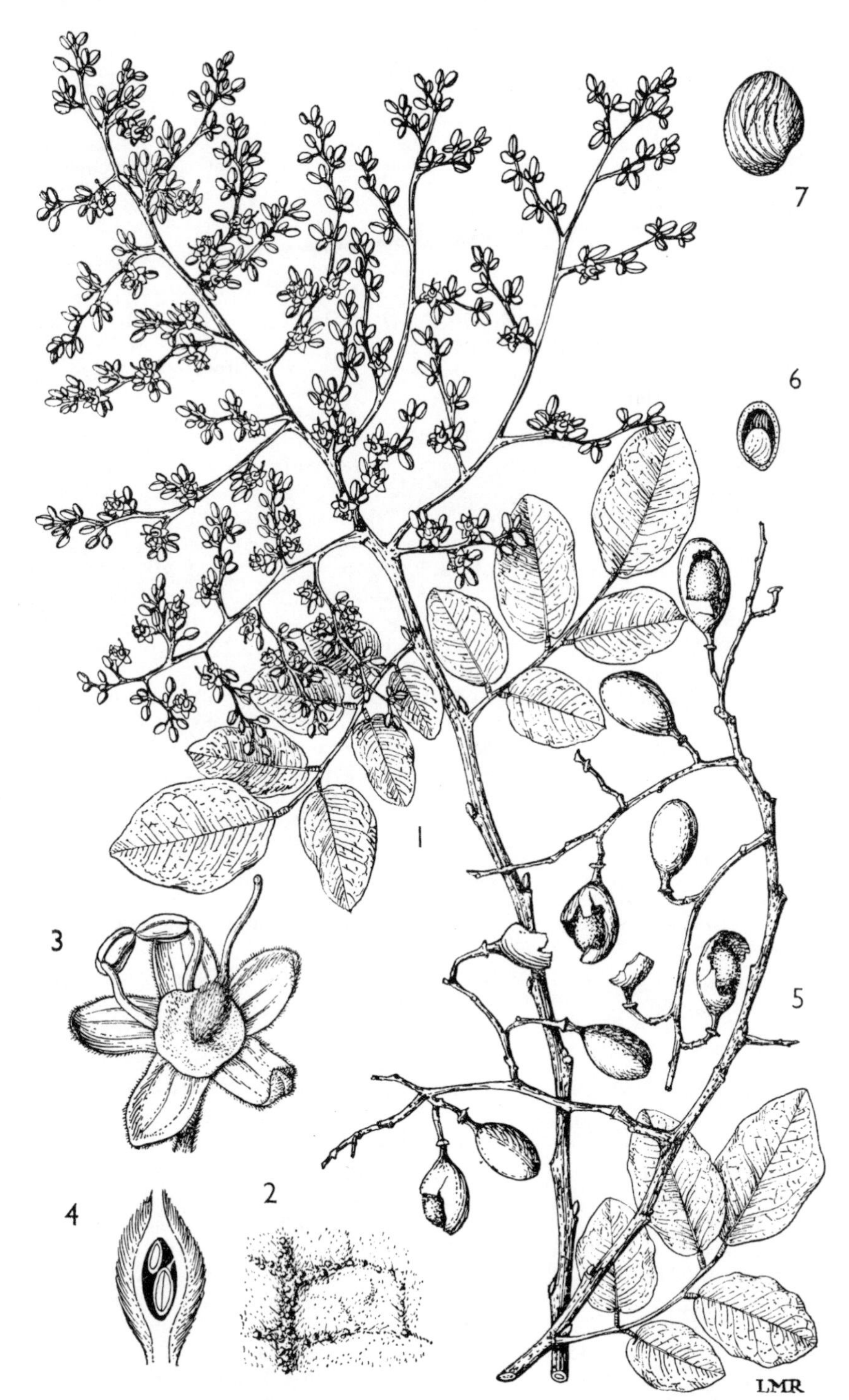

FIG. 17. *DIALIUM ORIENTALE*—**1**, part of flowering branch, × $\frac{2}{3}$; **2**, part of lower surface of leaflet, showing pustular venation, × 10; **3**, flower, × 6; **4**, ovary, longitudinal section, × 12; **5**, part of fruiting branch, × $\frac{2}{3}$; **6**, mesocarp opened to show seed inside, × $\frac{2}{3}$; **7**, seed, × 2. 1, 3, 4, from *Gisau* 63; 2, 5–7, from *Battiscombe* 802.

HAB. Lowland dry evergreen forest, riverine and swamp-forest, woodland; perhaps also in lowland rain-forest; 10–460 m.

SYN. [*D. schlechteri* sensu Steyaert in Bull. Soc. Roy. Bot. Belg. 84: 37 (1951), pro parte quoad *Busse* 2587, *non* Harms]

NOTE. *D. bipindense* Harms, a name which has been wrongly applied to *D. excelsum* in Uganda, is extremely close indeed to *D. holtzii*, but is rightly separated from that as a species on account of the rather smaller flowers with sepals 1·75–2 mm. long, stamen-filaments 0·75–1 mm. long, and anthers only 0·8 mm. long (not 1·75 mm. long as in *D. holtzii*). I am unable to confirm the distinction made by Steyaert (in Bull. Soc. Roy. Bot. Belg. 84: 34 (1951)) between the straight ovary of *D. holtzii* and the oblique ovary of *D. bipindense*.

3. **D. orientale** *Bak. f.* in J.B. 67 : 195 (1929); L.T.A.: 648 (1930); T.S.K.: 62 (1936); Steyaert in Bull. Soc. Roy. Bot. Belg. 84: 38 (1951); K.T.S.: 104 (1961). Type: Kenya, "Coast District", *Webber* 613 in *C.M.* 16436 (K, holo. !, EA, iso. !)

Tree or many-stemmed shrub 6–12(–18) m. high. Bark smooth, grey. Young branchlets puberulous or very shortly pubescent. Leaves: petiole and rhachis together 3·5–5(–7) cm. long; leaflets (5–)7–9, elliptic, with the lowest ones of the leaf ± ovate, (0·9–)1·5–4·5(–5·5) cm. long, 0·9–3·2 cm. wide, stiffly papery to subcoriaceous, obtuse to rounded often slightly emarginate not acuminate at apex, rounded to subcordate or less commonly ± cuneate at base, sparsely and inconspicuously pubescent (especially on midrib) to subglabrous; venation ± raised and reticulate on both surfaces. Flowers greenish-white, white or cream, in panicles up to 30 × 20 cm.; pedicels 1·5–2·5 mm. long, grey-puberulous. Sepals grey-puberulous outside, 2–2·5 mm. long. Petals 0. Stamens 2; filaments 1·5–2 mm. long; anthers 1·25–1·5 mm. long. Pods red to red-brown, obovoid-ellipsoid to subglobose, 1·3–1·8 cm. long, 1–1·3 cm. wide, puberulous outside. Seeds ± ellipsoid, 7–9 mm. long, 5–7 mm. wide, brown. Fig. 17, p. 105.

KENYA. Kilifi District: Mida, *Elliot* in *F.H.* 1489 in *C.M.* 16435 ! & Giriama Reserve, Bamba, Jan. 1938, *Dale* in *F.H.* 3898 !; Lamu District: Boni Forest, Milimani, 31 Dec. 1946, *J. Adamson* 342 in *Bally* 3836 !
TANGANYIKA. Tanga District: Kisosora, Jan. 1956, *Semsei* 2394 !
DISTR. **K7**; **T3**; not known elsewhere
HAB. Lowland dry evergreen forest, coastal evergreen bushland, grouped-tree grassland; near sea-level to 60 m.

SYN. *D. reticulatum* Burtt Davy & MacGregor in K.B. 1932: 261 (1932). Type: Kenya, Kilifi District, Mida, *Elliot* in *F.H.* 1489 in *C.M.* 16435 (K, holo. !, EA, iso. !)

12. TESSMANNIA

Harms in E.J. 45: 295, fig. 2 (1910); J. Léon. in B.J.B.B. 19: 384 (1949) & in Publ. I.N.E.A.C., Sér. Scient. 45: 35 (1950) & in Mém. 8°, Classe Sci., Acad. Roy. Belg. 30(2): 66 (1957)

Unarmed evergreen trees. Leaves simply pinnate; leaflets alternate, rarely opposite, usually emarginate at apex, marked with numerous pellucid gland-dots (which are visible at least before the leaflets have thickened with maturity), without any marginal swelling near base. Inflorescences of terminal or axillary racemes or panicles. Flowers ⚥, distichously arranged along the inflorescence-axes; bracteoles small, not enclosing the flower-buds, soon falling off. Sepals 4 (? rarely 5), very narrowly imbricate, almost valvate, fulvous-tomentose inside. Petals 5, subequal or one somewhat narrower than the rest, imbricate, clawed, with a crinkled lamina. Stamens 10; filaments tomentose below, one of them free, the rest connate below into a short sulcate tube; anthers dorsifixed, dehiscing by longitudinal slits. Ovary stipitate, usually tomentose but sometimes only so on the stipe; ovules several to numerous; style elongate, coiled when young, with a capitate stigma. Pods flattened, woody, indehiscent (or ? ultimately dehiscent into two valves), short, warted or smooth outside. Seeds 1–4, black or brown, exareolate, with a small basal aril.

There are 11 species in the genus, mainly in the rain-forest regions of central and west Africa. Since only two certain gatherings of the genus from East Africa are available, the present treatment is mainly based on the published works of Harms and Léonard. More material is wanted.

Leaflets 3–9 cm. long, mostly 1·4–3 cm. wide, gradually acuminate at apex, 5–8 per leaf . . . . 1. *T. burttii*
Leaflets 0·8–3 cm. long, 0·4–2 cm. wide, rounded or obtuse and emarginate at apex:
Leaflets 6–12 per leaf, 7–20 mm. wide; ovary and stipe tomentose; branchlets, petioles and leaf-rhachides hairy or glabrous . . . . . . 2. *T. martiniana*
Leaflets 16–26 per leaf, 4–6 mm. wide; ovary glabrous, on a villous stipe; branchlets, petioles and leaf-rhachides shortly hairy, later glabrescent . . 3. *T. densiflora*

1. **T. burttii** *Harms* in F.R. 43: 110 (1938); J. Léon. in Publ. I.N.E.A.C., Sér. Scient. 45: 37–40 (1950); F.F.N.R.: 128 (1962). Type: Zambia, Abercorn District, *B. D. Burtt* 6004 (K, holo.!)

Tree 4–12 m. high; bark smooth, grey to dark brown. Rhachis of leaf (together with petiole) (2–)4–10 cm. long; leaflets alternate, (2–)5–8, elliptic-lanceolate to lanceolate or narrowly ovate, 3–9·5 cm. long, 1·2–3(–3·8) cm. wide, gradually acuminate (acumen itself obtuse or emarginate), glabrous and with fine prominent reticulate venation on both surfaces (1–2 basal lateral nerves longer, more prominent and submarginal). Inflorescence of simple axillary or lateral racemes, single or few together; axis 1–7 cm. long, brown-pubescent as are (densely so) the pedicels and outside of sepals. Sepals 1·3–1·5 cm. long. Petals pink, with an apparently oblanceolate crinkled lamina. Ovary and stipe tomentose. Pods ± 4–5·5 cm. long, 2·5–2·8 cm. wide, densely brown-velvety-tomentose and with numerous rather small raised warts scattered over surface.

TANGANYIKA. Mpanda District: Niamanzi [Nyamnsi] R. 40 km. N. of Mpanda, July 1962, *Procter* 2115!
DISTR. **T4**; Zambia
HAB. "Riverine forest with *Monopetalanthus richardsiae*"; ± 1050 m.

NOTE. *Monopetalanthus richardsiae*, with which *T. burttii* grows, has a similar geographical range.

Wigg in *F.H.* 1888! from Kigoma District, Lugufu Forest (near Lugufu Station, Central Line), 17 July 1944, is very possibly young sterile growth of *T. burttii*. The leaflets are up to 10·5 cm. long. Two further gatherings have been made in Kigoma District, E. of Kokoti, Mihumo, 25 Dec. 1963, *Azuma* 1007! & Kabogo Mts., 4 Oct. 1962, *Toyoshima*! The first is of leafy twigs only, the second of two pods only. They are quite possibly conspecific, but there is no certainty. The leafy specimen has large leafy ovate-falcate stipules and on this character it might be *T. dewildemaniana* Harms from the Congo and Angola. Identification of such inadequate material is little more than guesswork. Properly collected specimens are wanted from this area.

2. **T. martiniana** *Harms* in E.J. 53: 463 (1915); L.T.A.: 707 (1930); T.T.C.L.: 107 (1949); J. Léon. in Publ. I.N.E.A.C., Sér. Scient. 45: 36–40 (1950). Type: Tanganyika, Rufiji District, Mohoro, *Martin* in *Holtz* 3131 (B, holo. †)

? Tree. Rhachis of leaf (together with petiole) 2–7 cm. long; leaflets alternate, 6–12, oblong to obovate-suborbicular, 0·8–3 cm. long, 7–20 mm. wide, rounded or obtuse and emarginate at apex, glabrous and (at least in var. *pauloi*) with fine prominent reticulate venation on both surfaces. Inflorescences of simple axillary racemes, or with 1–2 branches; axis 1–4·5 cm. long, densely brownish-velvety-pubescent as are the pedicels. Sepals ± 1 cm. long. Petals ± 2 cm. long, with an oblanceolate or oblong-oblanceolate lamina. Ovary and stipe tomentose. Fruit unknown.

var. **martiniana**

Branchlets, petioles and leaf-rhachides glabrous. Leaflets 9–12. Inflorescence-axis about 2–4·5 cm. long. Pedicels (together with receptacle) 8–10 mm. long.

TANGANYIKA. Rufiji District: Mohoro, Dec. 1912, *Martin* in *Holtz* 3131
DISTR. **T6**; not known elsewhere
HAB. Uncertain

var. **pauloi** *J. Léon.* in Mém. 8°, Classe Sci., Acad. Roy. Belg. 30(2): 68 (1957). Type: Tanganyika, Uzaramo District, *Paulo* 152 (BR, holo. !, EA, K, iso. !)

" Shrub or small tree." Branchlets, petioles and leaf-rhachides with short spreading pubescence. Leaflets 6–9. Inflorescence-axis ± 1 cm. long. Pedicels (together with receptacle) 3–4 mm. long.

TANGANYIKA. Uzaramo District: Kisiju, Sept. 1953, *Paulo* 152 !
DISTR. **T6**; not known elsewhere
HAB. On sea-shore

3. **T. densiflora** *Harms* in E.J. 53: 462 (1915); L.T.A.: 707 (1930); T.T.C.L.: 107 (1949); J. Léon. in Publ. I.N.E.A.C., Sér. Scient. 45: 36–40 (1950). Types: Tanganyika, Rufiji District, *Brulz* in *Holtz* 100 & *Martin* in *Holtz* 3130 (B, syn. †)

Tree 20–25 m. high, with an open crown; branchlets shortly hairy, later glabrescent (as are the petioles and leaf-rhachides). Rhachis of leaf (together with petiole) 3–6 cm. long; leaflets alternate to subopposite or opposite, 16–26, oblong, lanceolate, ovate-oblong or ovate-lanceolate, 1–1·5 cm. long, 4–6 mm. wide, obtuse and often slightly emarginate at apex, glabrous and reticulate on both surfaces. Inflorescences of axillary, dense-flowered racemes; axis 3–4 cm. long, densely velvety-hairy; pedicels (together with receptacle) up to 7 mm. long in young fruit; flowers red. Sepals ± 1·2 cm. long. Petals ± 2 cm. long. Ovary glabrous; stipe villous. Pods (not quite mature) elliptic-suborbicular or obliquely and broadly obovate, 3·3–4·5 cm. long, 2·5–2·8 cm. wide.

TANGANYIKA. Rufiji District: Kichi area, Aug. 1913, *Brulz* in *Holtz* 100 & near Mohoro Kichi Hills, Jan. 1913, *Martin* in *Holtz* 3130
DISTR. **T6**; not known elsewhere
HAB. Uncertain, said to be on sandy ground at 200–300 m.

13. BAIKIAEA

Benth. in G.P. 1: 581 (1865); J. Léon. in Mém. 8°, Classe Sci., Acad. Roy. Belg. 30(2): 72 (1957)

Unarmed evergreen (or rarely, in *B. plurijuga*, deciduous) trees. Leaves simply pari- or imparipinnate; leaflets usually alternate, sometimes opposite, without pellucid gland-dots, with a small ± marked swelling near the posticous margin of each leaflet close to the base*. Inflorescences of terminal or axillary racemes. Flowers ⚥, pedicellate, distichously arranged along the inflorescence-axes; bracteoles usually small, not enclosing the flower-buds,

* Considered by J. Léonard (l.c., 1957) to be a domatium. A small orifice to the swelling is usually visible. Compare *Afzelia* (p. 127) and *Intsia* (p. 129).

imbricate, almost valvate, fulvous-villous-tomentose inside, shortly and soon falling off. Sepals 4, the posticous one larger than the rest, very narrowly densely fuscous- or paler brown-tomentellous outside. Petals 5, 4 of them equal, the fifth narrower and usually differently coloured, all obovate, free, imbricate, crinkled around edges, villous along and near midrib. Stamens 10; filaments glabrous or villous below, one of them free, the rest connate at base into a short tube; anthers dorsifixed, dehiscing by longitudinal slits. Ovary stipitate, tomentose; ovules 1 to many; style elongate, glabrous, with an enlarged peltate depressed-subglobose stigma. Pods flattened, woody, dehiscing into two valves. Seeds large, exareolate, with thin and fragile or else hard testa.

Five species, all tropical African and all in the rain-forest region, except for *B. plurijuga* Harms in Zambia, Rhodesia, Angola and South West Africa.

Bracts 3–4 mm. long and 4–5 mm. wide; bracteoles 3–4·5 mm. long and 2–2·5 mm. wide . . . 1. *B. insignis*
Bracts 10–14 mm. long and 12–15 mm. wide; bracteoles 9–13 mm. long and 5–8 mm. wide . . . 2. *B. ghesquiereana*

1. **B. insignis** *Benth.* in Trans. Linn. Soc., Lond. 25: 314 (1865); L.T.A.: 704 (1930); J. Léon. in F.C.B. 3: 298, t. 23 (1952) & in Mém. 8°, Classe Sci., Acad. Roy. Belg. 30(2): 73 (1957). Type: Fernando Po, *Mann* 2342 (K, holo.!)

Tree 5–34 m. high; buttresses none or slight; bark grey or grey-brown, smooth or slightly fissured; branchlets and leaves glabrous (or rarely, and not in the Flora area, minutely pubescent). Leaves: petiole with rhachis up to 23(–30, *fide* F.C.B.) cm. long; petiolules 6–15 mm. long; leaflets (2–)3–8(–10, *fide* F.C.B.), coriaceous, ovate, elliptic, oblong-elliptic or oblong-lanceolate, up to 40 cm. long and 17 cm. wide, subacuminate to obtusely pointed at apex. Racemes 1·5–8·5(–12, *fide* F.C.B.) cm. long; bracts 3–4 mm. long, 4–5 mm. wide; bracteoles 3–4·5 × 2–2·5 mm. Sepals up to 11(–15, *fide* F.C.B.) cm. long. Petals up to 21 cm. long and 11 cm. wide, white or cream with the standard pale yellow. Stamen-filaments 6–16 cm. long, glabrous or villous below; anthers 1·8–2(–3, *fide* F.C.B.) cm. long. Ovary 2·5–3·8 cm. long. Style 5–9(–11, *fide* F.C.B.) cm. long. Pods (*fide* F.C.B.) 17–42(–60) cm. long, 5–12 cm. wide, densely brown-velvety. Seeds ellipsoid, 3–4·5 cm. long, 1·5–3 cm. wide, compressed, dark red, thin-shelled (description of seeds mostly from F.C.B.).

subsp. **minor** (*Oliv.*) *J. Léon.* in F.C.B. 3: 301 (1952) & in Mém. 8°, Classe Sci., Acad. Roy. Belg. 30(2): 73 (1957). Types: Congo Republic, without exact locality, *Smith* (BM, syn.) & *Burton* (K, syn.!)

Petiole with rhachis (1·5–)3·5–17 cm. long; petiolules 3–9(–10) mm. long; leaflets (3·5–)5·5–18·5(–23) cm. long, (1·5–)2·7–11 cm. wide. Sepals 3·5–8·5 cm. long. Petals 6–10(–11·5, *fide* F.C.B.) cm. long, 3·5–6(–7, *fide* F.C.B.) cm. wide. Stamen-filaments 6–11(–12, *fide* F.C.B.) cm. long. Fig. 18, p. 110.

UGANDA. Ankole District: Kagera R. banks between Nsongezi and Ruborogota, Oct. 1932, *Eggeling* 680 in *F.H.* 1054!; Busoga District: near Busia on Mjanji road, Oct. 1939, *Dale* U33!; Masaka District: Lake Nabugabo, Oct. 1932, *Eggeling* 591 in *F.H.* 966!

TANGANYIKA. Bukoba District: Kabirizi, Oct. 1931, *Haarer* 2250! & Minziro Forest, 5 Oct. 1953, *Willan* 32!; Mwanza District: Geita, Samina, July 1951, *Eggeling* 6259!; Buha District: Malagarasi ferry, June 1955, *Procter* 414!

DISTR. **U**2–4; **T**1, 4; ? Cameroun Republic (*fide* F.C.B.), Gabon, Cabinda, Congo Republic and Angola

HAB. Lowland rain-forest (? also in upland rain-forest at its lowest altitudes); 1140–1250 m.

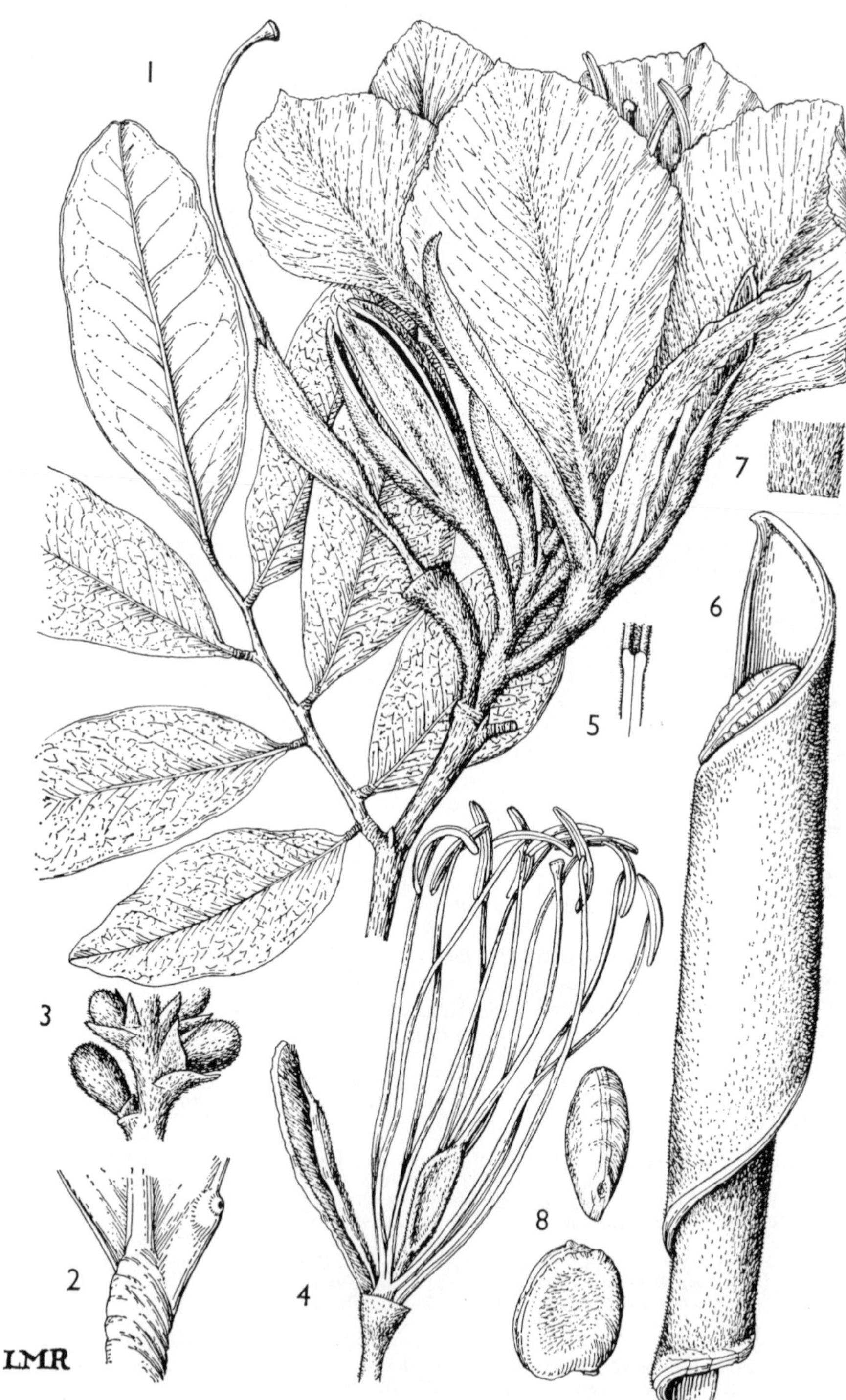

FIG. 18. *BAIKIAEA INSIGNIS* subsp. *MINOR*—**1,** part of branchlet showing inflorescence and leaf, × $\frac{2}{3}$; **2,** leaflet-base, lower surface, showing marginal gland, × 6; **3,** part of young inflorescence showing buds in two ranks, each bud with a bract and two bracteoles, × 2; **4,** flower, all parts removed except one sepal, ten stamens and ovary, × $\frac{2}{3}$; **5,** transverse section of upper suture of undehisced pod, × $\frac{2}{3}$; **6,** one valve of dehisced pod, × $\frac{2}{3}$; **7,** part of surface of pod, × 2; **8,** seed, side and basal views, × $\frac{2}{3}$. 1, from *Eggeling* 43 in *F.H.* 209 and *Haarer* 2250; 2, from *Willan* 32; 3, from *Mahon* 7; 4, from *Haarer* 2250; 5–8, from *Louis* 4378.

Syn. (of subsp.). *B. minor* Oliv., F.T.A. 2: 309 (1871); L.T.A.: 705 (1930); I.T.U., ed. 2: 57 (1952); Torre & Hillcoat in C.F.A. 2: 218 (1956)
B. eminii Taub. in P.O.A. C: 198 (1895); L.T.A.: 705 (1930); T.T.C.L.: 87 (1949); J. Léon. in B.J.B.B. 21: 129 (1951) & F.C.B. 3: 302 (1952). Types: Tanganyika, Bukoba, *Stuhlmann* 1012 (B, lecto. †, FHO, isolecto. !) & 1599 (B, syn. †, FHO, isosyn. !) & Bukoba District, Kafuro, *Stuhlmann* 1695a (B, syn. †, FHO, isosyn. !)

Note. Subsp. *insignis* differs from subsp. *minor* by having most organs, including the flowers, generally considerably larger. It occurs in Nigeria, Fernando Po, the Cameroun Republic, Gabon and the Congo Republic, but not in the Flora area.

Variation. Subsp. *minor* as it occurs in the Flora area is atypical of the subspecies in having a dark or blackish-brown indumentum over the outside of the calyx, ovary and pod, as in subsp. *insignis*. In typical subsp. *minor* it is a paler more rusty brown. Léonard has suggested (in F.C.B. 3: 302 (1952)) that subsp. *minor* in East Africa may be an eastern upland ecotype.

2. **B. ghesquiereana** *J. Léon.* in B.J.B.B. 21: 130, fig. 33 (1951). Type: Tanganyika, Kilwa District, *Busse* 3119 (BR, holo. !, B, BM, EA, iso. !)

Tree; branchlets and leaves glabrous. Leaves: petiole with rhachis 7–19 cm. long; petiolules 5–9 mm. long; leaflets 6–8, opposite or subopposite, coriaceous, elliptic, lanceolate, or sometimes ovate-elliptic, 7–19 cm. long, 3·2–6·8 cm. wide, gradually subacuminate at apex. Racemes (only young ones known): peduncle with axis up to 14 cm. long, like the bracts and bracteoles very densely dark-brown- or red-brown-villous outside; bracts 10–16 mm. long, 12–15 mm. wide; bracteoles 9–13 mm. long, 5–8 mm. wide. Buds very densely red-brown-villous outside. Open flowers and pods not yet known.

Tanganyika. Kilwa District: Matumbi Mts., Kibata, 8 July 1903, *Busse* 3119 & in *Peter* 48018 !
Distr. T8; not known elsewhere
Hab. Said to be in open " miombo " (*Brachystegia*) woodland; ± 300 m.

14. CYNOMETRA

L., Sp. Pl.: 382 (1753) & Gen. Pl., ed. 5: 466 (1754); J. Léon. in B.J.B.B. 21: 379 (1951) & in Mém. 8°, Classe Sci., Acad. Roy. Belg. 30(2): 94 (1957)

Unarmed, normally evergreen trees. Leaves simply pinnate; leaflets nearly always opposite*, in one to many pairs, without gland-dots. Inflorescence paniculate or (in East Africa ? only in 4, *C. brachyrrhachis* and 12, *C. filifera*) simply racemose. Flowers ⚥, spirally arranged along the inflorescence-axes; pedicels (in the Flora area, not necessarily elsewhere) jointed at apex and persistent after the flowers have fallen; bracteoles small, not enclosing the flower-buds. Sepals 4(–5), imbricate, glabrous inside, or with a few hairs near centre. Petals (4–)5, equal or nearly so. Stamens (8–)10(–12); filaments glabrous, free or almost so; anthers dehiscing by longitudinal slits. Ovary ± stipitate, ± pubescent or tomentose; ovules 1–4; style elongate; stigma minute. Pods either flattened, woody, and dehiscing elastically into 2 valves which are smooth or nearly so outside, or (typically for the genus, but not in indigenous species) indehiscent, thickened, and tubercled outside. Seeds 1–2(–4), with a very short funicle, thin-walled, exareolate.

A pantropical genus of perhaps 50–60 species, but figures uncertain.

C. cauliflora L., a native of tropical Asia, with a single pair of leaflets to each leaf, and small white or pink flowers in small racemes in congested masses on the trunk and

* The imperfectly known *C. brachyrrachis* Harms (see p. 114) and *C. filifera* Harms (p. 119) are the only possible exceptions in East Africa. They may, however, not be in their correct genus.

branches, has been recorded in cultivation in Tanganyika and Zanzibar (T.T.C.L.: 101 (1949); U.O.P.Z.: 224 (1949)).

The indigenous species are mostly imperfectly known and inadequately collected—out of the dozen species in only two, or possibly three—*C. alexandri* and *C. webberi*, and perhaps *C. suaheliensis*—are both flowers and mature pods known. It is likely that additional species will be discovered in the future.

Leaflets in one pair per leaf:
- Petiole 6–25 mm. long; inflorescence-axes glabrous or nearly so . . . . 1. *C. engleri*
- Petiole 2–3 mm. long; inflorescence-axes densely pubescent . . . . 2. *C. gillmanii*

Leaflets in 2–6 pairs per leaf; petiole 1–5(–10 in 3, *C. longipedicellata*) mm. long:
- Leaves with 4–6 pairs of leaflets; stipules linear-filiform, up to 2 cm. long or more; inflorescence racemose . . . . 12. *C. filifera*
- Leaves with 2–4 pairs of leaflets; stipules various (and for some species not known) but if linear then not more than 1 cm. long; inflorescence paniculate (except ? for 4, *C. brachyrrhachis*):
  - Pedicels 1·5–3·5 cm. long; leaflets in 2 pairs per leaf . . . . 3. *C. longipedicellata*
  - Pedicels up to 1·4 cm. long:
    - Terminal pair of leaflets (at least) gradually tapering or acuminate to an obtuse to acute apex*:
      - Inflorescence a short terminal raceme about 4–10 mm. long; pedicels (including receptacle) 4–7 mm. long, sparsely puberulous to subglabrous . . . . 4. *C. brachyrrhachis*
      - Inflorescence an axillary or terminal panicle 2–8 cm. long:
        - Pedicels glabrous, 6–14 mm. long; sepals glabrous outside, 4–5 mm. long; petals 5–6 mm. long . . . . 5. *C. alexandri*
        - Pedicels pubescent, 3–5 mm. long; sepals hairy, 2–3 mm. long; petals 3 mm. long . . . . 7. *C. ulugurensis*
    - Terminal pair of leaflets (as well as lateral leaflets) rounded or obtuse but not gradually tapering at apex:
      - Leaflets normally 3–4 pairs per leaf (occasional reduced leaves with 2 pairs only may also be present); sepals 2–3 mm. long . . . . 11. *C. webberi*
      - Leaflets always 2 pairs per leaf; sepals 3–4 mm. long:
        - Pedicels 1–2 mm. long, densely puberulous; leaf-rhachis very narrowly channelled above . . . . . 10. *C. sp. B*
        - Pedicels 3·5–12 mm. long, glabrous to slightly puberulous; leaf-rhachis broadly channelled above or almost winged:

* 6, *C. sp. A* comes into this group, but cannot at present be accounted for further since its inflorescence is still unknown. It occurs in the E. Usambara Mts., and is striking on account of its pods having a conspicuous apical beak.

The pedicels 7–12 mm. long, glabrous; young branchlets glabrous; panicle axes very minutely and inconspicuously puberulous (actual hairs only visible under a magnification of × 20 or more); stipe of ovary exserted for 1·5–2 mm. beyond mouth of hypanthium; petiolules 0·5–1 mm. long between proximal side of leaflet-base and leaf-rhachis; stipe of pod ± 1 cm. long . . 8. *C. greenwayi*

The pedicels 3·5–7 mm. long, glabrous to slightly puberulous; young branchlets puberulous to pubescent; panicle axes densely and obviously puberulous (hairs easily visible under a magnification of × 10); stipe of ovary exserted for 0·5 mm. beyond mouth of hypanthium; petiolules 0–0·5 mm. long between proximal side of leaflet-base and leaf-rhachis; stipe of pod ± 0·5 cm. long. . . 9. *C. suaheliensis*

1. **C. engleri** *Harms* in E.J. 38: 77, fig. 3 (1905); L.T.A.: 758 (1930); T.T.C.L.: 101 (1949); J. Léon. in B.J.B.B. 21: 388 (1951). Type: Tanganyika, Lushoto District, Sigi valley between Muheza and Longuza, *Engler* 379 & 391 (B, syn. †)

Evergreen tree 12–21 m. high (shrubby, 4–6 m. high, *fide* Engler, ex Harms). Leaves: stipules very quickly caducous, apparently not leafy; petiole 6–18 mm. long; leaflets 2, on petiolules 1–3 mm. long, ± obliquely elliptic, 6·5–14(–16, *fide* Harms) cm. long, 2·4–7·3 cm. wide, obtuse or somewhat emarginate at apex, glabrous; venation obscure or not apparent on both surfaces. Panicles 3–8 cm. long; axes glabrous or nearly so; pedicels 4–6 mm. long. Flowers small, white. Sepals 4, 2·5–3 mm. long. Petals 5, oblanceolate-oblong, 4–5 mm. long, 1–1·5 mm. wide. Pods unknown.

Tanganyika. Lushoto District: Sigi R., Longuza, 26 Oct. 1932, *Greenway* 3266! & 4 Oct. 1936, *Greenway* 4650! & Longuza, 29 Sept. 1936, *Greenway* 4633!
Distr. **T3**; not known elsewhere
Hab. Riverine forest; 100–300 m.

2. **C. gillmanii** *J. Léon.* in B.J.B.B. 21: 388, fig. 105 (1951). Type: Tanganyika, *Gillman* 1142 (K, holo.!, BR, EA, iso.!)

Tree, ? evergreen, 12–15 m. high. Leaves with petiole 2–3 mm. long; leaflets 2, on petiolules ± 0·5 mm. long, elliptic or obovate, 2–5·5 cm. long, 1–2·6 cm. wide, obtuse to rounded and sometimes emarginate at apex, glabrous. Panicles 0·8–1·5 cm. long, axes densely short-pubescent; pedicels 7–9 mm. long. Flowers small, white. Sepals 4, 3·5–4 mm. long. Petals 5, lanceolate or obovate, 3·5–6·5 mm. long, 1–1·5 mm. wide. Pods unknown.

Tanganyika. Kilwa District: Mkoe, 12 Dec. 1942, *Gillman* 1142!
Distr. **T8**; not known elsewhere
Hab. Insufficiently known; said to occur " on reddish sandy soil in Mkungwi–Mpande woodland "

Syn. *C. sp. No.* 14 sensu T.T.C.L.: 102 (1949)

3. **C. longipedicellata** *Harms* in E.J. 53: 460 (1915); L.T.A.: 758 (1930); T.T.C.L.: 101 (1949); J. Léon. in B.J.B.B. 21: 391 (1951). Types: Tanganyika, Lushoto District, Amani, Kwamkoro road, *Zimmermann* 1989 (B, syn. †, EA, isosyn. !) & Bomole, *Zimmermann* 1973 (B, syn. †)

Much branched evergreen tree up to 18 m. high, with broad rounded crown. Leaves: petiole with rhachis 2–5 cm. long; stipules persistent, foliaceous, obliquely ovate to ovate-oblong, 0·7–1·5 cm. long; petiole 3–10 mm. long; leaflets (2–)4, sessile, elliptic to ovate-elliptic or obovate-elliptic, 3–12 cm. long, 2–7 cm. wide, mostly obtuse at apex, not emarginate. Panicles compact; axes pubescent, up to 1–2·5 cm. long; pedicels unusually long (1·5–3·5 cm.), glabrous, jointed about 5 mm. below the hypanthium. Flowers white. Sepals 4, 4–7 mm. long. Petals ± 7–8 mm. long, but damaged ones only examined. Pods unknown.

TANGANYIKA. Lushoto District: Amani, 22 Sept. 1939, *Greenway* 5901 !
DISTR. **T3**; not known elsewhere
HAB. Lowland evergreen rain-forest; ± 900 m.

NOTE. Harms states in E.J. 53: 460 (1915) that a specimen of leaves only (*Holtz* 971) collected in the Pugu Hills near Dar es Salaam may perhaps belong to *C. longipedicellata*.

4. **C. brachyrrhachis** *Harms* in E.J. 53: 458 (1915); L.T.A.: 758 (1930); T.T.C.L.: 101 (1949); J. Léon. in B.J.B.B. 21: 398 (1951). Types: Tanganyika, Lushoto District, *Zimmermann* 2564 & *Grote* 3583 (both B, syn. †)

Tree; young branchlets shortly pubescent. Leaves: petiole 2–3 mm. long; rhachis with petiole 0·8–1·5 cm. long, pubescent or puberulous; leaflets 4, opposite, or sometimes subopposite or alternate, sessile, obliquely lanceolate, oblong-lanceolate, oblanceolate, ovate-lanceolate, or ovate, ± falcate-curved, 1–5 cm. long, 0·5–2 cm. wide, usually gradually acuminate and acute at apex, rarely shortly narrowed and obtuse, glabrous or subglabrous. Racemes terminal, short; axis ± 4–10 mm. long, puberulous; pedicels 4–7 mm. long (including hypanthium), sparsely puberulous to subglabrous. Sepals 4. Petals unknown. Pods unknown.

TANGANYIKA. Lushoto District: Amani, Bomole road, 1909, *Zimmermann* 2564 & Mar. 1912, *Grote* 3583
DISTR. **T3**; not known elsewhere
HAB. Uncertain, but presumably lowland rain-forest

NOTE. This species may well not be correctly placed generically. It might well, for instance, be a *Scorodophloeus*.

5. **C. alexandri*** *C. H. Wright* in Johnston, Uganda Prot. 1: 325 (1902); L.T.A.: 760 (1930); Eggeling & Harris, Fifteen Uganda Timbers (in Chalk, Burtt Davy & Hoyle, For. Trees Timb. Brit. Emp. 4): 23, fig. 3, t. 3 (1939); J. Léon. in B.J.B.B. 21: 390 (1951) & in F.C.B. 3: 316 (1952); I.T.U., ed. 2: 63, fig. 14, photo. 8 (1952). Type: Uganda, Toro District, E. of Ruwenzori, Kirurume R., *H. H. Johnston* (K, holo. !)

Flat-topped evergreen tree 10–50 m. high, branching rather low down, and with extensive thin buttresses near base; bark smooth but flaking, reddish-grey or light brown; young branchlets puberulous or shortly pubescent, or sometimes glabrous. Leaves: stipules ± persistent, linear, 3–6(–10) mm. long, 0·3–0·5(–0·75) mm. wide; petiole 2–4(–5) mm. long, puberulous; rhachis 0·5–3·5 cm. long, widely channelled or almost winged

* Probably named after Alexander Whyte, H. H. Johnston's professional collector, although C. H. Wright makes no mention of this in the published account of his new species, attributing the collection to " Alexander Johnston ".

above; leaflets 4(–6), sessile, obliquely lanceolate, elliptic-lanceolate, ovate-elliptic, or ovate, 1–10 cm. long, 0·5–3(–4) cm. wide, at least the larger terminal pair gradually tapering to an obtuse to subacute point, not emarginate, glabrous or nearly so. Panicles axillary and terminal, 2–6 cm. long; axes puberulous; pedicels 6–14 mm. long, glabrous, jointed below the flower. Flowers white or sometimes pink, sweetly scented. Sepals 4(–5 *fide* Léonard), 4–5 mm. long, glabrous outside. Petals 5, obovate-lanceolate, 5–6 mm. long. Pods smooth when mature, 5–10 cm. long, 3–5 cm. wide, rounded or apiculate at apex. Seeds ± 1·7–2 cm. in diameter.

UGANDA. Toro District: Semliki forests, 31 Oct. 1905, *Dawe* 636/525 !; Kigezi District: Maramagambo* Forest, Sept. 1936, *Eggeling* 3321 !; Entebbe, White Fathers Mission, July 1931, *Brasnett* 128 !
TANGANYIKA. Kigoma District: Uvinza, 29 Aug. 1950, *Bullock* 3232 ! & E. of Kokoti, Mihume, 25 Dec. 1963, *Azuma* 1005 !
DISTR. **U**2, ? 4; **T**4; Congo Republic
HAB. Lowland rain forest; 700–1220 m.

NOTE. *Cynometra alexandri*, known as Uganda Ironwood, produces a heavy timber of some importance. Its main area of distribution in Uganda is **U**2; the only specimen that I have seen from **U**4 is *Brasnett* 128, which is perhaps from a deliberately planted tree. The Tanganyika specimens, in fruit only or sterile, differ somewhat in aspect from the Uganda ones, but seemingly not in any significant way.

Various specimens from the Kivu Province of the Congo Republic have 8 or sometimes 10 leaflets, and probably represent a local race or variant; the pedicels of these may be up to 16 mm. long. This variant is to be sought in East Africa.

6. **C. sp. A**

Evergreen tree up to 40 m. high, with buttressed trunk and pale brown bark; young branchlets glabrous. Leaves: stipules apparently linear; petiole 3–5 mm. long, puberulous; rhachis 1·2–2·2 cm. long, channelled above; leaflets 4, sessile or nearly so, lanceolate to elliptic-lanceolate or ovate, 2–13 cm. long, 1·2–5 cm. wide, somewhat asymmetric at base, the terminal pair gradually tapering to a subacuminate obtuse point, glabrous. Flowers and inflorescences unknown. Pods compressed, almost smooth, ± 7–9 cm. long and 3–3·8 cm. wide, tapering at apex to a short but conspicuous beak continuing the line of the upper suture.

TANGANYIKA. Lushoto District: Amani–Maramba, 30 Oct. 1935, *Greenway* 4148 !
DISTR. **T**3; not known elsewhere
HAB. Lowland rain-forest; ± 900 m.

NOTE. This is apparently closely allied to *C. alexandri*, from which the markedly rostrate pods provide the only clear distinction at present. In view of this, of the very different geographical areas involved, and of the fact that the flowers of the Usambara tree are still unknown, it seems prudent to maintain the two as distinct, at least for the time being.

Other sterile specimens from the E. Usambara Mts., *Peter* K 280, 48742 (K 300), 48741 (K 301), 48124 (K 412), 48906 (K 556), 48618 (K 593), and K 1084, are perhaps conspecific. They show stipules up to 7 × 1 mm., leaf-rhachides to 4 cm. long, and leaflets up to 16 × 5·5 cm. It is quite possible that the foliage of these specimens, and of *Greenway* 4148, may be from juvenile or sucker shoots. If this is so, then the possibility that this group of specimens represents juvenile states of *Cynometra sp. B* (p. 117) must be borne in mind. If this were proved, then, of course, the present plant would be more closely related to (though specifically distinct from) *C. suaheliensis*.

There is also another possibility to be considered: except for the non-foliaceous stipules, the very short petioles about 4 mm. long, and the more reduced lower pair of leaflets, there is a strong resemblance to *Julbernardia magnistipulata* (Harms) Troupin (see p. 146), and flowers may possibly show that to be the correct genus for these at present rather baffling specimens. The undescribed name *Berlinia obliqua* A. Peter (see T.T.C.L.: 89) was used by Peter for various specimens, some of them *Cynometra sp. A*, other true *Julbernardia magnistipulata*.

Peter 48128 (B !) and 49379 (B !), both from the E. Usambara Mts., the former from

* Sheet labelled Malabigambo (a forest of which name is to be found in Masaka District) but cited as Kigezi District, Marabagambo, in I.T.U., ed. 2: 63 (1952).

Amani, the latter from Sangarawe, are similar to the specimens already cited but have ovate foliaceous stipules. They may be separated from *Julbernardia magnistipulata* by *inter alia* the puberulous branchlets and leaf-rhachides.

These unsolved problems illustrate the urgent need for more good collecting in the E. Usambara Mts.

7. **C. ulugurensis** *Harms* in E.J. 53: 461 (1915); L.T.A.: 758 (1930); T.T.C.L.: 101 (1949); J. Léon. in B.J.B.B. 21: 399 (1951). Type: Tanganyika, Morogoro District, *Rupprecht* in *Holtz* 3100 (B, holo. †)

Tall tree; branchlets glabrous, or minutely puberulous when young. Leaves: rhachis, together with the 1–2·5 mm. long petiole, 0·8–1·5 cm. long, puberulous; leaflets 4, sessile, obliquely lanceolate, oblong-lanceolate, or ovate, 0·8–4 cm. long, 0·5–1·5 cm. wide, acute, or more usually ± gradually and often acutely acuminate at apex, glabrous or nearly so. Panicles 5–8 cm. long, composed of 1–3 cm. long racemes; axes and pedicels pubescent, the latter 3–5 mm. long. Sepals 4, 2–3 mm. long, hairy. Petals 5, oblong or lanceolate-oblong, 3 mm. long. Pods unknown.

TANGANYIKA. Morogoro District: Kimboza Forest Reserve, Feb. 1913, *Rupprecht* in *Holtz* 3100
DISTR. **T6**; not known elsewhere
HAB. Uncertain, presumably in forest; ± 660 m.

8. **C. greenwayi** *Brenan* in K.B. 17: 209, fig. 1 (1963). Type: Kenya, Kilifi District, *Greenway* 10440 (K, holo. !, BR, S, iso. !)

Tree up to about 10 m. high, with flaking grey-brown bark; young branchlets glabrous. Leaves: stipules very small and very soon falling away; petiole 1–3 mm. long, puberulous above; rhachis 0·5–2 cm. long, channelled above; leaflets 4, with petiolules 0·5–1 mm. long below the proximal side of the leaflet-base, ± coriaceous, obovate or obovate-elliptic, 1·1–10·5 cm. long, 0·8–6·5 cm. wide, rounded or obtuse but not gradually tapering at apex, glabrous. Panicles axillary or falsely terminal, 2·5–7 cm. long; axes very minutely and inconspicuously puberulous (actual hairs only visible with a magnification of at least × 20); pedicels 7–12 mm. long, glabrous, jointed at top below flower. Flowers white, sweetly scented. Sepals 4, 4·5 mm. long. Petals 4–5, obovate-oblanceolate, 5·5–6 mm. long. Immature pods ± 5·7 cm. long and 2·8 cm. wide, on a stipe ± 1 cm. long, and with an apical beak ± 0·5 cm. long.

KENYA. Kilifi District: SE. of Gedi, 4 Jan. 1962, *Greenway* 10440 !
DISTR. **K7**; not known elsewhere
HAB. "Common with *Ehretia* in sand-dunes with scattered coral-rock outcrops"; 8 m.

9. **C. suaheliensis** (*Taub.*) *Bak. f.*, L.T.A.: 759 (1930); T.T.C.L.: 101 (1949); J. Léon. in B.J.B.B. 21: 391 (1951), pro parte, excl. syn. *C. sp. No.* 13 et spec. *Greenway* 2956; K.T.S.: 103 (1961). Types: Kenya, Mombasa, *Wakefield* (B, syn. †, K, isosyn.!) & Tanganyika, Pangani, *Stuhlmann* 365 (B, syn. †)

Evergreen shrub or tree 5–15 m. high; bark smooth, red-grey (*Graham* U.767 in *F.H.* 2238) or rough (*Tanner* 3596); young branchlets puberulous or shortly pubescent. Leaves: stipules not apparent (? absent); petiole 2–3 mm. long, puberulous; rhachis 0·4–1·3 cm. long (–2·7 cm. on juvenile shoots), broadly channelled or almost winged above; leaflets 4, with glabrous or puberulous petiolules, asymmetrically obovate-elliptic or elliptic, 0·7–8 cm. long, 0·5–4·4 cm. wide (larger on juvenile shoots), mostly rounded at apex, not gradually tapering, glabrous. Racemes or panicles terminal, 2–9 cm. long; axes densely puberulous; pedicels 3·5–7 mm. long, glabrous to slightly puberulous, jointed at top below flower. Flowers white. Sepals 4, 3–3·5 mm. long. Petals 5, oblong-lanceolate, ± 3·5–5·5 mm. long. Pods smooth, 4–6 cm. long, 2–3 cm. wide, on a stipe ± 0·5 cm. long, shortly (± 2–3 mm.) beaked at apex.

KENYA. Kwale District: Kinango, Jan. 1930, *R. M. Graham* U.767 in *F.H.* 2238!; Kilifi District: Sabaki R. 6 km. N. of Malindi, 11 Nov. 1961, *Polhill & Paulo* 737! & 40 km. NW. of Malindi, Marafa, 21 Nov. 1961, *Polhill & Paulo* 824!
TANGANYIKA. Pangani District: Bushiri, 30 Nov. 1950, *Faulkner* Kew No. 699! & Pangani, Sakura, Mwera, 10 July 1957, *Tanner* 3596!
DISTR. **K7**; **T3**; not known elsewhere
HAB. ? Lowland dry evergreen forest, riverine forest, coastal evergreen bushland; sea-level–150 m.

SYN. *Theodora suaheliensis* Taub. in P.O.A. C: 198 (1895)
Schotia suaheliensis (Taub.) Harms in V.E. 3(1): 454, fig. 248 (1915)
Cynometra sp. sensu T.S.K.: 64 (1936)

NOTE. *Holtz* 6955, from Uzaramo District, Pugu Hills, 29 Nov. 1903 (EA!), represents either a very marked variant of *C. suaheliensis* or a distinct species. It differs from *C. suaheliensis* in having pedicels densely and shortly pubescent, in the two pairs of leaflets on each leaf being subequal in size and the terminal pair less obovate. More and better material is much needed.

10. **C. sp. B**

Evergreen tree 21 m. high with " slender crown " and drooping branches; young branches puberulous. Leaves: stipules soon caducous, linear-subulate, ± 5–7 mm. long and 0·75 mm. wide; petiole ± 3·5–5 mm. long, puberulous; rhachis 1·2–2·4 cm. long, very narrowly channelled above making it almost terete; leaflets 4, with puberulous petiolules, narrowly and somewhat asymmetrically obovate to elliptic, 2·3–11 cm. long, 1·2–5 cm. wide, rounded or obtuse and sometimes slightly emarginate at apex, subglabrous. Panicles terminal and axillary, 4–8 cm. long; axes densely puberulous; pedicels 1–2 cm. long, clothed like the panicle-axes, jointed below the flower. Sepals 4, 4 mm. long. Petals oblanceolate-elliptic, 5·5–6·5 mm. long. Pods unknown.

TANGANYIKA. Lushoto District: Amani, 29 Apr. 1932, *Greenway* 2956!
DISTR. **T3**; not known elsewhere
HAB. Lowland rain forest; 900 m.

SYN. *C. sp. No.* 13 sensu T.T.C.L.: 101 (1949)
[*C. suaheliensis* sensu J. Léon. in B.J.B.B. 21: 391 (1951) pro parte, quoad syn. *C. sp. No.* 13 et spec. *Greenway* 2956, *non* (Taub.) Bak. f.]

NOTE. I cannot follow J. Léonard in considering this to be *C. suaheliensis*, since *Greenway* 2956 differs in: 1, the very narrowly channelled, almost terete leaf-rhachis; 2, the larger leaflets; and 3, the very short pedicels. The leaflets are mostly damaged and may not show their real shape satisfactorily. It is possible that the terminal pair may have more tapering apices than the specimen indicates. See note under *C. sp. A* (p. 115).

11. **C. webberi** *Bak. f.* in J.B. 67: 196 (1929); L.T.A.: 761 (1930); T.S.K.: 64 (1936); T.T.C.L.: 101 (1949); J. Léon. in B.J.B.B. 21: 392 (1951); K.T.S.: 103 (1961). Type: Kenya, coastal forests, *Webber* 603 (K, holo.!, EA, iso.!)

Evergreen tree 4·5–18 m. high, with bushy crown, buttressed base to trunk, and smooth grey bark; young branchlets shortly pubescent to puberulous. Leaves: stipules not seen; petiole 1–3 mm. long, puberulous; rhachis 0·8–3(–4) cm. long, conspicuously channelled above; leaflets normally 6–8 (although occasional reduced leaves with only 4 may be present also), elliptic to obovate-elliptic or oblong-elliptic, 0·9–3·2(–4·2) cm. long, 0·4–1·2(–2·5) cm. wide, rounded to obtuse or sometimes slightly emarginate at apex, not gradually tapering, glabrous or almost so. Panicles mostly terminal, 2–14 cm. long; pedicels 1·5–10(–12, *fide* Léonard) mm. long, pubescent like

FIG. 19. *CYNOMETRA WEBBERI*—**1,** branchlet with leaves and inflorescences, × $\frac{2}{3}$; **2,** part of leaf-rhachis showing basal parts of leaflets, × 4; **3,** part of inflorescence, showing persistent pedicels, × $\frac{2}{3}$; **4,** flower-bud, showing one bract and two bracteoles, × 8; **5,** flower, showing jointed pedicel, × 4; **6,** receptacle cut longitudinally, showing attachment of stipe of ovary, also gynoecium itself, × 8; **7,** pod, before dehiscence, × $\frac{2}{3}$; **8,** valve of dehisced pod, × $\frac{2}{3}$; **9,** upper suture of pod, × $\frac{2}{3}$; **10,** seed, × $\frac{2}{3}$. 1–6, from *Drummond & Hemsley* 4257; 7, from *Tanner* 3983; 8–10, from *Graham* in *F.H.* 2168.

the panicle-axes, jointed below the flower. Flowers sweetly scented. Sepals 4, green, 2–3 mm. long. Petals 5, white, 3·5–4·5 mm. long. Pods smooth, 4–6 cm. long, 2·6–3·3 cm. wide, apiculate. Fig. 19.

KENYA. Kwale District: 5 km. S. of Mazeras, Mwachi, 10 Sept. 1953, *Drummond & Hemsley* 4257 !; Kilifi District: Arabuko, Oct. 1929, *R. M. Graham* B. 699 in *F.H.* 2168 ! & 40 km. NW. of Malindi, Marafa, 22 Nov. 1961, *Polhill & Paulo* 838 !
TANGANYIKA. Tanga District: Mtotohovu, 8 Dec. 1935, *Greenway* 4241 ! & Steinbruch Gorge, 3 Nov. 1957, *Faulkner* 2091 !; Uzaramo District: 88 km. W. of Dar es Salaam on Morogoro road, 26 Aug. 1951, *Trapnell* 2170 !
DISTR. **K7**; **T3**, 6; not known elsewhere
HAB. Lowland dry evergreen forest, ? evergreen woodland, deciduous woodland (*Brachystegia*); near sea-level to ± 150 m.

12. **C. filifera** *Harms* in N.B.G.B. 13: 414 (1936); T.T.C.L.: 100 (1949); J. Léon. in B.J.B.B. 21: 398 (1951). Type: Tanganyika, Lindi District, *Schlieben* 5795 (B, holo. †, BM, BR, iso. !)

Tree up to about 3–5 m. high; young branchlets densely pubescent. Leaves sessile or subsessile; stipules linear-filiform, up to 2 cm. or more long; rhachis pubescent, not winged, 3–6 cm. long; leaflets 8–12, alternate or opposite, ovate or oblong, 1–3 cm. long, 0·8–1·5 cm. wide, ± obtuse or obtusely subacuminate at apex, not emarginate, glabrous. Peduncle together with axis of flowering raceme 3–6 cm. long, hairy; pedicels 1–1·4 cm. long, ± hairy in lower part, jointed a little below flower. Flowers white, red outside. Sepals 4, 4–5 mm. long. Petals 4, obovate-oblong or oblong, 4 mm. long. Pods unknown.

TANGANYIKA. Lindi District: 20 km. from Lindi, Mlinguru, 26 Dec. 1934, *Schlieben* 5795 ! & Lindi Creek, *Gillman* 1164 !
DISTR. **T8**; not known elsewhere
HAB. Uncertain, said to occur in " bush-land " at about 100 m. by Schlieben, and in coastal thicket on brown loam or limestone by Gillman

NOTE. The generic position is doubtful. More material, including fruits, is desired. The racemose inflorescences, strobiliform when young, and the definitely alternate arrangement of at least most of the leaflets combine to make *C. filifera* very anomalous among the other East African species of *Cynometra*. The petiolules are not twisted, however, and the basal parts of the pedicels persist after the flowers have fallen, as in the other *Cynometra* species.

15. **ZENKERELLA**

Taub. in E. & P. Pf. 3(3): 386 (1894); J. Léon. in B.J.B.B. 21: 408 (1951) & in Mém. 8°, Classe Sci., Acad. Roy. Belg. 30(2): 94 (1957)

Podogynium Taub. in E.J. 23: 173 (1896)

Unarmed evergreen trees. Leaves simple, without pellucid gland-dots, glabrous or almost so. Inflorescence of axillary racemes borne singly, their axes (in East African species) so short that the flowers appear to be fasciculate. Flowers ⚥, white or flesh-pink, spirally arranged along the inflorescence-axes; pedicels not jointed at apex; bracteoles small, not enclosing the flower-buds. Sepals 4(–5), imbricate, glabrous. Petals 5, equal or nearly so. Stamens 10; filaments glabrous, free or shortly connate at base; anthers dehiscing by longitudinal slits. Ovary stipitate, glabrous to pubescent; stipe shortly adnate along one side to the hypanthium-tube; ovules 1–3; style elongate. Pods stipitate, compressed, boat-shaped, splitting along lower and partially at least along upper suture into 2 coriaceous or subcoriaceous transversely venose valves. Seeds 1 per pod, large, with a thin wrinkled (when dry—? fleshy when fresh) wall, apparently exareolate.

Six to seven species in tropical Africa, all forest dwellers. None of the East African species is yet adequately known, and further good flowering and fruiting specimens of all would be welcome.

Stipules not seen, apparently very quickly caducous; branchlets glabrous to puberulous; ovary glabrous to puberulous:
- Branchlets and petioles glabrous; ovary glabrous or not:
  - Ovary glabrous . . . . 1. *Z. capparidacea*
  - Ovary puberulous . . . . 2. *Z. grotei*
- Branchlets and petioles puberulous, later glabrescent; ovary glabrous . . . . . . 3. *Z. schliebenii*

Stipules persistent, foliaceous, large, mostly more than 1 cm. long; branchlets pubescent; ovary glabrous. . . . . 4. *Z. egregia*

1. **Z. capparidacea** (*Taub.*) *J. Léon.* in B.J.B.B. 21: 413 (1951). Type: Tanganyika, Morogoro District, *Stuhlmann* 8851 (B, holo. †)

Small tree 8 m. high; branchlets glabrous. Leaves: stipules not seen, apparently very quickly caducous; petiole 4–6(–8, *fide* Taubert) mm. long, glabrous; lamina oblong-elliptic or possibly obovate-elliptic, 8–12(–14, *fide* Taubert) cm. long, 2·7–4·5(–6·5, *fide* Taubert) cm. wide, acuminate at apex, rounded to cuneate-rounded at base. Axis of raceme 0–2 mm. long; pedicels glabrous, 10–15 mm. long. Flower-buds pale pink. Sepals 4·5–5 mm. long. Petals ± 6 mm. long (only damaged ones seen). Ovary glabrous. Pods unknown.

TANGANYIKA. Morogoro District: Uluguru Mts., Bondwa Hill, 23 Mar. 1953, *Drummond & Hemsley* 1756!
DISTR. T6; not known elsewhere
HAB. Upland rain-forest; ± 1900 m.

SYN. *Podogynium capparidaceum* Taub. in E.J. 23: 173 (1896)
Cynometra capparidacea (Taub.) Harms in E. & P. Pf., Nachtr. 4 zu 3(3): 124 (1914) & in E.J. 53: 459 (1915) & in V.E. 3(1): 435 (1915); L.T.A.: 757 (1930); T.T.C.L.: 100 (1949)

NOTE. The distinctions between this species, *Z. grotei* and *Z. schliebenii* are small. It may well be that the two latter are really only geographical races or subspecies of *Z. capparidacea*, but the available material is, I consider, still insufficient to justify drastic alterations in the taxonomy.

2. **Z. grotei** (*Harms*) *J. Léon.* in B.J.B.B. 21: 415 (1951). Type: Tanganyika, Lushoto District, *Grote* 3803 (B, syn. †, B, EA, iso. !) & 5637 (B, syn. †, EA, iso. !)

Small tree with thin trunk, tending to gain support from other trees (*Drummond & Hemsley* 2596); tree to 25–30 m., with rounded crown (*Grote* 3803, *Greenway* 4676); branchlets glabrous. Leaves: stipules not seen, apparently very quickly caducous; petiole 2·5–6 mm. long, glabrous; lamina oblong-elliptic or elliptic, 3·3–15(–18) cm. long, 1·3–5·5(–8) cm. wide, obtusely pointed to acute or acuminate at apex, rounded to rounded-cuneate at base, glabrous. Axis of raceme 0–3 mm. long; pedicels 5–10 mm. long, puberulous (*Drummond & Hemsley* 2596) or glabrous and 1 to nearly 2 cm. long (*Grote* 3803). Flowers white. Sepals 5–6 mm. long. Petals 7–8 mm. long, 3–4 mm. wide. Ovary puberulous or shortly pubescent. Pods elliptic or suborbicular, somewhat asymmetric, 4–5·5 cm. long, 2–4 cm. wide. Seeds ± 2–3 cm. in diameter. Fig. 20/7–10.

TANGANYIKA. Lushoto District: Amani, *Grote* 3803 in *Peter* 48000! & 48002! & Kwamkoro road, Oct. 1913, *Grote* 5637 & Shagai Forest, near Sunga, 17 May 1953, *Drummond & Hemsley* 2596! & Sangarawe, 14 Oct. 1936, *Greenway* 4676!
DISTR. T3; not known elsewhere
HAB. Upland rain-forest; 970–2000 m.

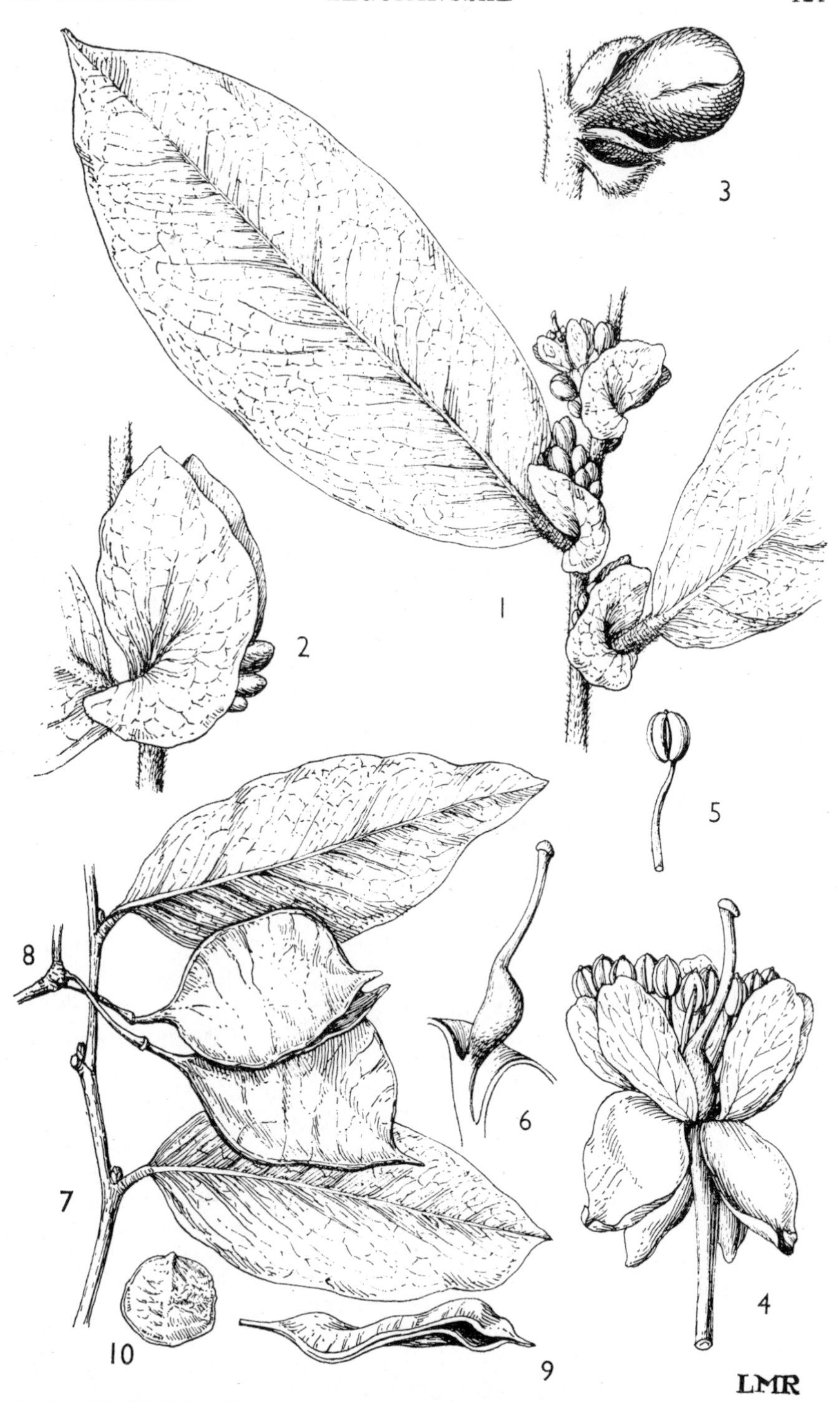

FIG. 20. *ZENKERELLA EGREGIA*—**1**, part of branchlet with leaves and inflorescences, × ⅔; **2**, stipules, × 2; **3**, flower-bud showing one bract and two bracteoles, × 6; **4**, flower, × 3; **5**, stamen, × 3; **6**, receptacle cut longitudinally, showing attachment of stipe of ovary, also gynoecium itself, × 3. *Z. GROTEI*—**7**, part of branchlet with leaves, × ⅔; **8**, pods before dehiscence, × ⅔; **9**, upper suture of pod, × ⅔; **10**, seed, × ⅔. 1, 4–6, from *Greenway* 2520; 2, 3, from *Peter* 48426; 7, from *Drummond & Hemsley* 2596; 8–10, from *Willan* 586.

SYN. *Cynometra grotei* Harms in E.J. 53: 459 (1915); L.T.A.: 757 (1930); T.T.C.L: 100 (1949)

3. **Z. schliebenii** (*Harms*) *J. Léon.* in B.J.B.B. 21: 413, fig. 111 (1951) & in Mém. 8°, Classe Sci., Acad. Roy. Belg. 30(2): 95 (1957). Type: Tanganyika, Morogoro District, Uluguru Mts., *Schlieben* 3854 (B, holo. †, BR, iso. !)

Tree 6–20 m. high; branchlets puberulous, glabrescent. Leaves: stipules very caducous; petiole 3–5 mm. long, puberulous; lamina oblong-lanceolate, elliptic, or oblanceolate, 3–14 cm. long, 1–4(–5·5) cm. wide, obtusely subacuminate at apex, obtuse to rounded at base. Axis of raceme very short or scarcely developed; pedicels glabrous, ± 8–15 mm. long. Flowers whitish-pink. Sepals 6–7 mm. long. Petals 9–11 mm. long, oblanceolate. Pods ± semicircular (upper suture straight, lower curved), 4·2–5 cm. long, 2·7–3·4 cm. wide, brown. Seeds ± 3–3·7 cm. long, 2–2·5 cm. wide, and 0·7–0·9 cm. thick.

TANGANYIKA. Morogoro District: Uluguru Mts., 8 Nov. 1932, *Schlieben* 2929! & *Schlieben* 3452! & above Morningside, May 1953, *Paulo* 64!
DISTR. T6; not known elsewhere
HAB. Upland rain-forest; 1430–1780 m.

SYN. *Cynometra schliebenii* Harms in N.B.G.B. 12: 84 (1934); T.T.C.L.: 100 (1949)

4. **Z. egregia** *J. Léon.* in B.J.B.B. 21: 414 (1951) & in Mém. 8°, Classe Sci., Acad. Roy. Belg. 30(2): 95 (1957). Type: Tanganyika, Lushoto District, Mangubu-Kiwanda, *Greenway* 4683 (K, holo. !, BR, EA, FHO, iso.!)

Tree 10–18 m. high; branchlets pubescent. Leaves: stipules obliquely reniform, 1–3(–4) cm. long, 0·5–2 cm. wide, foliaceous, persistent; lamina oblong-elliptic to obovate-elliptic, (6–)11·5–25 cm. long, (2·3–)4·2–9·5 cm. wide, acuminate at apex, widely rounded to subcordate and slightly asymmetric at base. Axis of raceme ± 0·5–1 cm. long; pedicels glabrous, 6–10 mm. long (to 16 mm., *fide* Léonard). Sepals 6–7 mm. long. Petals 6–10 mm. long, 2–3 mm. wide. Ovary glabrous. Fig. 20/1–6, p. 121.

TANGANYIKA. Lushoto District: Kwamtili, 5 Dec. 1917, *Peter* K610! & Sigi R., Longuza, 26 Oct. 1932, *Greenway* 3267!; Morogoro District: near Kimboza on Mikese–Kisaki road, 4 Sept. 1930, *Greenway* 2520!
DISTR. T3, 6; not known elsewhere
HAB. Lowland rain-forest, riverine and swamp forest; 270–400 m.

SYN. *Cynometra egregia* Hora & Greenway, Check-lists For. Trees & Shrubs Brit. Emp. 5(1) (Tanganyika Terr.): 70 (1940), *nom. nud.*; T.T.C.L.: 100 (1949)

16. SCORODOPHLOEUS

Harms in E.J. 30: 77 (1901); J. Léon. in B.J.B.B. 21: 418 (1951) & in Mém. 8°, Classe Sci., Acad. Roy. Belg. 30(2): 102 (1957)

Unarmed evergreen trees. Leaves simply pinnate; leaflets alternate, or some of them subopposite to opposite, 3–20, without gland-dots. Inflorescence racemose. Flowers ⚥, spirally arranged along the inflorescence-axis; pedicels not jointed; bracteoles very small, not enclosing the flower-buds. Sepals 4, imbricate, glabrous inside or with a few short hairs near central part and at apex. Petals 5, subequal. Stamens 10; filaments free, glabrous; anthers dehiscing by longitudinal slits. Ovary stipitate, pubescent near margins; stipe of the ovary adnate along one side to the wall of the subcylindric-turbinate hypanthium; ovules 2; style elongate; stigma very small, capitate. Pods flattened, woody, dehiscing elastically into 2 valves which are smooth or nearly so outside. Seeds large, compressed, with thin wrinkled (when dry—? fleshy when fresh) wall, exareolate.

A genus of 2 species in tropical Africa.

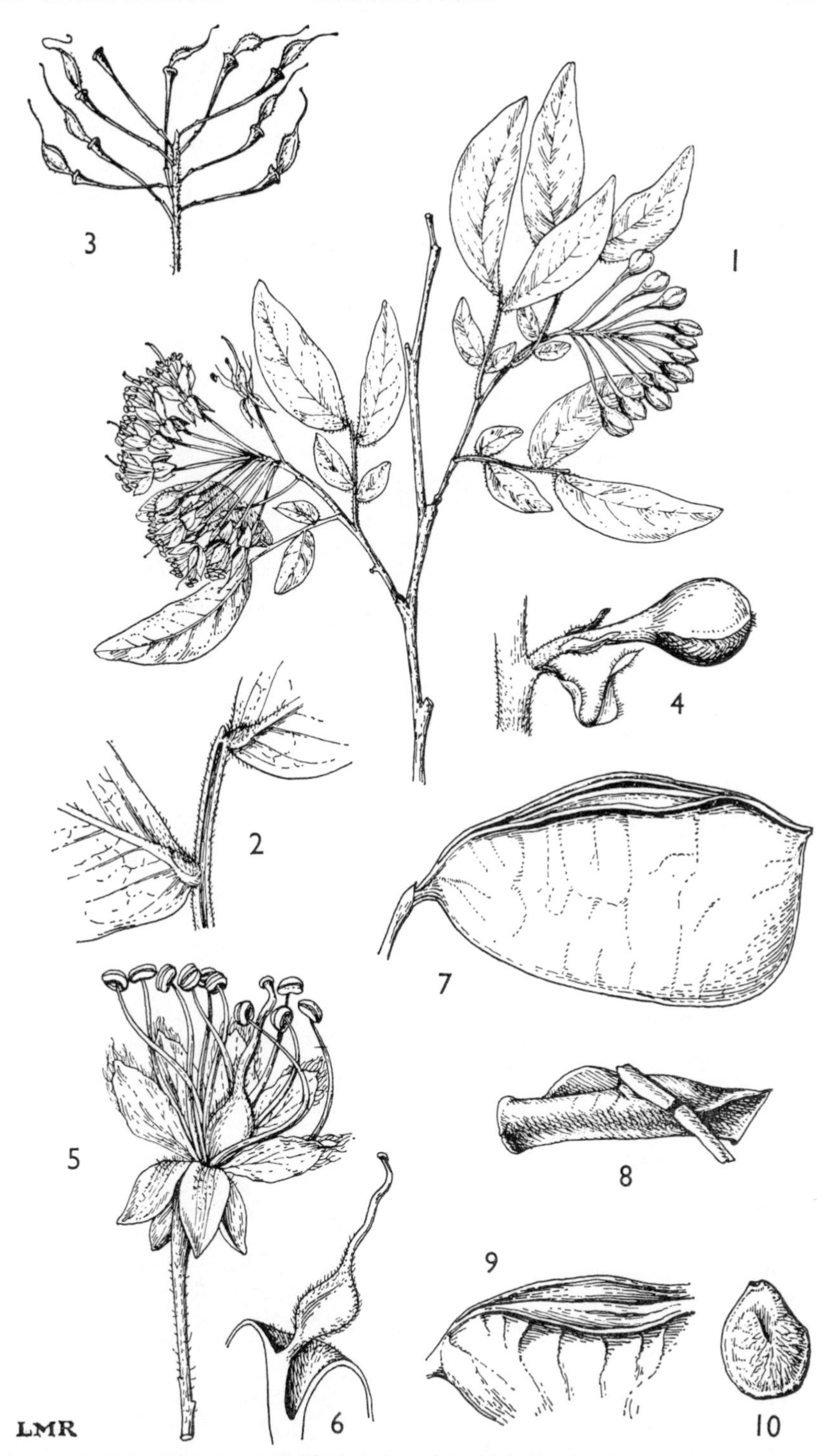

Fig. 21. *SCORODOPHLOEUS FISCHERI*—**1**, branchlets with leaves and inflorescences, × $\frac{2}{3}$; **2**, part of leaf-rhachis, showing basal parts of leaflets, × 4; **3**, part of inflorescence showing young fruits, × $\frac{2}{3}$; **4**, flower-bud showing bract and two bracteoles, × 8; **5**, flower, showing jointed pedicel, × 4; **6**, receptacle cut longitudinally, showing attachment of stipe of ovary and also gynoecium itself, × 8; **7**, pod before dehiscence, × $\frac{2}{3}$; **8**, valve of dehisced pod, × $\frac{2}{3}$; **9**, upper suture of pod, × $\frac{2}{3}$; **10**, seed, × $\frac{2}{3}$. 1, 2, 5, 6, from *Eggeling* 6707; 3, from *Milne-Redhead & Taylor* 7346; 4, from *Semsei* 2969; 7, from *Holst* 3578; 8–10, from *Tanner* 3597.

S. fischeri (*Taub.*) *J. Léon.* in B.J.B.B. 21 : 419 (1951); K.T.S. : 107 (1961). Type: Tanganyika, near Pangani, Kiwanda, *Fischer* 144 (B, holo. †, BM, drawing !)

Tree 6–25 m. high; bark smooth, grey or grey-brown. Leaves: petiole 3–4 mm. long; leaflets 3–5, mostly alternate, the terminal one largest, the rest decreasing in size downwards, glabrous or with some inconspicuous pubescence on midrib and margins; terminal leaflet asymmetrically rhombic-elliptic to rhombic-lanceolate, narrowed to an obtuse to subacute sometimes minutely emarginate apex, 3–11 cm. long, 1·4–5·5 cm. wide; lower leaflets elliptic to ovate, down to 1 cm. long and 0·7 cm. wide. Racemes terminal and lateral; axis ± 1·5–2 cm. long; pedicels 0·5–2·1 cm. long (excluding hypanthium), glabrous or sparingly and inconspicuously pubescent. Flowers strongly honey-scented. Sepals reflexed, pink-tinged, 5–6 mm. long. Petals white, 6·5–10 mm. long, 2–4 mm. wide. Pods obliquely obovate-elliptic, apiculate at apex, rounded at base, 5·5–7 cm. long, 3·3–4·5 cm. wide, with their upper suture transversely flattened. Seeds ellipsoid, 2–2·5 cm. long, 1·4–1·9 cm. wide, 5–6 mm. thick. Fig. 21, p. 123.

KENYA. Mombasa, Nov. 1884, *Wakefield* !
TANGANYIKA. Lushoto District: Daluni–Mashewa, 26 Oct. 1935, *Greenway* 4132 !; Morogoro District: Turiani, Sept. 1953, *Eggeling* 6707 !; Lindi District: Sudi, 12 Dec. 1942, *Gillman* 1135 !
ZANZIBAR. Zanzibar I., Kituoni, 30 Oct. 1932, *Vaughan* 2008 !
DISTR. **K7**; **T3**, 6, 8; **Z**; not known elsewhere
HAB. Lowland dry evergreen forest, riverine forest, and sometimes in wooded grassland; 30–670 m.

SYN. *Theodora fischeri* Taub. in P.O.A. C: 198 (1895)
Schotia fischeri (Taub.) Harms in V.E. 3(1): 453 (1915)
Cynometra fischeri (Taub.) Bak. f., L.T.A.: 758 (1930)
Cynometra sp. No. 15 sensu T.T.C.L.: 102 (1949)

NOTE. The branchlets, petioles, and leaf-rhachides vary from glabrous or nearly so to rather densely pubescent. The former appears commoner than the latter (12 and 8 gatherings respectively). The pubescent variant has so far been found only in **T3** and **T6**, and no less than five of its eight gatherings are from Handeni District. There may well be a case for naming these variants, but so little material is available from **T8** and Kenya that it seems best to delay this.

17. AFZELIA

Sm. in Trans. Linn. Soc. 4: 221 (1798), *nom. conserv.*; J. Léon. in Reinwardtia 1: 61–66 (1950) & in Mém. 8°, Classe Sci., Acad. Roy. Belg. 30(2): 106 (1957)

Unarmed evergreen or sometimes deciduous trees. Leaves simply and normally paripinnate; stipules with their basal parts connate into a persistent intrapetiolar scale, and with their upper parts free and caducous; petiolules twisted; leaflets opposite or subopposite, without translucent gland-dots, but normally with a small dot-like gland at proximal side of leaflet base either on lower surface in angle between margin and midrib or on margin itself*. Inflorescences simply racemose, or of racemes grouped into panicles; flowers spirally arranged along the inflorescence-axes; pedicels jointed at base; bracteoles large, well-developed, concavo-convex, almost completely concealing the young flower-buds, one bracteole overlapping the other by its margins; both bracteoles caducous before the flower opens. Hypanthium ± elongate. Sepals 4, imbricate, unequal (2 outer, 2 inner). Petal 1, large, clawed, the others rudimentary or absent. Stamens normally 7 fertile and 2 staminodes. Ovary stipitate, the stipe adnate to the hypanthium; style long; stigma small, ± capitate; ovules many. Pods dehiscing into 2 thickly

* Compare the perhaps similar marginal swelling in *Baikiaea* (p. 110), and similar ones on the surface in *Intsia* (p. 129).

woody valves. Seeds embedded in white pith, large, thick, hard, furnished with a basal brightly coloured aril.

A genus of 13 species, six in Malesia, the others in tropical Africa.

Leaflets (mature) rounded, or sometimes obtuse, and often emarginate at apex, not at all acuminate; inflorescence simply racemose or once-forked; hypanthium 1·1–2·5 cm. long; aril cup-shaped; pods straight . . . . 2. *A. quanzensis*

Leaflets apiculate or shortly acuminate at apex; inflorescence normally paniculate, rarely and perhaps casually simply racemose; hypanthium 0·3–4·5 cm. long; aril cup-shaped or bilobed; pods straight or curved:

Hypanthium 0·3–0·6 cm. long; large petal 1·1–2 cm. long; other rudimentary petals present; pods straight; aril cup-shaped, not bilobed. . . . 1. *A. africana*

Hypanthium 1·5–4·5 cm. long; petal 3·5–6 cm. long; other rudimentary petals absent; pods curved-reniform; aril markedly bilobed . . . 3. *A. bipindensis*

1. **A. africana** *Pers.*, Syn. Pl. 1: 455 (1805); Harms in V.E. 3(1): 458, fig. 251 (1915); L.T.A.: 700 (1930); Chalk, Burtt Davy, Desch & Hoyle, Twenty W. Afr. Timber Trees (For. Trees & Timb. Brit. Emp. 2): 15, fig. 2 & photos. (1933); J. Léon. in Reinwardtia 1: 64 (1950) & in F.C.B. 3: 355, fig. 27/C (1952); I.T.U., ed. 2: 55, fig. 12 (1952); F.W.T.A., ed. 2, 1: 459, fig. 149 (1958). Type: uncertain

Tree 6–30 m. high, with flat or rounded spreading crown; bark grey to dark brown, scaly; branchlets glabrous. Leaves: rhachis with petiole 4–32 cm. long or more; leaflets (mature) 2–5(–6) pairs, petiolulate, elliptic to ovate-elliptic, mostly 5–15 cm. long, 2·8–6·9(–8·5) cm. wide, obtusely pointed to ± acuminate at apex. Inflorescences paniculate, with racemose branches 3–13 cm. long, rarely (? casually and by reduction) inflorescence simply racemose. Flowers sweetly scented, with hypanthium 0·3–0·6 cm. long. Sepals shortly and densely velvety-pubescent or -puberulous outside, broadly obovate-elliptic to rotund, 2 outer 6–9 mm. long, 5–7 mm. wide, 2 inner 7–11 mm. long, 6–10 mm. wide. Large petal 1·1–2 cm. long, white or greenish-white, with pinkish-crimson streak down centre, with a rather long claw suddenly widened into a bilobed lamina 0·7–1·1 cm. wide. Stamens 7 fertile, with ± pubescent filaments. Style ± pubescent below. Pods straight, 10·5–20·5 cm. long, 5·5–8·5 cm. wide. Seeds black, ellipsoid or oblong-ellipsoid, 1·6–3 cm. long, 1·1–2·1 cm. wide, with an orange cup-shaped basal aril.

UGANDA. W. Nile District: Metuli, Dec. 1931, *Brasnett* 306! & Payida, escarpment foot, 21 Mar. 1945, *Greenway & Eggeling* 7238!; Acholi District: N. of Kitgum, Jan. 1937, *Sangster* 223!

DISTR. U1, 2 (*fide* I.T.U., ed. 2: 57); from Senegal eastwards to the Sudan Republic and Uganda, including the Congo Republic

HAB. Wooded grassland; 1220–1370 m.

SYN. *Intsia africana* (Pers.) O. Kuntze, Rev. Gen. Pl. 1: 192 (1891)

2. **A. quanzensis*** *Welw.* in Ann. Conselho Ultram. 1858: 586 (1859);

* In spite of the " correction " by various authors, including J. Léonard in F.C.B. 3: 354 (1952), of the spelling of the epithet to *cuanzensis*, there is no evidence that " *quanzensis* " was an unintentional orthographic error. Welwitsch used the initial " qu " repeatedly, and Quanza is the version used in Stieler's Hand-Atlas (1882) and also, with Cuanza as a synonym, in Justus Perthes' Specialkarte von Afrika (1893). See also Chalk, Burtt Davy & Desch, Some E. Afr. Conif. and Legum.: 26–28 (1932).

L.T.A.: 701 (1930); Burtt Davy, Fl. Pl. Ferns Transv. 2: 327, fig. 53 (1932); Chalk, Burtt Davy & Desch, Some E. Afr. Conif. and Legum. (For. Trees & Timb. Brit. Emp. 1): 26, fig. (1932); T.S.K.: 63 (1936); T.T.C.L.: 87 (1949); U.O.P.Z.: 109 (1949), pro parte, excl. fig. rami floriferi (= *Intsia bijuga*); J. Léon. in Reinwardtia 1: 64 (1950) & in F.C.B. 3: 354, fig. 27/B (1952) (as "*cuanzensis*"); Torre & Hillcoat in C.F.A. 2: 215 (1956) (as "*cuanzensis*"); Roti-Michelozzi in Webbia 13: 142 (1957); K.T.S.: 96, fig. 18 (1961) (as "*cuanzensis*"); F.F.N.R.: 98, fig. 21/J (1962). Type: Angola, Cuanza Norte, R. Cuanza, between Sansamanda and Quisonde, *Welwitsch* 594 (LISU, holo., BM, K, iso.!)

Tree 1·5–35 m. high, with very spreading crown; bark grey to pale brown, reticulate, or coming off in large flakes leaving yellowish-brown patches; branchlets pubescent, puberulous, or glabrous. Leaves: rhachis with petiole 6–32 cm. long; leaflets (mature) (3–)4–6(–7 or, *fide* Roti-Michelozzi, to 9) pairs, petiolulate, ovate-elliptic, oblong-elliptic, or elliptic, 2·3–9(–12) cm. long, 1·5–6·0(–7·2) cm. wide, rounded, or sometimes obtuse, and often emarginate at apex. Inflorescences erect, of simple or once-forked racemes; axis 2–7·5 cm. long. Flowers very sweetly scented, with hypanthium 1·1–2·5 cm. long. Sepals shortly pubescent or puberulous outside, outer 2 elliptic, 0·9–1·7 cm. long, 0·7–1·3 cm. wide, inner 2 obovate-spathulate, 1·7–2·5 cm. long, 0·9–1·8 cm. wide. Large petal upwardly-turned, 2·5–4·5 cm. long, ± pubescent and green outside, red inside, with a rather long claw suddenly widened into a deeply bilobed lamina 2·2–3·1(–3·8) cm. wide. Stamens 7(–9) fertile, with glabrous to pubescent filaments mostly green like the ovary. Style pubescent or glabrous. Pods straight, 7–23 cm. long, 4·5–8·3 cm. wide. Seeds black, oblong-ellipsoid or ellipsoid, 2–3·4 cm. long, 0·9–1·7 cm. wide, with an orange, red, or vermilion, cup-shaped basal aril. Fig. 22.

KENYA. Kwale District: Shimba Hills, *Battiscombe* 35!; Mombasa, Apr. 1876, *Hildebrandt* 1967!; Kilifi District, 6 Mar. 1945, *Jeffery* 115!

TANGANYIKA. Shinyanga, *Koritschoner* 1712!; Pangani, Madanga, Kumbamtoni, 27 Apr. 1956, *Tanner* 2798!; Ufipa District: Milepa, 22 Oct. 1935, *Michelmore* 1093!

ZANZIBAR. Zanzibar I.: recorded in U.O.P.Z.: 109 (1949); Pemba I., Ras Mkumbuu, 22 Dec. 1930, *Greenway* 2772!

DISTR. **K**7; **T**1–8; **Z**; **P**; Congo Republic, Somali Republic (S.) (*fide* Chiovenda, Fl. Somala 2: 179, figs 105–109 (1932)), Mozambique, Malawi, Zambia, Rhodesia, Angola, Botswana, the Transvaal and Swaziland*

HAB. Thickets, woodland, lowland dry evergreen forest; 0–1340 m.

SYN. *Intsia quanzensis* (Welw.) O. Kuntze, Rev. Gen. Pl. 1: 192 (1891); P.O.A. C: 199 (1895)

[*Afzelia africana* sensu Chalk, Burtt Davy & Desch, Some E. Afr. Conif. and Legum. (For. Trees & Timb. Brit. Emp. 1), pro parte quoad fig. (p. 27) (1932); T.T.C.L.: 87 (1949), pro parte quoad distrib., spec. *Burtt* 4747, et notam Burttianam, *non* Pers.]

NOTE. According to F.C.B. 3: 354 (1952), the leaflets may be in up to 10 pairs, and up to 14 cm. long, and the racemes up to 10 cm. long. I have not been able to confirm these. The number of leaflets may be taken from the field-notes to *Welwitsch* 628, quoted in Hiern, Cat. Afr. Pl. Welw. 1: 300 (1896).

There are nearly always 7 fertile stamens and 2 staminodes. Very rarely there are deviations: *Hildebrandt* 1967!, from Mombasa, apparently has 9 fertile stamens and, according to *Welwitsch*, there may sometimes be only five.

A. quanzensis is composed in East Africa of two fairly well-marked variants:—

1. With petiolules and leaf-bases quite glabrous. Confined to altitudes between sea-level and 500 m., with (so far!) one exception, *Hornby* 340, collected at Mpwapwa at about 1070 m.

2. With some short spreading hairs on leaf-rhachis, petiolules and leaflet-bases; sometimes hairs on petiolule only. Altitude-range (as far as at present known) 1053–1835 m. The isotype at Kew of *A. quanzensis* (*Welwitsch* 594) corresponds with this variant.

* In C.F.A. 2: 215 (1956), said to occur in Natal. I have not seen specimens from there.

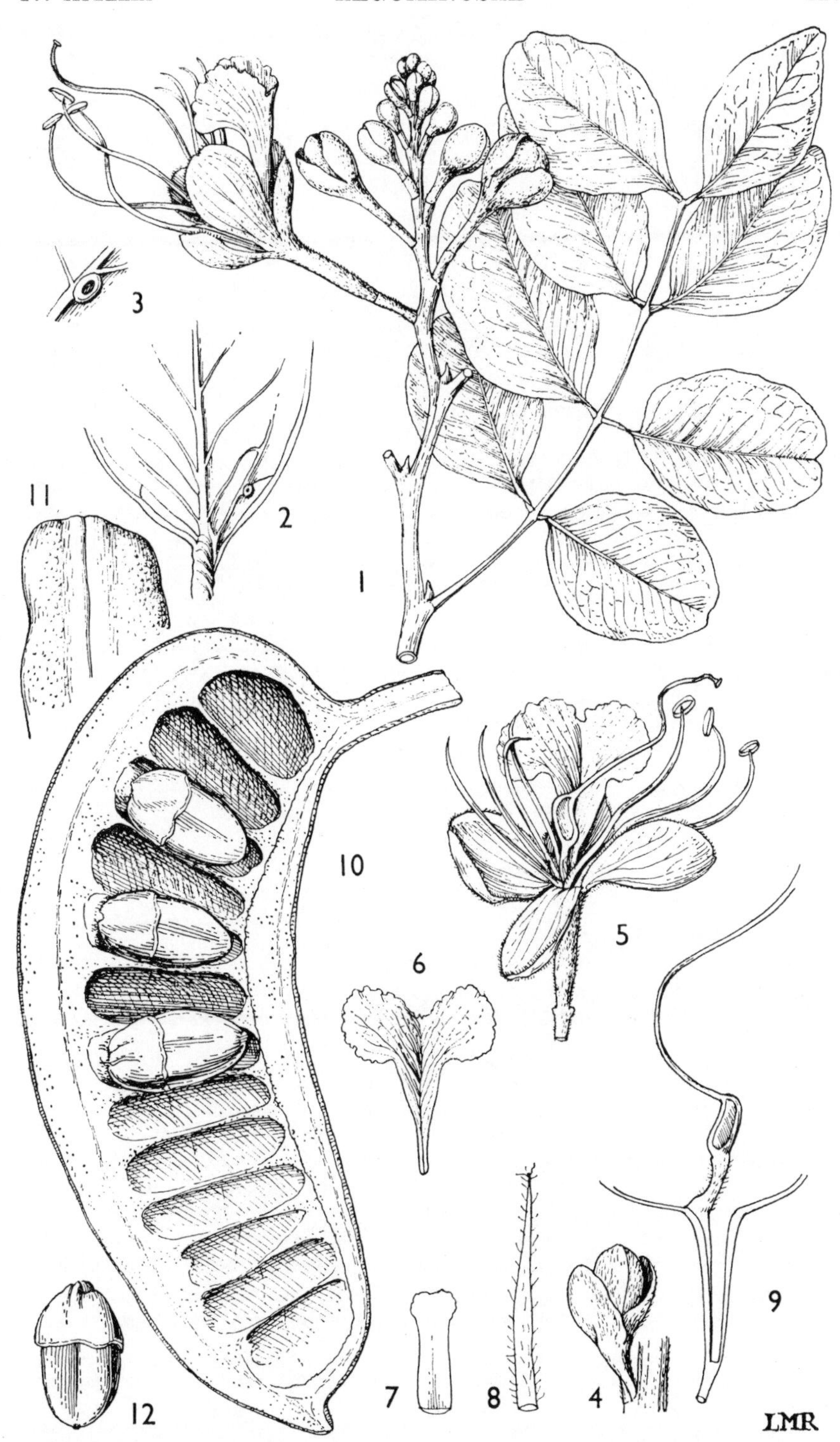

FIG. 22. *AFZELIA QUANZENSIS*—**1,** part of branchlet, showing leaf and inflorescence, × $\frac{2}{3}$; **2,** base of leaflet, lower surface, showing gland, × 4; **3,** gland, × 8; **4,** flower-bud with overlapping bracteoles, × 2; **5,** flower, × $\frac{2}{3}$; **6,** large petal, × $\frac{2}{3}$; **7,** one of four small petals, × 8; **8,** one of two staminodes, × 8; **9,** longitudinal section of hypanthium showing adnate stipe of ovary, × 1; **10,** pod, dehisced, × $\frac{2}{3}$; **11,** part of pod showing suture, × $\frac{2}{3}$; **12,** seed, showing aril, × $\frac{2}{3}$. 1–4, from *Hornby* 340; 5–9, from *Milne-Redhead & Taylor* 7061; 10–12, from *Richards* 6348.

Although these variants are valid in East Africa, they do not appear to be always so elsewhere. Thus, *Sandwith* 41 (Zambia, Quien Sabe, 1190 m.) shows mostly glabrous foliage.

3. **A. bipindensis** *Harms* in E.J. 49 : 426 (1913) ; L.T.A. : 700 (1930) ; Chalk, Burtt Davy & Desch, Some E. Afr. Conif. and Legum. (For. Trees & Timb. Brit. Emp. 1) : 28, 33 (1932) ; Chalk, Burtt Davy, Desch & Hoyle, Twenty W. Afr. Timber Trees (For. Trees & Timb. Brit. Emp. 2) : 21, fig. 3 & photos. (1933) ; J. Léon. in Reinwardtia 1 : 65 (1950) & in F.C.B. 3 : 355, t. 26, fig. 27/D (1952) ; Torre & Hillcoat in C.F.A. 2 : 216 (1956) ; F.W.T.A., ed. 2, 1 : 459 (1958). Type : Cameroun Republic, Bipindi, *Zenker* 3738 (B, holo. †, K, iso. !)

Tall tree, 18–40 m. high ; bark usually reddish-brown and scaly (but said on *Sangster* 538 to be " grey, fairly smooth ") ; young branchlets glabrous. Leaves : rhachis with petiole 9–32 cm. long ; leaflets (4–)5–7(–8, *fide* F.C.B.) pairs, petiolulate, oblong to oblong-elliptic, 5–13(–20, *fide* F.C.B.) cm. long, 2·6–5·8(–7, *fide* F.C.B.) cm. wide, ± apiculate or shortly acuminate at apex. Inflorescence paniculate. Flowers sweetly scented, with hypanthium 1·5–4·5 cm. long. Sepals densely tomentellous outside, outer 2 elliptic, 1–1·7 cm. long, 0·8–1·5 cm. wide, inner 2 obovate-elliptic, 1·4–2 cm. long, 0·9–1·6 cm. wide. Petal upwardly turned, 3·5–6 cm. long, with a long red-purple claw, and a crumpled ± bilobed lamina 3–5 cm. wide which is white at first and then turns pink. Stamens 7 fertile, with pubescent filaments, red like the style ; staminodes 2, elongate, 0·5–2 cm. long. Style very sparingly pubescent. Pods curved-reniform, (8–)11–19·5 cm. long and 5·5–8 cm. wide (*fide* F.C.B.). Seeds black, ovoid-ellipsoid, 2·5–3·3(–4·5, *fide* F.C.B.) cm. long, 1–1·7(–2, *fide* F.C.B.) cm. wide, with a deeply bilobed orange aril, one at least of whose lobes reaches to about the middle or near the top of the seed.

UGANDA. Toro District : Bwamba, Kidongo, Aug. 1937, *Eggeling* 3374 ! & Bwamba, Jan. 1939, *Sangster* 494 ! & May 1939, *Sangster* 538 !
DISTR. U2 ; Nigeria, Cameroun Republic, Gabon, Congo Republic, Central African Republic and Angola
HAB. Lowland rain-forest ; ± 900 m.

SYN. [*Afzelia bella* sensu I.T.U., ed. 2 : 57 (1952), *non* Harms]

18. INTSIA

Thou., Gen. Nov. Madag. : 22 (1808) ; Meijer Drees in Bull. Jard. Bot. Buitenz., sér. 3, 17 : 87 (1938)

Unarmed evergreen trees. Leaves simply paripinnate ; stipules connate into a persistent intrapetiolar scale ; petiolules twisted ; leaflets opposite or subopposite, without translucent gland-dots, but glandular as in *Afzelia* on the leaflet surface on one or both sides of the base. Inflorescences simply racemose to paniculate ; flowers spirally arranged along the inflorescence-axes ; pedicels jointed at base; bracteoles ± well-developed and concavo-convex, ± concealing the young flower-buds ; both bracteoles caducous before the flower opens. Hypanthium ± elongate. Sepals and petals as in *Afzelia*. Stamens 3 fertile and 4–7 staminodes (*fide* Meijer Drees). Ovary as in *Afzelia*. Pods indehiscent or ultimately ± dehiscent ; valves woody, but much thinner than in *Afzelia*, and ± venose outside. Seeds large, hard, exarillate.

Nine species, in tropical Asia and on the coasts and islands of the Indian and Pacific Oceans. Very close to *Afzelia*, differing mainly in the androecium and the exarillate seeds.

I. bijuga (*Colebr.*) *O. Kuntze*, Rev. Gen. 1 : 192 (1891) ; Meijer Drees in Bull. Jard. Bot. Buitenz., sér. 3, 17 : 89 (1938) ; T.T.C.L. : 103 (1949). Type :

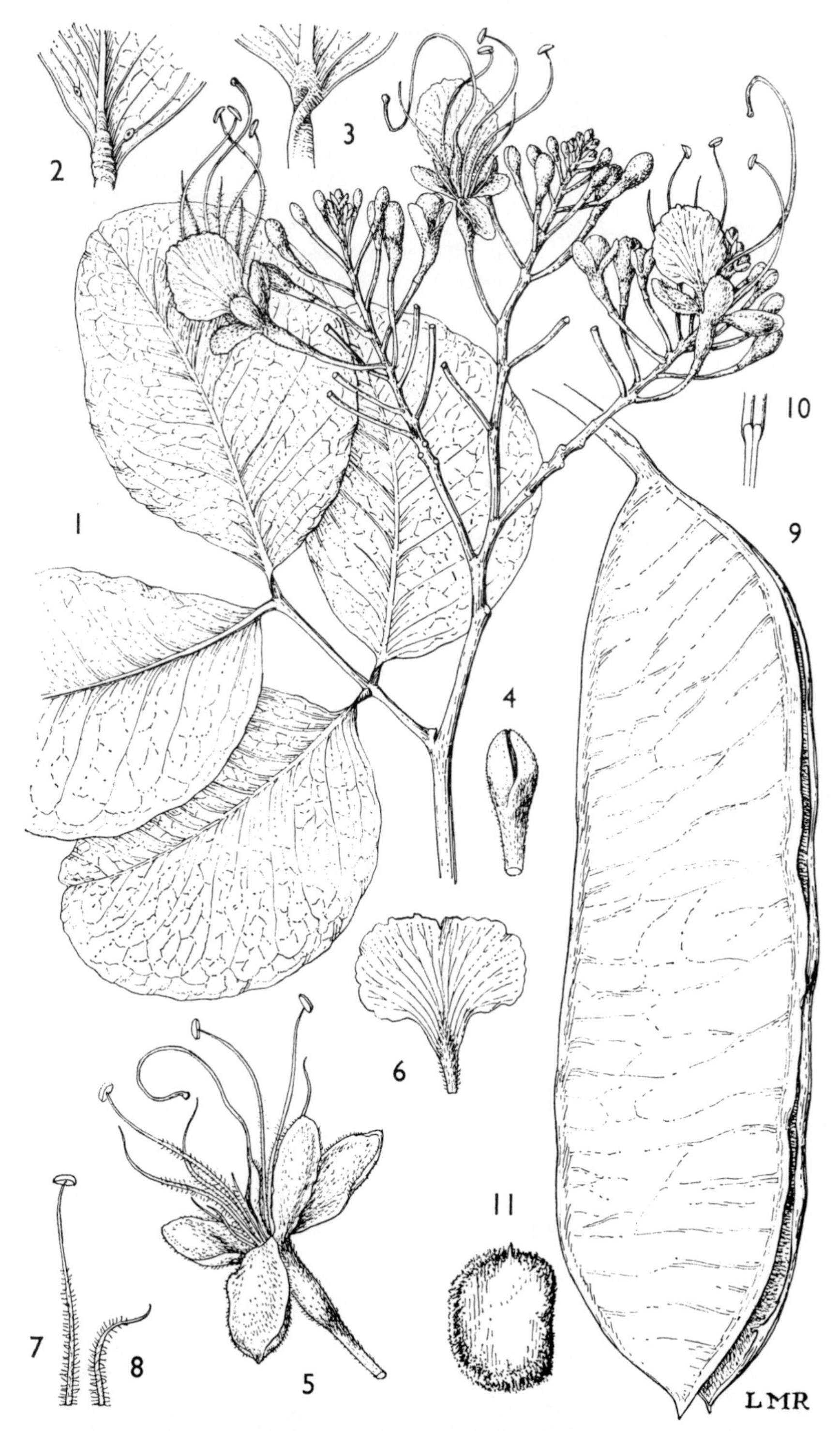

Fig. 23. *INTSIA BIJUGA*—**1**, part of branchlet showing inflorescence and leaf, × $\frac{2}{3}$; **2**, leaflet-base (lower surface) showing glands, × 2; **3**, leaflet-base, upper surface, showing twisted petiolule, × 2; **4**, flower-bud showing overlapping bracteoles, × 4; **5**, flower with petal removed, × 1; **6**, petal, × 1; **7**, fertile stamen, × 1; **8**, staminode, × 1; **9**, pod, × $\frac{2}{3}$; **10**, transverse section of upper suture of undehisced pod, × $\frac{2}{3}$; **11**, seed, × $\frac{2}{3}$. 1–4, from *A. C. Smith* 1383 (Fiji); 5–8, from *Greenway* 2707; 9–11, from *Maitland* 471.

cultivated at the Calcutta Botanic Garden; no certainly authentic specimen seen, but *Herb. Wallich* 5823A! is from there.

Tree 5–40 m. high, buttressed when old; bark pale grey; young branchlets glabrous or nearly so. Leaves glabrous (in East Africa); petiole with rhachis (1·5–)3·5–10·5 cm. long; leaflets in (1–)2, very rarely 3, pairs, asymmetrically ovate to elliptic or rotund, 4–16 cm. long, 2·8–11 cm. wide, rounded to obtuse or obtusely pointed at apex, venose. Inflorescence a corymbose panicle 3–10 cm. long and 4–12 cm. wide. Sepals 6–16 mm. long. Petal 1, white or pink, 1·3–3·3 cm. long, long-clawed, with a crisped lamina 1–3·4 cm. wide. Stamens red. Pods 10–28 cm. long, 4–7·2 cm. wide. Seeds 2·2–3·5 cm. long, 1·5–3·2 cm. wide, ± covered with detachable rusty scurf. Fig. 23, p. 129.

ZANZIBAR. Pemba I., SW. of Verani, 11 Dec. 1930, *Greenway* 2707!; SW. of Pemba, Panza I., 13 Feb. 1929, *Greenway* 1402!; W. of Pemba, Makongwe I., 13 Feb. 1929, *Greenway* 1414!

DISTR. **P**; coasts and islands of the Indian and Pacific Oceans. *I. bijuga* is cultivated in Tanganyika and in Zanzibar.

HAB. On and near the sea-shore

SYN. *Macrolobium bijugum* Colebr. in Trans. Linn. Soc. 12: 359, t. 17 (1819)
Afzelia bijuga (Colebr.) A. Gray, Bot. U.S. Expl. Exped. 1: 467, t. 51 (1854); U.O.P.Z.: 107 (1949), *nom. illegit., non Afzelia bijuga* (Willd.) Spreng. (1827)
[*A quanzensis* sensu U.O.P.Z.: 109 (1949), pro parte, quoad icon. rami floriferi tantum, *non* Welw.]

NOTE. The petiolules and leaflet midribs are said to be sometimes pubescent in the East Indies, but not in East Africa. A. Gray (l.c.) illustrated two smaller lateral petals in addition to the normal single one, but I have not observed any lateral ones myself.
For an account of the dispersal of *I. bijuga*, whose seeds, owing to the lightness of the cotyledons, are widely dispersed by ocean currents, see Ridley, Dispersal of Plants Throughout the World: 280–281 (1930).

19. DANIELLIA

J. J. Benn. in Pharm. Journ. 14: 252 (1855); J. Léon., Étude Botanique des Copaliers du Congo Belge (Publ. I.N.E.A.C., Sér. Scient. 45): 90 (1950) & in Mém. 8°, Classe Sci., Acad. Roy. Belg. 30(2): 110 (1957)

Paradaniellia Rolfe in K.B. 1912: 96 (1912)

Unarmed evergreen or deciduous trees. Leaves simply paripinnate; stipules intrapetiolar, connate into a structure lanceolate in outline, enfolding the axillary bud, open on the abaxial side, and very soon caducous*; petiolules not twisted; leaflets opposite or subopposite, with pellucid gland-dots (sometimes restricted to certain parts of the leaflet, such as near the margins or midrib or base). Inflorescence paniculate**; ultimate branches racemose; flowers spirally arranged; pedicels jointed at base; bracteoles large, well-developed, concavo-convex, ± completely concealing the young flower-buds, one bracteole overlapping the other by its margins, both caducous before the flower opens. Hypanthium elongate. Sepals 4, imbricate (2 outer, 2 inner). Petals either (2–)3 large or medium + 2(–3) very small, or 1 large or medium + 4 very small. Stamens 10, free, or 9 of them shortly connate at base. Ovary stipitate, the stipe adnate to the hypanthium; style long; stigma small. Pods dehiscing into 2 valves; endocarp coriaceous, curling up and separating from the stiffly papery to thinly woody exocarp; seeds solitary, large, with smooth hard testa, affixed near the distal end of the pod, dis-

* Compare the stipules of *Ficus*, as acutely noted by Bennett himself. In some species of *Daniellia*, e.g. *D. ogea* (Harms) Rolfe and *D. pynaertii* De Wild., the stipules may be very elongate, up to 8–9 cm.

**Sometimes (not in East Africa) the main axis of the apparent panicle continues as a leafy shoot, and then the inflorescences proper might perhaps be more justly considered as the lateral racemose branches.

FIG. 24. *DANIELLIA OLIVERI*—**1,** leaf with stipule; **2,** longitudinal section of stem with petiole base, and stipule round bud; **3,** stipule (side facing stem); **4,** inflorescence; **5,** one of four sepals; **6,** one large petal; **7,** one of four small petals; **8,** one of ten stamens; **9,** ovary; **10,** dehisced pods; **11,** inner lining (endocarp) of dehisced pod before it is shed. All × $\frac{2}{3}$. 1, from *Brasnett* 308 and *Deighton* 4217; 2, 3, from *Chipp* 623; 4–9, from *Brasnett* 308; 10, 11, from *Elliott* 14.

persed together with and while still remaining attached to the exocarp of one valve of the pod; funicle rather long, expanded at its end into a small aril.

There are eight species in this genus, all tropical African, widespread, but absent from eastern tropical Africa except for Uganda and the Sudan Republic.

D. oliveri (*Rolfe*) *Hutch. & Dalz.*, F.W.T.A. 1: 341, fig. 131/b (July 1928) & in K.B. 1928: 382 (Oct. 1928); L.T.A.: 697 (1930); J. Léon. in Publ. I.N.E.A.C., Sér. Scient. 45: 118, fig. 15 (1950) & in F.C.B. 3: 348, fig. 26 (1952); I.T.U., ed. 2: 63, fig. 15, photos. 9/a & b (1952). Types: several from West Africa, including *Heudelot* 364 from Senegal, indicated by Léonard as the type (? lectotype) and *Barter* 978 from Nigeria, the basis of Hook., Ic. Pl., t. 2406 (K, syn. !)

Deciduous tree 9–25(–45) m. high; bark pale grey, scaly, with crimson slash; crown obconical, flat-topped. Leaves: rhachis with petiole 17–50 cm. long; leaflets 4–10 pairs, ovate-elliptic to ovate-oblong, 3·5–18 cm. long, 2·5–9·5 cm. wide, asymmetric at base, ± acuminate at apex, pellucid-dotted all over, glabrous, or pubescent and becoming glabrous; petiolules 0·5–1·5 cm. long. Panicles 6–25 × 5–37 cm. Flowers white, fragrant. Sepals elliptic, 1·2–1·7 cm. long, 0·8–1·1(–1·5 when spread out) cm. wide. Petals: 1 oblong-elliptic, 0·8–1·2 cm. long, 0·3–0·45 cm. wide; 4 very small and inconspicuous. Stamen-filaments exserted, 3–3·5 cm. long, free, glabrous. Pods 6–9·5 cm. long, 3–4·5 cm. wide. Seeds obovoid or ellipsoid, 2–2·6 cm. long, 1·3–1·9 cm. wide. Fig. 24, p. 131.

UGANDA. W. Nile District: Adumi Camp, nearly 1 km. from Congo border, Dec. 1931, *Brasnett* 308! & Metuli, 25 Nov. 1941, *A. S. Thomas* 4063! & Aringa, 5 km. W. of Ladonga, 25 Nov. 1957, *Langdale-Brown* 2384!

DISTR. U1; widespread in tropical Africa north of the Equator from Senegal eastwards to the Sudan Republic and southwards to the Cameroun Republic, Congo Republic and Uganda

HAB. Wooded grassland, with *Butyrospermum paradoxum*; 1060–1530 m.

SYN. *Paradaniellia oliveri* Rolfe in K.B. 1912: 96 (1912)

NOTE. This tree yields an oleo-resin known as West African gum copal.
The description of this species by J. Léonard in F.C.B. 3: 348 (1952) gives ranges of measurements for certain organs that I have not so far been able to confirm (they are certainly not thereby to be taken as untrustworthy, however): stipules to 18 cm. long, sepals to 2·1 cm. long, larger petal to 1·6 × 0·6 cm., pods to 5 cm. wide.

20. TRACHYLOBIUM

Hayne in Flora 10: 743 (1827); J. Léon. Étude Botanique des Copaliers du Congo Belge (Publ. I.N.E.A.C., Sér. Scient. 45): 125 (1950) & in Mém. 8°, Classe Sci., Acad. Roy. Belg. 30(2): 115 (1957)

Unarmed evergreen tree. Leaves with a single pair of leaflets; stipules lateral, free, very small, very quickly falling off; leaflets opposite, with numerous pellucid gland-dots. Inflorescence paniculate; flowers spirally arranged along the racemose ultimate branches; pedicels jointed at base; bracteoles large, well-developed, concavo-convex, ± completely concealing the very young flower-buds, one bracteole overlapping the other by its margins, both caducous before the flower opens. Receptacle short. Sepals 4, imbricate (2 outer, 2 inner). Petals 5: upper 3 large, clawed, lower 2 very small, or occasionally all 5 well-developed and subequal. Stamens 10, free. Ovary stipitate; ovules 4; style long; stigma small, capitate. Pod indehiscent, thick, woody, ± resinous-warted, with a pithy endocarp, 1–3-seeded. Seeds hard, ± ellipsoid.

A single species of the coast of east tropical Africa, Madagascar, Mauritius and the Seychelles.

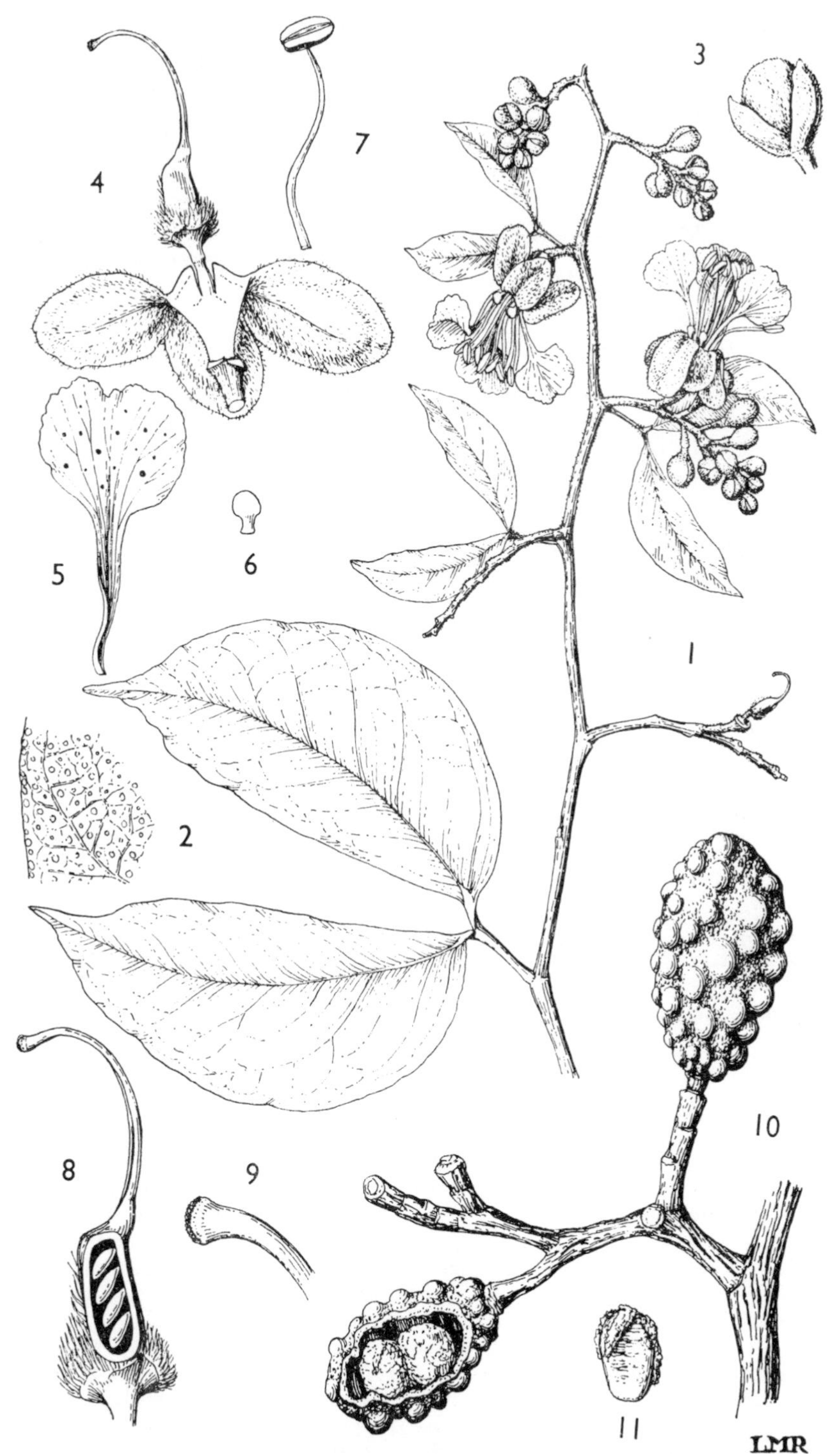

FIG. 25. *TRACHYLOBIUM VERRUCOSUM*—**1**, part of branchlet showing leaf and inflorescence, × ⅔; **2**, part of surface of leaflet showing glands, × 4; **3**, flower-bud showing overlapping bracteoles, × 3; **4**, section through hypanthium showing ovary and three of four sepals, × 2; **5**, one of three large upper petals, × 2; **6**, one of two minute lower petals, × 2; **7**, one of ten free stamens, × 2; **8**, ovary showing ovules, × 3; **9**, stigma, × 8; **10**, pods, × ⅔; **11**, seed, × ⅔. 1–9, from *Trump* 100; 10, 11, from *Wallace* 832.

T. verrucosum (*Gaertn.*) *Oliv.*, F.T.A. 2: 311 (1871); L.T.A.: 737 (1930); Greenway in E. Afr. Agric. Journ. 6: 250 (1941); T.T.C.L.: 107 (1949); U.O.P.Z.: 475 (1949); J. Léon. in Publ. I.N.E.A.C., Sér. Scient. 45: 126 (1950) & in Mém. 8°, Classe Sci., Acad. Roy. Belg. 30(2): 116 (1957); K.T.S.: 109, fig. 22 (1961). Type: Gaertner's description and figure of *Hymenaea verrucosa*, of which probably no specimen was kept by him.

Tree 6–24 m. high (–40 m., *fide* Léonard, l.c., 1950); bark grey or greyish-white, smooth or rough; young branchlets puberulous to glabrous. Leaflets ovate to elliptic or oblong-elliptic, asymmetric, 3·5–12 cm. long, 2–5·7 cm. wide (smaller ones often present with the inflorescence, larger (to 15 × 7·4 cm.) on juvenile shoots), coriaceous, glabrous or subglabrous, obtusely pointed to shortly subacuminate at apex. Panicle up to 35 × 25 cm., but often much smaller; ultimate branches and pedicels densely and shortly whitish-pubescent. Flowers white. Sepals 7–11 mm. long, densely appressed-pubescent outside and silvery-silky inside. Larger petals ± 1·5–2 cm. long; lamina ± 8–10 mm. in diameter. Pods ovoid-oblong, ellipsoid-oblong, or obovoid, 2·5–5 cm. long, 1·5–3 cm. wide. Seeds 1·3–1·8 × 0·9–1·2 cm. Fig. 25, p. 133.

KENYA. Kilifi District: Arabuko, Mar.–Apr. 1929, *R. M. Graham* B263 in *F.H.* Ox. 289 & in *C.M.* 13951! & Gedi, 30 Mar. 1954, *Trump* 100!
TANGANYIKA. Tanga District: Kwale to Tanga, 31 Jan. 1939, *Greenway* 5840!; Rufiji District: Mafia I., 23 Mar. 1933, *Wallace* 832!; Newala District: Makonde Plateau, 19 July 1941, *Gillman* 1018!
ZANZIBAR. ? Zanzibar I., *Kirk*!
DISTR. **K7**; **T3**, 6, 8; **Z**; Mozambique, Madagascar, Mauritius and the Seychelles
HAB. On or near the coast, in lowland dry evergreen forest, woodland and coastal evergreen bushland; 15–240 m.

SYN. *Hymenaea verrucosa* Gaertn., Fruct. 2: 306, t. 139/7 (1791)
Trachylobium hornemannianum Hayne in Flora 1827: 744 (1827) & Arzneyk. Gewächse 11, t. 18 (1830); Oliv., F.T.A. 2: 311 (1871); P.O.A. B: 305, 414 (1895). Type: Mauritius, *Hornemann* (whereabouts doubtful, ? C)
T. mossambicense Klotzsch in Peters, Reise Mossamb., Bot. 1: 21, t. 2 (1861); Kirk in J.L.S. 11: 1 (1869). Type: Mozambique, Querimba (Quisanga), *Peters* (B, holo. †, K, iso.!)

NOTE. This tree is tapped for gum copal (Zanzibar copal), and fossil or semi-fossil copal is dug up from places where it grew formerly.

In Madagascar and Mauritius the flowers may have all five petals well-developed, or only three, a variation that is probably genetic. I have observed only the latter sort of flower in East African specimens.

21. OXYSTIGMA

Harms in E. & P. Pf., Nachtr. zu 3(3): 195 (1897); J. Léon. in Bull. Inst. Roy. Col. Belge 21: 747 (1950) & in Mém. 8°, Classe Sci., Acad. Roy. Belg. 30(2): 128 (1957)

Unarmed evergreen trees. Leaves alternate, pari- or imparipinnate, with 1–9 opposite or alternate leaflets with numerous pellucid gland-dots which may be obscured in thick leaflets; petiolules not twisted; stipules small, very caducous. Inflorescence of spiciform racemes aggregated into panicles; flowers spirally arranged; bracteoles very small, not concealing the young flower-buds, persistent. Sepals 5(–6, *fide* Léonard), imbricate (2 smaller outer, 3 somewhat larger inner). Petals 0. Stamens (8–, *fide* Léonard) 10; filaments free, pubescent below. Disc present between stamens and ovary, small, cushion-like. Ovary sessile, pubescent; ovule 1; style elongate, glabrous, tapering to a minute terminal stigma. Fruits indehiscent, ± compressed, asymmetric, non-stipitate, with or without a proximal wing; pericarp usually with longitudinal veins running from apex of fruit. Seeds solitary, large, ruminate or foveolate, resinous.

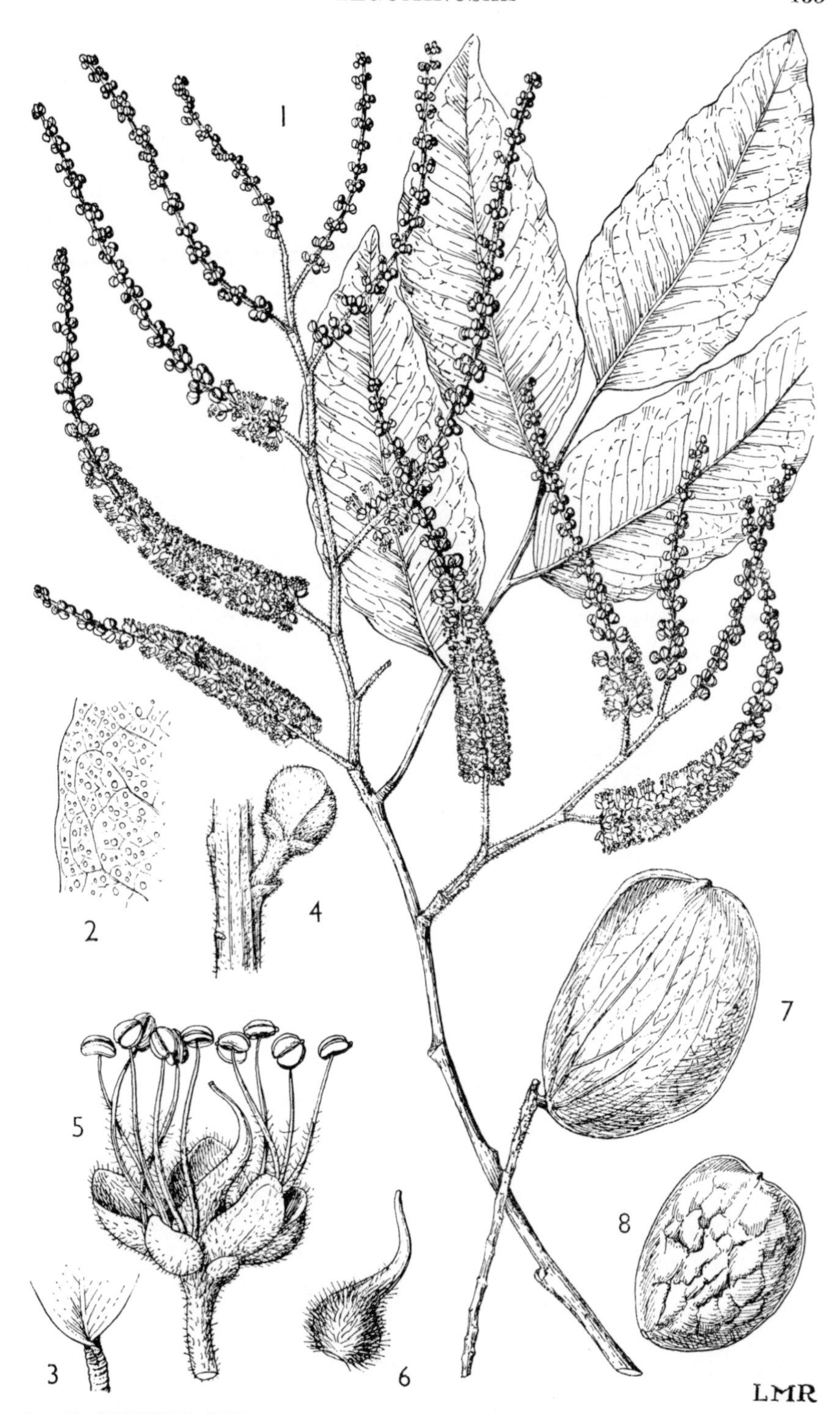

Fig. 26. *OXYSTIGMA MSOO*—**1,** branchlet with leaf and inflorescences, × $\frac{2}{3}$; **2,** part of leaflet, showing glands, × 2; **3,** leaflet-base, × 2; **4,** flower-bud with bract and two bracteoles, × 8; **5,** flower, × 8; **6,** ovary, × 8; **7,** pod, × $\frac{2}{3}$; **8,** seed, × $\frac{2}{3}$. 1–6 from *Bancroft* 8; 7, 8, from *Drummond & Hemsley* 1313.

Five species, in tropical Africa, one in East Africa and all the others in the central and western rain-forest regions.

The Asiatic genus *Kingiodendron* Harms is very closely related indeed to *Oxystigma*, and may perhaps not prove sufficiently distinct to be maintained. Both genera were first published simultaneously.

O. msoo *Harms* in F.R. 13 : 417 (1914); L.T.A. : 774 (1930); T.T.C.L. : 105 (1949); J. Léon. in Bull. Inst. Roy. Col. Belge 21 : 748 (1950). Type : Tanganyika, Moshi District, Rau Forest, *Deininger* (B, holo. †)

Tree up to 30–40 m. high, with grey slightly flaky bark; young branchlets glabrous. Leaves glabrous; rhachis 4–14 cm. long (to 19 cm. in juvenile foliage); leaflets mostly 5–7, alternate, ovate-elliptic, elliptic or oblong, 4–14 cm. long, 1·7–7 cm. wide, ± acuminate. Main and lateral axes of panicle shortly pubescent; spiciform branches 7–17·5 cm. long. Sepals 1·5–2·5 mm. long. Stamens 5–6 mm. long. Fruits obovate-elliptic, 4–6 cm. long, 2·5–4 cm. wide, longitudinally veined, proximally winged; winged part rigid, up to ± 2·5 cm. long. Fig. 26, p. 135.

TANGANYIKA. Moshi District: Rau Forest, *Lewis* 3 ! & 29 Jan. 1936, *Greenway* 4530 ! & 2 Apr. 1936, *Bancroft* 8 ! & 25 Feb. 1953, *Drummond & Hemsley* 1313 !
DISTR. **T2**; not known elsewhere (but see note below)
HAB. Ground-water forest; 640–1260 m.

NOTE. *Zimmermann* 3257 (EA !), from rain-forest at Amani, is an *Oxystigma*, and may well be *O. msoo*, but the specimen is a very poor one. Effort should be made to find this tree again at Amani. If the record is confirmed, then **T3** must be added to the distribution given above.

22. GUIBOURTIA

J. J. Benn. in J.L.S. 1 : 149 (1857); J. Léon. in B.J.B.B. 19 : 400 (1949) & Étude Botanique des Copaliers du Congo Belge (Publ. I.N.E.A.C., Sér. Scient. 45) : 67 (1950) & in Mém. 8°, Classe Sci., Acad. Roy. Belg. 30(2) : 137 (1957)

Unarmed evergreen trees. Leaves with a single pair of leaflets, or rarely (not in East Africa) with a single leaflet; stipules free, usually small and caducous; leaflets opposite, asymmetric, usually with numerous pellucid gland-dots. Inflorescence paniculate; flowers sessile or pedicellate, spirally arranged along the spicate or racemose ultimate branches; bracteoles small, persistent or falling before the flowers open. Sepals 4, imbricate (2 outer, 2 inner). Petals 0. Stamens (8–)10(–12, *fide* Léonard); filaments free, glabrous. Ovary stipitate or sessile; ovules 2–4; style elongate, ending in a capitate stigma. Pods indehiscent or (not in East Africa) dehiscent along one suture, compressed, thick or thin. Seeds solitary, large.

Thirteen species in tropical Africa, and three (one of them doubtfully a *Guibourtia*) in the West Indies and South America.

G. schliebenii (*Harms*) *J. Léon.* in B.J.B.B. 19 : 404 (1949) & in Publ. I.N.E.A.C., Sér. Scient. 45 : 70, 78 (1950) & Mém. 8°, Classe Sci., Acad. Roy. Belg. 30(2) : 141 (1957). Type : Tanganyika, Lindi District, Lake Lutamba, *Schlieben* 6123 (B, holo. †, BM, BR, EA, K, iso. !)

Tree 6–20 m. high; young branchlets puberulous to shortly pubescent. Leaves: petiole 4–7 mm. long; leaflets ovate-falcate with outer margins convex and inner nearly straight, 1·7–5 cm. long, 0·8–2·5 cm. wide, ± obtusely acuminate, prominently venose both sides, glabrous except near base, ± translucently dotted. Panicles up to ± 8 cm. long and wide; branches to ± 4 cm. long, puberulous to shortly pubescent. Flower-buds sessile or nearly so, globose, with very caducous bracteoles. Flowers white. Inner

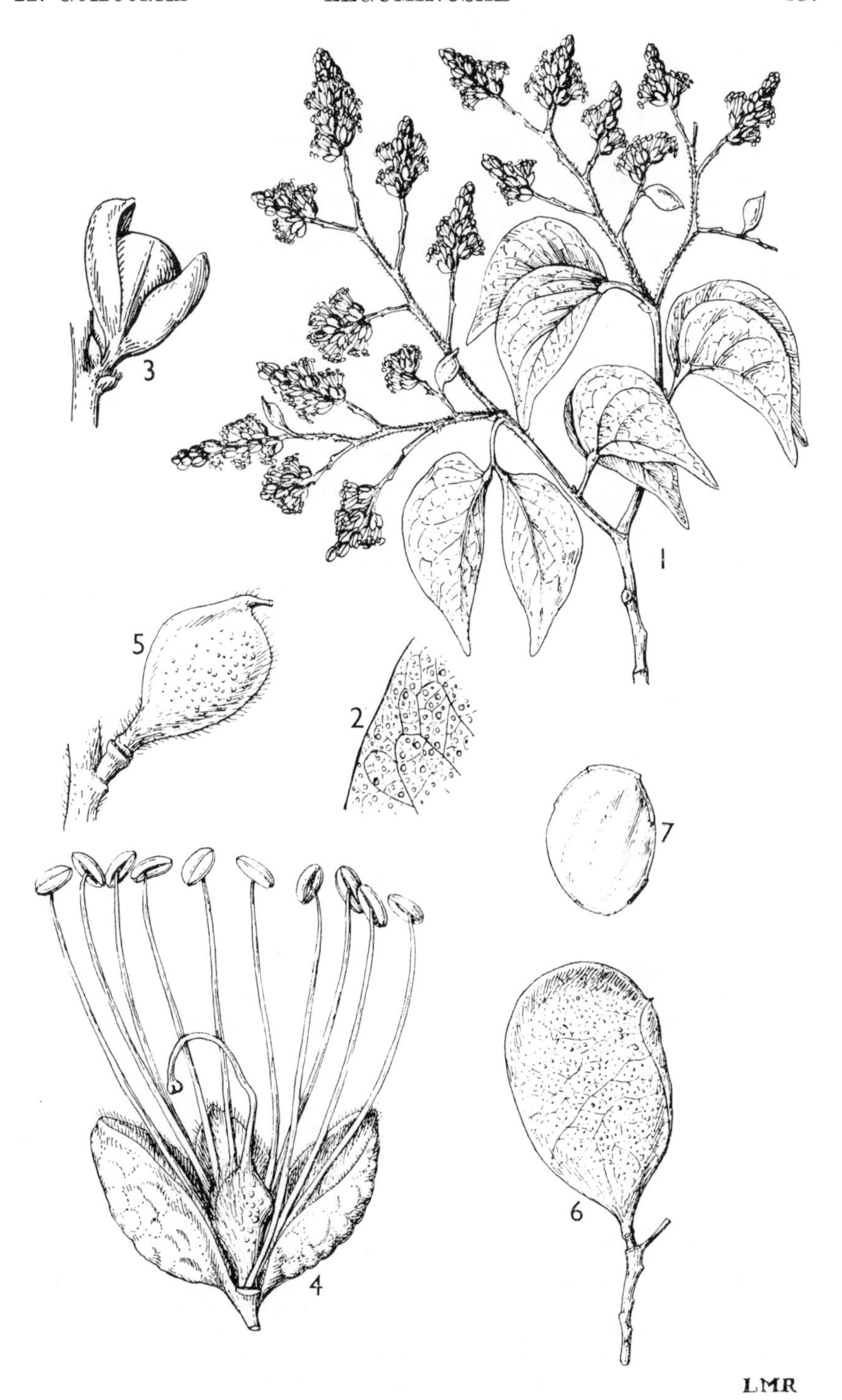

Fig. 27. *GUIBOURTIA SCHLIEBENII*—**1,** branchlets, showing leaves and inflorescences, × ⅔; **2,** part of leaflet-surface, showing glands, × 2; **3,** young flower-bud, showing bracteoles, × 6; **4,** flower, with one sepal removed, × 8; **5,** young pod, × 4; **6,** mature pod, × 1; **7,** seed, × 1. 1–5, from *Allen* 112; 6, 7, from *Barbosa* 2195.

sepals ± 3 mm. long, outer ± 4 mm. long. Stamens ± 5–7 mm. long. Pods indehiscent, flattened, brown, obovate-elliptic, asymmetric, 3–3·5 cm. long, 2–2·3 cm. wide, glabrous, reticulate-venose. Fig. 27, p. 137.

TANGANYIKA. Lindi District: Lake Lutamba, 17 Mar. 1935, *Schlieben* 6123! & Sudi, 25 May 1943, *Gillman* 1458!
DISTR. **T8**; Mozambique
HAB. Imperfectly known: *Schlieben* 6123 from near lake-shore at 200 m. alt., *Gillman* 1458 from " woodland on reddish-orange sands "

SYN. *Copaifera schliebenii* Harms in N.B.G.B. 13: 415 (1936); T.T.C.L.: 99 (1949)

NOTE. *Ede & Amani* 1!, from **T6**, Ulanga District, Kilombero valley, is very like *G. schliebenii* except for having petioles 0·7–1·2 cm. long, and leaflets 5·5–7 cm. long and 2·5–3·5 cm. wide. This may well be merely a robust state of *G. schliebenii*, but the specimen is poor and shows only a leafy shoot with young fruits. More is needed from this district. *Ede & Amani* 1 comes from a large tree 21–25 m. high, common as an emergent in " semi-evergreen ground-water thickets " at about 270 m. alt.

23. ISOBERLINIA

Craib & Stapf in K.B., Addit. Ser. 9: 266 (1911) & in K.B. 1912: 93 (1912); Duvign. in Inst. Roy. Col. Belge, Bull. Séances 21(2): 432–3 (1950); Troupin in B.J.B.B. 20: 302 (1950); Hauman in Inst. Roy. Col. Belge, Bull. Séances 23(2): 477 (1952); J. Léon. in Mém. 8°, Classe Sci., Acad. Roy. Belg. 30(2): 173 (1957)

Unarmed trees. Leaves paripinnate; stipules intrapetiolar, connate, or free in their upper part; leaflets opposite or subopposite, in 2–5 pairs, petiolulate, ± unequal-sided at base, without translucent gland-dots. Inflorescence normally paniculate; branches racemose or paniculate; bracteoles 2, well-developed, valvate, completely enclosing the flower-buds, persistent. Hypanthium tubular, 2·5–8 mm. long. Sepals (4–)5(–7), subequal. Petals 5(–6); usually all well-developed, but the upper one always somewhat longer or wider than the others and often of a different shape. Stamens 10 (–14); filaments free, exserted. Ovary sessile to shortly stipitate, and inserted on the tube of the hypanthium; ovules 4–8. Pod dehiscing elastically or explosively into 2 flattened smooth or sometimes obliquely nerved woody valves; upper suture not at all winged. Seeds compressed, ± obovate-elliptic, with thin testa and short funicle.

An exclusively tropical African genus of five species.

Upper (posticous) petal bilobed or deeply emarginate; ultimate branches of panicle short, 2·5–8 cm. long; inflorescence comparatively dense; bracts subtending pedicels 2–4 mm. long:
  Primary lateral nerves 7–10 each side of the midrib in mature leaflets; pedicels of open flowers 10–16 mm. long; rain-forest tree to 46 m. high . . . . 1. *I. scheffleri*
  Primary lateral nerves (8–)11–16 on each side of the midrib in mature leaflets; pedicels of open flowers usually only 3–5 mm. long, but sometimes elongate, and up to 15 mm.; woodland tree to ± 17 m. high. 2. *I. angolensis*
Upper (posticous) petal rounded or scarcely emarginate at apex; ultimate branches of panicle often elongate, 3·5–25 cm. long; inflorescence comparatively lax to rather dense; bracts subtending pedicels 2–8 mm. long:
  Leaflets glabrous or nearly so to minutely and sparsely puberulous on the surface beneath; nerves glabrous to sparsely and inconspicuously pubescent; upper

(posticous) petal usually 4·5–5 mm. wide; inflorescence rather lax to rather dense, with tomentellous to puberulous or glabrescent axes; ultimate branches of inflorescence 3·5–8(–18) cm. long; bracts subtending pedicels 2–3·5 mm. long . . . . . . 3. *I. doka*

Leaflets ± pubescent to tomentose on the surface beneath; nerves ± densely pubescent to tomentose; upper (posticous) petal 6–10 mm. wide; inflorescence lax, with tomentose axes; ultimate branches of inflorescence 5–25 cm. long; bracts subtending pedicels 4–7 mm. long . . . . . . 4. *I. tomentosa*

1. **I. scheffleri** (*Harms*) *Greenway* in K.B. 1937: 416 (1937); T.T.C.L.: 104 (1949); Troupin in B.J.B.B. 20: 306 (1950); Brenan in K.B. 17: 220 (1963). Type: Tanganyika, Lushoto District, Derema, *Scheffler* 201 (B, holo.†, BM, EA, K, iso. !)

Tall evergreen tree 30–46 m. high; trunk cylindrical, with small buttresses; bark smooth or nearly so, grey to greyish-black. Leaves: upper part of stipules free, thin, lanceolate, caducous, or apparently absent; petiole with rhachis 4–14·5 cm. long; leaflets 2–4 pairs, elliptic to ovate-elliptic, sometimes slightly obovate-elliptic, 5–18 cm. long, 3–9·5 cm. wide (larger on juvenile shoots), pubescent on midrib and lateral nerves beneath, with the surface beneath usually sparsely minutely and inconspicuously appressed-puberulous; primary lateral nerves ± 7–10 on each side of the midrib in mature leaflets. Panicles terminal, rounded, much branched, dense; ultimate racemose branches short, 2·5–6 cm. long, shortly brown-tomentose or -tomentellous. Bracts small, 2–3 mm. long, falling while the flower-buds are very young indeed. Pedicels of open flowers 10–16 mm. long. Bracteoles brown, 8·5–13 mm. long, 6–10 mm. wide. Sepals (4–)5. Petals white; upper one broadly obovate or obtriangular, ± bilobed at apex, 9–12 mm. long, 8–11·5 mm. wide; 4(–5) smaller, 4–10 mm. long, 1·5–4 mm. wide. Stamens 10–11. Pods large, 15–31 cm. long, 5·5–9·3 cm. wide, brown, flat, woody, tomentose, ± obliquely nerved.

TANGANYIKA. Lushoto District: Sigi, 5 Jan. 1933, *Greenway* 3319 ! & Amani, 18 Jan. 1940, *Greenway* 5913 ! & Balangai Forest, 23 Feb. 1951, *Hughes* 57 !

DISTR. **T3**, ? 6 (see below); not known elsewhere

HAB. Lowland and upland rain-forest, common and locally dominant; 460–1680 m.

SYN. *Berlinia scheffleri* Harms in E.J. 30: 83 (1901); L.T.A.: 688 (1930)
Westia scheffleri (Harms) Macbr. in Contrib. Gray Herb., n.s. 59: 21 (1919)

NOTE. *Semsei* 846 ! (EA, K) from Morogoro District, Kimboza Forest Reserve, " tall tree, in fruit and common ", July 1952, and *Wigg* 44 ! (EA), from S. Nguru in the same district, are probably *I. scheffleri*. The specimens consist only of leaves and a pod-fragment and leaves alone respectively, so that better material, with flowers, is needed to confirm the occurrence of the species in **T6**.

2. **I. angolensis** (*Benth.*) *Hoyle & Brenan* in K.B. 4: 78 (1949); Troupin in B.J.B.B. 20: 304 (1950); Hauman in F.C.B. 3: 378, fig. 31 (1952); Torre & Hillcoat in C.F.A. 2: 204 (1956); J. Léon. in Mém. 8°, Classe Sci., Acad. Roy. Belg. 30(2): 175, 176 (1957); F.F.N.R.: 125 (1962), saltem pro parte; Brenan in K.B. 17: 220 (1963). Type: Angola, Cuanza Norte, between Pungo Andongo and Candumbo, *Welwitsch* 568 (LISU, syn., BM, isosyn. !)

Tree 1–12(–17) m. high; bark grey to brown, fissured. Leaves: stipules connate into an intrapetiolar scale with upper parts (? always) absent; petiole with rhachis 6·5–25 cm. long; leaflets 3–4 (rarely –5) pairs, ovate, elliptic, lanceolate or oblong, (4–)7–19 cm. long, 2–7·8 cm. wide (larger on coppice or

FIG. 28. *ISOBERLINIA ANGOLENSIS* var. *LASIOCALYX*—**1,** branchlet with leaf and inflorescence, × $\frac{2}{3}$; **2,** flower-bud and bract, × 2; **3,** bracteoles and calyx, × 1; **4,** flower with bracteoles removed, × 2; **5,** upper petal, × 2; **6,** pod, × $\frac{2}{3}$; **7,** upper suture of pod cut transversely, × $\frac{2}{3}$; **8,** seed, × $\frac{2}{3}$. *I. TOMENTOSA* —**9,** bud and bract, × 2; **10,** upper petal, × 2. 1–5, from *Stolz* 1957; 6, from *Greenway* 3626; 7, 8, from *Procter* 704; 9, 10, from *Richards* 1796.

juvenile shoots), with variable indumentum (see below); primary lateral nerves (8–)11–16 on each side of the midrib in mature leaflets. Panicles terminal, much branched, ± dense; ultimate racemose branches short, 3–8 cm. long, ± shortly brown-tomentose. Bracts 3–4 mm. long. Pedicels of open flowers usually 3–4 mm. long, sometimes longer and up to 15 mm. Bracteoles brown to greenish-brown, 10·5–16 mm. long, 6–9 mm. wide. Sepals 5. Petals white; upper one obovate or oblong-obovate, bilobed or deeply emarginate at apex, 9–12 mm. long, 6·5–8 mm. wide; 4 smaller, 7–12 mm. long, 3–5 mm. wide. Stamens 10. Pods large, 19–35 cm. long, 6–8 cm. wide, rusty-tomentose to glabrescent. Fig. 28/1–8.

KEY TO INTRASPECIFIC VARIANTS

Leaflets glabrous or sparsely pubescent on surface beneath, the nerves being sometimes more densely pubescent:
  Hypanthium glabrous outside . . . . . var. **angolensis**
  Hypanthium ± pubescent outside . . . . . var. **lasiocalyx**
Leaflets ± densely pubescent to shortly tomentose all over lower surface; hypanthium ± pubescent outside, rarely glabrous . . . . var. **niembaënsis**

var. **angolensis**; Brenan in K.B. 17: 221 (1963)

TANGANYIKA. Buha District: S. of Malagarasi on Kibondo–Kasulu road, 21 Mar. 1954, *S. W. G. Smith* 857! & near Mkuti R., Feb. 1955, *Procter* 368!; Mbeya District: Mbozi, 9 Apr. 1932, *R. M. Davies* 603!
DISTR. **T4**, 7; Congo Republic, Sudan Republic (*Myers* 10812!), Zambia and Angola; ? in Cameroun Republic
HAB. Deciduous woodland; ± 800–900 m.

SYN. *Berlinia angolensis* Benth. in Trans. Linn. Soc. 25: 310 (1865); Oliv., F.T.A. 2: 296 (1871); L.T.A.: 687 (1930)
Westia angolensis (Benth.) Macbr. in Contrib. Gray Herb., n.s. 59: 21 (1919)

var. **lasiocalyx** *Hoyle & Brenan* in K.B. 4: 78 (1949); T.T.C.L.: 103 (1949); Troupin in B.J.B.B. 20: 306 (1950); Hauman in F.C.B. 3: 380 (1952); Brenan in K.B. 17: 222 (1963). Type: Malawi, Nyika Plateau, *Whyte* (K, holo.!). Fig. 28/1–8.

TANGANYIKA. Kigoma District: Kigoma–Machaso, 18 Feb. 1926, *Peter* 37001!; Iringa District: Ulete, 21 Feb. 1932, *St. Clair-Thompson* 515!; Rungwe District: Mulinda Forest, 27 Mar. 1913, *Stolz* 1957! & Tukuyu–Massoka [? Masoko] road, 23 Mar. 1932, *St. Clair-Thompson* 1304!
DISTR. **T4**, 7; Congo Republic, Malawi, Zambia and Angola
HAB. Deciduous woodland, sometimes dominant or co-dominant; 900–1740 m.

SYN. *Berlinia densiflora* Bak. in K.B. 1897: 269 (1897); L.T.A.: 687 (1930). Type as for *Isoberlinia angolensis* var. *lasiocalyx*
B. stolzii Harms in E.J. 53: 465 (1915); L.T.A.: 687 (1930). Types: Tanganyika, Rungwe District, Mulinda Forest, *Stolz* 1472 (B, syn. †) & 1957 (B, syn. †, EA, K, isosyn.!)
Isoberlinia densiflora (Bak.) Milne-Redh. in K.B. 1937: 415 (1937)
[*I. tomentosa* sensu C.F.A. 2: 205 (1956), *non* (Harms) Craib & Stapf]

var. **niembaënsis** (*De Wild.*) *Brenan* in K.B. 17: 222 (1963). Type: Congo Republic, Niemba Camp, *Delevoy* 289 (BR, holo.!)

TANGANYIKA. Mwanza District: Geita, 7 June, 1937, *B. D. Burtt* 6573!; Kahama District: between Kahama and Ushirombo, *B. D. Burtt* 5494!; Tabora District: Urambo, 6 Oct. 1949, *Bally* 7532! & Kaliuwa, 18 Oct. 1949, *Shabani* 28!
DISTR. **T1**, 4; Congo Republic, Malawi, Zambia and Angola
HAB. Deciduous woodland; 1160–1370 m.

SYN. *Berlinia niembaënsis* De Wild., Pl. Bequaert. 3: 145 (1925); L.T.A.: 689 (1930)
Isoberlinia niembaënsis (De Wild.) Duvign. in Inst. Roy. Col. Belge, Bull. Séances 21(2): 434 (1950); Troupin in B.J.B.B. 20: 308 (1950); Hauman in F.C.B. 3: 382 (1952)

NOTE. The specimens from Mwanza and Kahama Districts are sterile, and some allowance therefore must be made. It is strange that var. *niembaënsis* has not been found in **T7**, although it occurs to the south of the Flora area.

The specimens referred to as possible hybrids of *I. angolensis* and *I. tomentosa* by Hoyle & Brenan in K.B. 4: 78 (1949) are *I. angolensis* var. *niembaënsis*.

3. **I. doka** *Craib & Stapf* in K.B., Addit. Ser. 9: 267 (1911) & in K.B.1912: 94 (1912); Troupin in B.J.B.B. 20: 303 (1950); Hauman in F.C.B. 3: 380 (1952); I.T.U., ed. 2: 66, photo. 10 (1952); J. Léon. in Mém. 8°, Classe Sci., Acad. Roy. Belg. 30(2): 178 (1957); F.W.T.A., ed. 2, 1: 468 (1958); Brenan in K.B. 17: 223 (1963). Types: Nigeria, Katagum, *Dalziel* 364 & Zaria, *Dudgeon* 9 (both K, syn.!)

Tree 10–18 m. high or more. Leaves: upper part of stipules free, lanceolate, to ± 2 cm. long, caducous, or apparently absent; petiole with rhachis 11–24 cm. long; leaflets 3–4 pairs, ovate to elliptic, 6–18 cm. long, 3·3–13 cm. wide, glabrous or sparsely and inconspicuously pubescent on midrib and lateral nerves beneath, with the surface glabrous or nearly so to minutely and sparsely puberulous; primary lateral nerves 6–11 on each side of the midrib. Panicles rather lax to rather dense; ultimate racemose branches 3·5–8(–18) cm. long, tomentellous to puberulous or glabrescent. Bracts 2–3·5 mm. long. Pedicels of open flowers 2–5 mm. long. Bracteoles 9–12 mm. long, 6–9 mm. wide, fawn. Sepals 5, white. Petals white; upper one oblong-elliptic, 8–12 mm. long, 4·5–5 mm. wide, rounded or slightly emarginate but not bilobed at apex; 4 smaller 6–12 mm. long, 3–4 mm. wide. Stamens 10. Pods 15–30 cm. long, 5–7 cm. wide, with brown indumentum partially rubbing off with maturity. Seeds 2·5–3·3 × 1·8–2·5 cm.

UGANDA. W. Nile District: Ladonga, Dec. 1944, *Eggeling* 5550! & hills near Ulepi, Mar. 1935, *Eggeling* 1701! & Uippi escarpment ridge, 21 Mar. 1945, *Greenway & Eggeling* 7239!

DISTR. U1; from Guinée Republic in the W. to the Sudan Republic in the E., occurring in the N. part of the Congo Republic, but not S. of the Equator

HAB. Deciduous woodland; ± 1220 m.

NOTE. The Uganda specimens here referred to *I. doka* are somewhat atypical in having sparse pubescence on the midrib and lateral nerves beneath, in this way differing from glabrous-nerved West African *I. doka* and showing some approach to *I. tomentosa*. Whether this is a regional variant of *I. doka* or the result of introgression from *I. tomentosa* can hardly be decided on the limited evidence available.

4. **I. tomentosa** (*Harms*) *Craib & Stapf* in K.B. 1912: 93 (1912); T.T.C.L.: 104 (1949); Troupin in B.J.B.B. 20: 307 (1950); Hauman in F.C.B. 3: 381 (1952); J. Léon. in Mém. 8°, Classe Sci., Acad. Roy. Belg. 30(2): 175, 176 (1957); Brenan in K.B. 17: 224 (1963). Type: Tanganyika, Rungwe District, Umuamba, Lake Likaba, *Goetze* 1315 (B, holo. †)

Tree 3–12 m. high, with rounded or spreading crown. Leaves: stipules often with foliaceous ovate-cordate auricles up to ± 2·5–4 cm. long and 2–2·5 cm. wide; petiole with rhachis 15–27 cm. long; leaflets 3–4(–5) pairs, ovate-elliptic, elliptic, or less commonly oblong-elliptic, (7–)10–25 cm. long, (4–)4·8–12·7 cm. wide (larger—to ± 40 × 20 cm.—on coppice or juvenile shoots), ± densely pubescent or tomentose at least on midrib and lateral nerves beneath and sometimes all over; primary lateral nerves (9–)12–14 on each side of the midrib in mature leaflets. Panicles lax; ultimate racemose branches 5–25 cm. long, dark brown-tomentose. Bracts 4–8 mm. long. Pedicels of open flower 4–13 mm. long. Bracteoles brown, 11–20 mm. long, 10–12 mm. wide. Sepals 5–6. Petals white; upper one obovate or obovate-elliptic, rounded not bilobed at apex, 12–15 mm. long, 6–10 mm. wide; 4–5 smaller, 10–13 mm. long, 3·5–5 mm. wide. Stamens 10(–14). Pods large, 15–30 cm. long, 5·4–8·5 cm. wide, with dense rusty-brown tomentum partially rubbing off with maturity. Seeds 2·5–3·2 × 1·8–3 cm. Fig. 28/9, 10, p. 140.

TANGANYIKA. Buha District: Murungu–Malagarasi ferry, 25 Aug. 1950, *Bullock* 3210!; Mpanda District: Kabungu, 22 July 1948, *Semsei* in *F.H.* 2488!; Rungwe District: Masukulu, 4 Nov. 1912, *Stolz* 1641!

DISTR. T4, 7; Guinée Republic to the Cameroun Republic, also in the Congo Republic, Sudan Republic, Malawi and Zambia

HAB. Deciduous woodland; 610–1680 m.

SYN. *Berlinia tomentosa* Harms in E.J. 30: 321 (1901); L.T.A.: 689 (1930)
Westia tomentosa (Harms) Macbr. in Contrib. Gray Herb., n.s., 59: 21 (1919)

NOTE. So far, all specimens of *Isoberlinia* with foliaceous auricles to the stipules prove to be *I. tomentosa* but these appendages are by no means present on all specimens of *I. tomentosa*. Observers in the field are asked to test the validity and constancy of this distinction.

See note under *I. doka* (p. 142).

24. BERLINIA

Hook. f. & Benth. in Hook., Niger Flora: 326 (1849); Duvign. in Inst. Roy. Col. Belge, Bull. Séances 21(2): 432–3 (1950); Troupin in B.J.B.B. 20: 298 (1950), pro parte; Hauman in Inst. Roy. Col. Belge, Bull. Séances 23: 476 (1952); J. Léon. in Mém. 8°, Classe Sci., Acad. Roy. Belg. 30(2): 180 (1957)

Unarmed trees. Leaves paripinnate; stipules intrapetiolar, connate into a scale (at least on mature shoots); leaflets opposite or subopposite, in (1–)2–5 pairs, petiolulate, equal or ± unequal-sided at base, with or without translucent dots. Flowers in racemes or panicles of racemes; bracteoles 2, well-developed, valvate, completely enclosing the flower-buds, persistent. Hypanthium tubular, 7–20 mm. long. Sepals 5, subequal. Petals 5; either the upper one relatively very large and long-clawed with the four other petals much reduced, or (not in the Flora area) all 5 large and subequal. Stamens 10; filaments exserted, 9 of them shortly connate at base, the other one free. Ovary shortly stipitate, inserted on the tube of the hypanthium; ovules 2–8. Pod dehiscing elastically into 2 flattened woody valves which are obliquely transversely nerved when mature; upper suture narrowly winged on each side (see fig. 29/7). Seeds compressed, ± obovate-elliptic to elliptic or quadrate, with thin testa and short funicle.

About 15 species, mostly in the forest areas of central and western tropical Africa.

B. orientalis *Brenan* in K.B. 17: 211, fig. 2 (1963). Type: Mozambique mouth of Msalo [Msalu] R., *Allen* 35 (K, holo.!)

Tree 8–18 m. high, with glabrous branchlets. Leaflets in 2–4 pairs, ovate, elliptic or elliptic-lanceolate, 4·5–16·5 cm. long, 2·5–7 cm. wide, glabrous or nearly so, obtuse or obtusely subacuminate at apex. Racemes 3–11 cm. long, ± aggregated into terminal panicles. Pedicels 1·5–3 cm. long, tomentellous, as are the obovate 2–3 cm. long and 1–1·4 cm. wide bracteoles; the latter tomentellous inside also. Hypanthium 0·8–1·2 cm. long, glabrous or nearly so outside. Upper petal 3·5–4·5 cm. long, with a lamina 2–2·5 cm. long and 2·5–3·8 cm. wide which is white flushed with green in the central part; other petals 1–1·7 cm. long, with linear laminae 7–13 mm. long and 1–1·5 mm. wide. Pods large, pale brown. Fig. 29/1–6, p. 144.

TANGANYIKA. Masasi, Nyengedi, 23 Mar. 1943, *Gillman* 1231!; Newala District: Kitama, 12 Dec. 1942, *Gillman* 1067!; Mikindani District: about 7 km. inland from Mtwara, Nov. 1953, *Eggeling* 6741!

DISTR. T8 (but see note below); Mozambique

HAB. Woodland and thicket; 60–450 m.

SYN. [*B. auriculata* sensu T.T.C.L.: 89 (1949), *non* Benth.]

FIG. 29. *BERLINIA ORIENTALIS*—**1,** branchlet with leaf and inflorescence, × $\frac{2}{3}$; **2,** part of bracteole to show indumentum, × 1; **3,** flower showing bracteoles, calyx and ovary, other parts removed, × $\frac{2}{3}$; **4,** one of four smaller petals, × 2; **5,** stamen, × $\frac{2}{3}$; **6,** immature pod, × $\frac{2}{3}$. 1, 4, 5, from *Eggeling* 6741; 2, 3, from *Mendonça* 1005; 6, from *Gomes e Sousa* 4575.

NOTE. *Pitt* [*Pitt-Schenkel*] 948 !, from **Tanganyika, Bukoba District, 1933,** appears to be conspecific. It seems highly improbable that *B. orientalis* should occur as a native in Bukoba District, and unlikely, considering its native range, (though not impossible) that it should have been introduced there. Some mistake in labelling seems to have occurred, as the alleged collector, Mr. Pitt, is convinced that this specimen was not in fact collected by him.

25. JULBERNARDIA

Pellegr. in Boissiera 7: 297 (1943); Troupin in B.J.B.B. 20: 309 (1950), pro parte; Hauman in Bull. Inst. Roy. Col. Belge 23: 477 (1952); J. Léon. in Mém. 8°, Classe Sci., Acad. Roy. Belg. 30(2): 188 (1957)

Pseudoberlinia Duvign. in Bull. Inst. Roy. Col. Belge 21: 431 (1950); Hauman in F.C.B. 3: 402 (1952)

[*Isoberlinia* sensu auct. afr. mult., pro parte, *non* Craib & Stapf]

Unarmed, often evergreen trees. Leaves paripinnate, with leaflets in one to many pairs; stipules intrapetiolar, always connate below, bicuspidate above or with 2 relatively large foliaceous lobes simulating free stipules, ± persistent or very quickly falling off; lobes sometimes somewhat auriculate at base; petiolules usually twisted; leaflets opposite, ± markedly asymmetric at base, with venation prominent on both surfaces; translucent dots often present but sometimes absent (and sometimes inconstant in one species). Flowers in terminal usually much-branched panicles; bracteoles 2, well-developed, valvate, completely enclosing the flower-buds, persistent, keeled down back. Hypanthium absent. Sepals 5, well-developed, equal or nearly so in length, ciliate. Petals 5, equal or nearly so in length, all subequal and well-developed, 1 large and 4 small, or all small. Stamens 10, fertile; 9 filaments shortly connate below, the tenth free. Ovary densely pubescent, very shortly stipitate; stipe free; ovules few (up to 5); style elongate; stigma capitate, abruptly enlarged. Pods compressed, 1–5-seeded, elastically dehiscing into 2 woody valves; valves obliquely transversely nerved (nerves often obscure when pods are ripe); upper suture usually with a flange-like wing or ridge projecting laterally. Seeds compressed, without areoles, with a hard testa, borne on short funicles.

About 8 species, all in tropical Africa.

Leaflets in 1–3 pairs, with glabrous margins; stipules with ± persistent green foliaceous lobes; pedicels of open flowers 0·4–1·5 cm. long; petals very unequal in length, one much longer than the other four:
  Leaflets in 2–3 pairs . . . . 1. *J. magnistipulata*
  Leaflets in one pair only . . . . . . 2. *J. unijugata*
Leaflets normally in 3–6 pairs, usually with markedly pubescent margins (at least when young)—margins sometimes glabrous in 3, *J. paniculata*; stipules not foliaceous, very quickly falling; pedicels of open flowers 0·2–0·6 cm. long; petals equal in length or one only somewhat longer than the others:
  Pairs of leaflets (2–)3–4; sepals broadly obovate-spathulate, contiguous or slightly imbricate, 3·5–4·5 mm. wide; largest petal ± rotund or triangular, ± 6 mm. long; bracts much shorter than the bracteoles . . . . . . 3. *J. paniculata*
  Pairs of leaflets (2–)4–6(–8); sepals oblong, not or only slightly wider above than below, not

contiguous, 1·5 mm. wide; largest petal ovate, 6·5–9 mm. long; bracts usually half as long to more than as long as the bracteoles (occasionally only one-third as long) . . . . 4. *J. globiflora*

1. **J. magnistipulata** (*Harms*) *Troupin* in B.J.B.B. 20: 314 (1950); J. Léon. in Mém. 8°, Classe Sci., Acad. Roy. Belg. 30(2): 192 (1957); K.T.S.: 105 (1961). Type: Tanganyika, Lushoto District, Amani, *Zimmermann* 2003 (B, syn. †) & *Grote* 3437 (B, syn. †, EA, isosyn. !)

Evergreen tree, rarely a bush, (2·5–)10–25 m. high, with smooth grey or grey-brown bark; young branchlets glabrous. Leaves: stipule-lobes large, foliaceous, ± persistent, 0·4–4·7 cm. long, 0·25–3 cm. wide; petiole with rhachis 3·8–9 cm. long; leaflets in (1–)2–3 pairs, elliptic, oblong-elliptic or obovate-elliptic, or ovate-lanceolate, 6–11·3(–15) cm. long, 2·3–5(–7·8) cm. wide, obtuse, glabrous (including margins). Panicles up to ± 20 cm. long and wide, brown-tomentellous; pedicels of open flowers 4–7 mm. long. Flowers cream to white. Bracteoles 8–9 mm. long and wide, brown-tomentellous or -tomentose outside. Sepals imbricate, rotund to suborbicular, 3·5–5 mm. long, 3–4 mm. wide. Petals very unequal; larger one suborbicular, shortly clawed, 7 mm. long, 6 mm. wide; others ovate or oblanceolate, 2–5 mm. long, 1–1·5 mm. wide. Pods obovate-oblong or oblong, glabrescent to glabrous when mature, 5·5–14 cm. long, 2·5–3·5(–5) cm. wide. Seeds dark brown, 1·6–2·3 × 1·3–1·8 cm.

KENYA. Kwale District: about 3 km. on Kinango–Mariakani road, 29 Aug. 1953, *Drummond & Hemsley* 4039 !; Mombasa District: Port Tudor, *Sulemani* in *F.H.* 3236 !; Kilifi District: 42 km. NW. of Malindi, Marafa, 20 Nov. 1961, *Polhill & Paulo* 803 !

TANGANYIKA. Lushoto District: Sigi R., 4 Nov. 1936, *Greenway* 4710 ! & Lushoto–Mombo road, Sept. 1953, *Eggeling* 6698 !; Pangani District: Bushiri, 30 Sept. 1950, *Faulkner* 682 !

DISTR. **K7**; **T3**; not known elsewhere

HAB. Lowland rain-forest, riverine forest, coastal evergreen bushland and coastal *Brachystegia* woodland; 0–1150 m.

SYN. *Berlinia magnistipulata* Harms in N.B.G.B. 8: 148 (1922); L.T.A.: 691 (1930); T.S.K.: 63 (1936)
Isoberlinia magnistipulata (Harms) Milne-Redh. in K.B. 1937: 415 (1937); T.T.C.L.: 104 (1949)

NOTE. There is at Kew a Kenya specimen said to have been collected on the Mau (*G. S. Baker* 308 !). This is so far outside the otherwise known area of the species that, unless it is confirmed, an error in labelling must be strongly suspected.

The *nomen nudum Berlinia obliqua* A. Peter is mentioned in T.T.C.L.: 89. The specimen cited, *B. D. Burtt* 274 (EA !) is *Julbernardia magnistipulata*. *Berlinia obliqua* appears on numerous specimens collected by Peter. Some of them are *J. magnistipulata*, others are *Cynometra sp. A* (see p. 115).

The pods of *J. magnistipulata* are usually about 3–3·5 cm. wide. *Wigg* in *F.H.* 1029, from Tanganyika, Lushoto District, Mombo–Lushoto road, has them about 4–5 cm. wide. There is insufficient evidence to indicate if this variant is worth taxonomic recognition, and for the present I am regarding it merely as an extreme of *J. magnistipulata*.

2. **J. unijugata** *J. Léon.* in Mém. 8°, Classe Sci., Acad. Roy. Belg. 30(2): 193 (1957). Type: Tanganyika, Buha District, Mkuti R., *Procter* 365 (EA, holo. !, K, iso. !)

Evergreen tree 2–18 m. high, nearly always forked at 3–6 m., with smooth brown or light grey bark flaking off in large plates; young branchlets glabrous. Leaves: stipule-lobes foliaceous, ± persistent, semi-ovate, 0·5–1·2 cm. long, 0·2–0·5 cm. wide; leaflets in one pair only, ovate to oblong-lanceolate or lanceolate, 3·3–11 cm. long, 1·5–5·1 cm. wide, obtusely pointed or subacuminate at apex, glabrous (including margins). Panicles up to ±

20 cm. long and wide, shortly blackish-brown-tomentose; pedicels of open flowers 1–1·5 cm. long. Flowers white inside. Bracteoles 11–15 mm. long, 8–12 mm. wide. Sepals imbricate, elliptic to ovate-elliptic, 5·5–6·5 mm. long, 3·5–5 mm. wide. Petals very unequal; larger one ovate-elliptic, shortly clawed, 7–9 mm. long, 6–7 mm. wide; others very small, ovate-triangular, 1·5–4·5 mm. long, 0·5–1·5 mm. wide. Pods brown-tomentose, 8–15 (or more) cm. long, 3–4·5 cm. wide. Seeds dark brown, 1·9–2·2 × 1·4–1·7 cm.

TANGANYIKA. Kigoma District: Uvinza, 29 Aug. 1950, *Bullock* 3241! & Feb. 1956, *Procter* 422! & 29 Jan. 1957, *F. G. Smith* 1375! & 104 km. S. of Kigoma, Mugombazi, *Harley* 9493!

HAB. Riverine forest, ? lowland rain-forest; 1080–1580 m.

DISTR. **T4**; not known elsewhere

NOTE. Immediately distinguished from the other species of the genus in East Africa by the leaflets being always in one pair.

3. **J. paniculata** (*Benth.*) *Troupin* in B.J.B.B. 20: 316 (1950); Torre & Hillcoat in C.F.A. 2: 211 (1956); J. Léon. in Mém. 8°, Classe Sci., Acad. Roy. Belg. 30(2): 192, 195, 196 (1957); F.F.N.R.: 125 (1962). Types: Angola, Huila, *Welwitsch* 581 (LISU, lecto., BM, K, isolecto.!) & 582 (BM, syn.!)

Evergreen, flat-topped tree 2–20 m. high, with smooth whitish or rough dark grey bark; young branchlets pubescent to tomentellous, glabrescent. Leaves: stipules intrapetiolar, ± 5–7 mm. long, connate below, bicuspidate above, non-foliaceous, very caducous; petiole with rhachis (3–)8·5–22(–28) cm. long; leaflets in (2–)3–4 pairs, oblong-elliptic to ovate-oblong, sometimes somewhat obovate-oblong, (2–)6·5–13(–17) cm. long, (1·3–)2·5–6·3(–10) cm. wide, obtuse or rounded at apex, asymmetrically cuneate-attenuate (or upper margin rounded) at base, ± pubescent beneath, sometimes pubescent above, rarely glabrous on both surfaces; margins fringed with whitish pubescence at least when young, fringe rarely absent. Panicles to ± 15–25 × 15–30 cm., brown-tomentellous; bracts 2–3 mm. long, much shorter than the bracteoles; pedicels of open flowers 2–5 mm. long. Bracteoles 9–12 mm. long, 8–10 mm. wide. Sepals broadly obovate, contiguous or slightly imbricate, 4–5 mm. long, 3·5–4·5 mm. wide. Petals white; larger one ± rotund or triangular, shortly clawed, 6 mm. long, 5 mm. wide; others spathulate, 5·5–7 mm. long, with claw 3–4 mm. long gradually widened into the 2–3 mm. long and 1·5–2·5 mm. wide lamina. Pods obovate-oblong or oblong, 5–10·5 cm. long, 2–3·6 cm. wide, brown-tomentose. Seeds 1·4–1·7 × 1·3–1·6 cm. Fig. 30/12, p. 148.

TANGANYIKA. Mpanda District: Kabungu, May 1953, *F. G. Smith* 800!; Ufipa District: Sumbawanga, Aug. 1957, comm. *Shepherd* H53/57!; Mbeya District: Mbozi, 29 Aug. 1933, *Greenway* 3627!

DISTR. **T4, 7**; Congo Republic, Mozambique, Zambia, Malawi and Angola

HAB. Deciduous woodland, usually occurring with *Brachystegia floribunda*, common and locally dominant; 1150–1550 m.

SYN. *Berlinia paniculata* Benth. in Trans. Linn. Soc. 25: 311 (1865); L.T.A.: 687 (1930)
Westia paniculata (Benth.) Macbr. in Contrib. Gray Herb. 59: 21 (1919)
Isoberlinia paniculata (Benth.) Greenway in K.B. 1928: 203 (1928); T.T.C.L.: 104 (1949)
Pseudoberlinia paniculata (Benth.) Duvign. in Bull. Inst. Roy. Col. Belge 21: 434 (1950); Hauman in F.C.B. 3: 403 (1952)

NOTE. The panicles are characteristically borne upright above the leaves.

4. **J. globiflora** (*Benth.*) *Troupin* in B.J.B.B. 20: 314 (1950); J. Léon. in Mém. 8°, Classe Sci., Acad. Roy. Belg. 30(2): 192, 195 (1957); F.F.N.R.: 125 (1962). Type: Malawi, Shire Highlands, *Buchanan* 138 (K, holo.!)

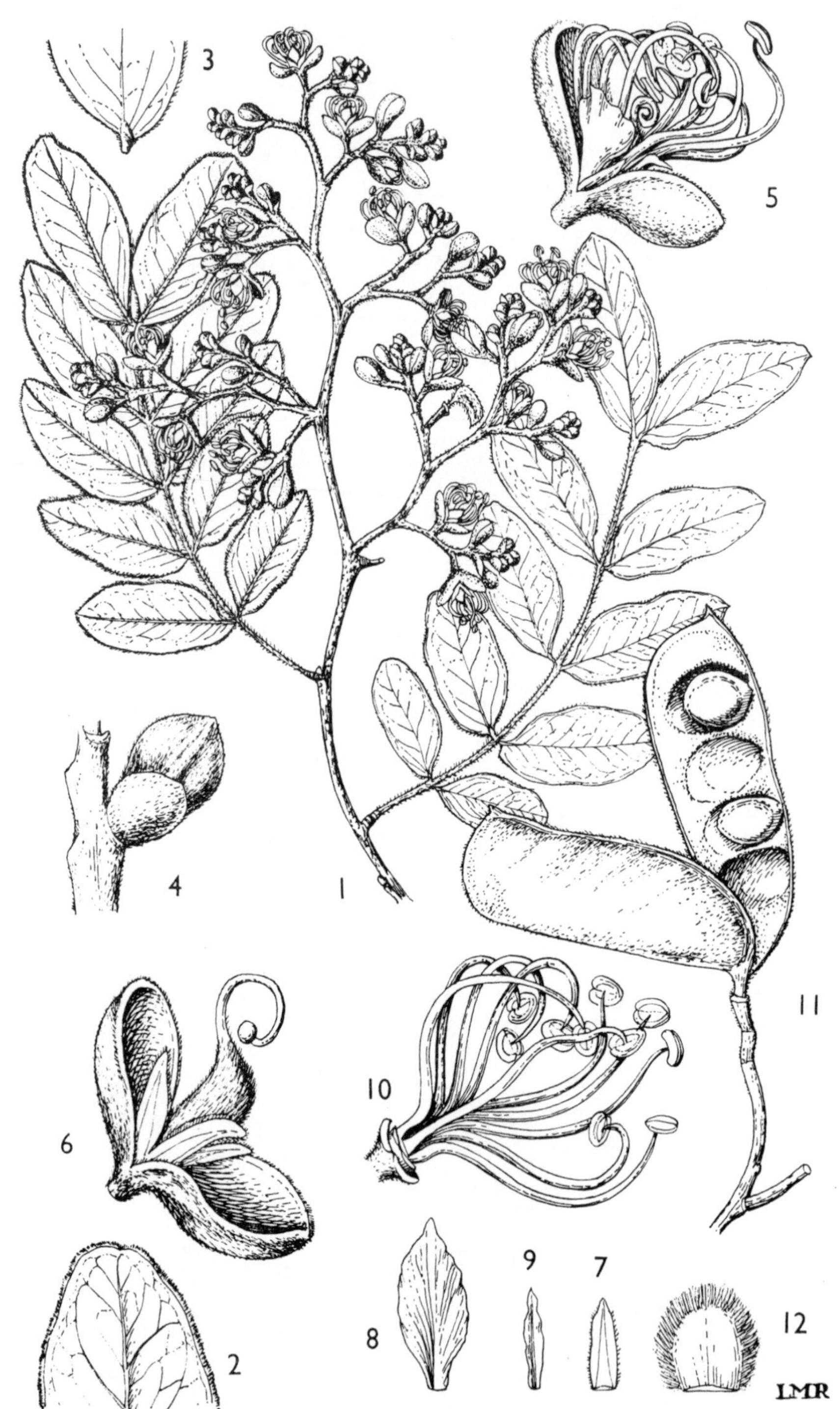

FIG. 30. *JULBERNARDIA GLOBIFLORA*—**1,** branchlet with leaves and inflorescence, × $\frac{2}{3}$; **2,** apical part of leaflet showing pubescent margin, × 4; **3,** leaflet-base, × 2; **4,** flower-bud with bract, × 4; **5,** flower, × 3; **6,** flower showing bracteoles, calyx and ovary (petals and stamens removed), × 3; **7,** sepal, × 3; **8,** larger petal, × 3; **9,** lateral petal, × 3; **10,** stamens, nine of them connate below, one free, × 3; **11,** pod dehisced, × $\frac{2}{3}$. *J. PANICULATA*—**12,** sepal, × 3. 1–10 from *Welch* 290; 11, from *Soil Cons. Dept.* O.F.C. 2; 12, from *Duff* 137/33.

Usually a tree 5–15 m. high, with flat or rounded spreading crown, but sometimes shrubby or flowering from coppice shoots 0·3–2 m. high; bark rough or smooth when young, grey; young branchlets puberulous to pubescent or tomentose, soon glabrescent. Leaves: stipules intrapetiolar, about 3–5 mm. long, connate below, bicuspidate above, non-foliaceous, very caducous; petiole with rhachis (3–)5–20 cm. long; leaflets in (2–)4–6(–8) pairs, narrowly oblong-elliptic or oblong-lanceolate, or sometimes narrowly obovate-oblong, (1–)2–8·5(–11·5) cm. long, (0·6–)1–3·3(–5·6) cm. wide, obtuse or rounded and sometimes slightly emarginate at apex; margins fringed with whitish pubescence. Panicles up to ±30 cm. long and wide, brown-tomentellous or shortly tomentose; bracts 2–10 mm. long, usually half as long to more than as long as the bracteoles (occasionally only one-third as long); pedicels of open flowers 2–6 mm. long. Bracteoles 7–10 mm. long, 6–9 mm. wide. Sepals oblong, not or only slightly wider above than below, non-contiguous, 2·5–4·5 mm. long, 1·5 mm. wide. Petals white; larger one ovate, shortly clawed, 6·5–9 mm. long, 4–5 mm. wide; others oblanceolate to oblanceolate-spathulate or linear, 3–8 mm. long, 0·5–3 mm. wide. Pods obovate-oblong or oblong, 4–9 cm. long, 2–3·2 cm. wide, brown-tomentose. Seeds dark brown, 1·3–1·6 × 1–1·5 cm. Fig. 30/1–11.

TANGANYIKA. Handeni District: Handeni–Bagamoyo road, Sept. 1950, *Semsei* 549!; Mpwapwa, *Hornby* 352!; Morogoro, near swimming-pool, 11 Mar. 1953, *Drummond & Hemsley* 1481!; 19 km. E. of Songea, 28 Mar. 1956, *Milne-Redhead & Taylor* 9364!

DISTR. **T**1, 3–8; Congo Republic, Mozambique, Zambia, Rhodesia, Malawi and Botswana

HAB. Deciduous woodland; ecologically a very important species in the Flora area, ranging throughout the *Brachystegia* areas in Tanganyika, probably the commonest species, and often dominant; usually growing with *Brachystegia spiciformis* and forming tsetse-fly (*Glossina morsitans*) habitat (T.T.C.L.: 104); 490–1830 m.

SYN. *Brachystegia globiflora* Benth. in Hook., Ic. Pl. 14: 43 (1881)
Berlinia eminii Taub. in P.O.A. C: 199 (1895). Types: Tanganyika, Tschaja (? Lake Chaya in Dodoma District)–Karagwe, *Stuhlmann* 498 & Tabora District, Igonda, *Boehm* 157a (both B, syn. †)
Berlinia globiflora (Benth.) Harms in V.E. 3(1): 472 (1915); L.T.A.: 689 (1930)
Westia eminii (Taub.) Macbr. in Contrib. Gray Herb. 59: 21 (1919)
Isoberlinia globiflora (Benth.) Greenway in K.B. 1928: 203 (1928); T.T.C.L.: 104 (1949)
Pseudoberlinia globiflora (Benth.) Duvign. in Bull. Inst. Roy. Col. Belge 21: 434 (1950); Hauman in F.C.B. 3: 405 (1952)

26. ENGLERODENDRON

Harms in E.J. 40: 27 (1907); J. Léon. in Mém. 8°, Classe Sci., Acad. Roy. Belg. 30(2): 199 (1957)

Unarmed evergreen trees. Leaves paripinnate; stipules apparently connate into a minute intrapetiolar scale; petiolules not twisted; leaflets opposite or subopposite, in 2–5 pairs, petiolulate, equal-sided at base, without translucent gland-dots. Flowers in fairly lax irregular terminal panicles; bracteoles 2, well-developed, valvate, completely enclosing the flower-buds, persistent. Receptacle cup-shaped. Sepals 6(–7, *fide* Harms), the posticous one rather broader than the rest and 2-toothed at apex. Petals 6–7, subequal. Stamens 12–14: 6–7 exserted, well-developed, fertile, alternating with 6–7 much smaller included ones with very small anthers*. Ovary inserted at base of hypanthium, shortly stipitate (stipe not projecting beyond receptacle), tomentellous; ovules 4–5(–6, *fide* Harms); style elongate, glabrous above; stigma small, capitate, scarcely wider than style. Pod short, dehiscing into 2 flattened woody valves marked with faint obliquely transverse nerves;

* The anthers on the included stamens in a single flower may be apparently all rudimentary and sterile, or all rather larger, opening and containing pollen. Harms (l.c.) states that the exserted stamens may be 8 and the included ones 5, but I cannot confirm this.

FIG. 31. *ENGLERODENDRON USAMBARENSE*—**1**, branchlet with leaves and inflorescence, × $\frac{2}{3}$; **2**, flower, × 2; **3**, calyx, × 2; **4**, sepal, × 4; **5**, petal, × 4; **6**, part of androecium, showing large and small stamens alternating, two of larger only partly shown, × 2; **7**, ovary, × 2; **8**, dehisced pod, × $\frac{2}{3}$; **9**, seed, × $\frac{2}{3}$. 1, from *Greenway* 4670; 2–7, from *Greenway* 1061; 8, 9, from *Greenway* 6112.

upper suture not winged. Seeds 1–2, large, on short funicles (fully mature ones not seen).

A single species, apparently confined to the Flora area.

E. usambarense *Harms* in E.J. 40: 28, fig. 2 (1907); L.T.A.: 692 (1930); T.T.C.L.: 102 (1949). Type: Tanganyika, Lushoto District, between Amani and Bomole, *Engler* 3436 (B, holo. † & iso. (in *Peter* 48011)!)

Shrub or tree up to 25 m. high; bark smooth, grey-brown to dark brown; branchlets brown-puberulous to shortly brown-tomentellous, soon glabrescent. Leaves: petiole with rhachis 4–12·5(–15) cm. long; leaflets elliptic to lanceolate, 2·5–10·8(–12·5) cm. long, 1–4·9(–6) cm. wide, narrowly acuminate, ± pubescent or puberulous on midrib beneath, more sparingly so on lateral nerves and surface. Flowers white. Pedicels ± 6–11 mm. long, dark-brown-tomentellous; bracteoles ± 7–12 mm. long and 7–8 mm. wide, dark-brown-tomentellous outside, more shortly so inside. Hypanthium 4–7 mm. long, patchily puberulous to subglabrous. Petals oblong-oblanceolate, 10–12(–15, *fide* Harms) mm. long, 3–5 mm. wide, pubescent on claw. Pods obliquely rotund to oblong-elliptic, 4–9·5(–12·5, *fide Semsei* 3014) cm. long, 3·7–5 cm. wide, densely brown-tomentellous or shortly tomentose. Seeds ± 3·7 × 3·1 cm., compressed. Fig. 31.

TANGANYIKA. Lushoto District: Amani, 21 Dec. 1928, *Greenway* 1061! & Sangarawe, 19 Jan. 1937, *Greenway* 4860! & Kwamkoro, 12 Nov. 1959, *Semsei* 2942!
DISTR. **T3**; not known elsewhere
HAB. Lowland rain-forest; 760–1000 m.

NOTE. There is a certain amount of variation in the indumentum. The two extremes are:
a. Pubescence on midrib and lateral nerves of the leaflets beneath inconspicuous, appressed; hairs mostly straight.
b. Pubescence on midrib and lateral nerves of the leaflets beneath conspicuous, coarser and spreading; hairs mostly crisped. The indumentum on the petiolules, rhachides, petioles and young branchlets is also somewhat denser and more spreading in b.
Extreme (a) is exemplified by *Greenway* 4670!, from Tanganyika, Lushoto District, Tongwe–Mlinga, (b) by *Semsei* 2833!, from Tanganyika, Lushoto District, Amani West Forest Reserve. I do not see any satisfactory reason for giving formal recognition to these two extremes, since there do not appear to be other characters correlated, and intermediates occur, of which *Greenway* 1061 and 4860, cited above, are examples. The isotype, *Engler* 3436, is similar to *Greenway* 1061.

27. TAMARINDUS

L., Sp. Pl.: 34 (1753) & Gen. Pl., ed. 5: 20 (1754); J. Léon. in Mém. 8°, Classe Sci., Acad. Roy. Belg. 30(2): 200 (1957)

Unarmed evergreen tree. Leaves paripinnate; stipules free, ± asymmetrically lanceolate, very quickly falling off (usually to be seen only with the very youngest leaves); leaflets opposite, in rather numerous pairs, almost sessile, asymmetric at base; translucent gland-dots absent. Flowers in lax terminal and lateral racemes; bracteoles 2, well-developed, valvate, completely enclosing the young flower-buds but quickly falling off before the buds are full-sized. Hypanthium shortly elongate-turbinate. Sepals 4, imbricate. Petals: upper 3 well-developed; lower 2 minute, setiform, below the staminal band. Stamen-filaments connate to about half-way into a pubescent band terminating in 3 upcurved anther-bearing filaments alternating with 5 sterile teeth (1–2 of these rarely elongated into short filaments). Ovary ± pubescent, long-stipitate, with the stipe adnate to one side of the hypanthium; ovules 8–14; style elongate, gradually enlarged into the capitate stigma. Pods indehiscent, with a dry outer shell and pulpy inner layer. Seeds ± compressed, with a continuous-margined areole on each face.

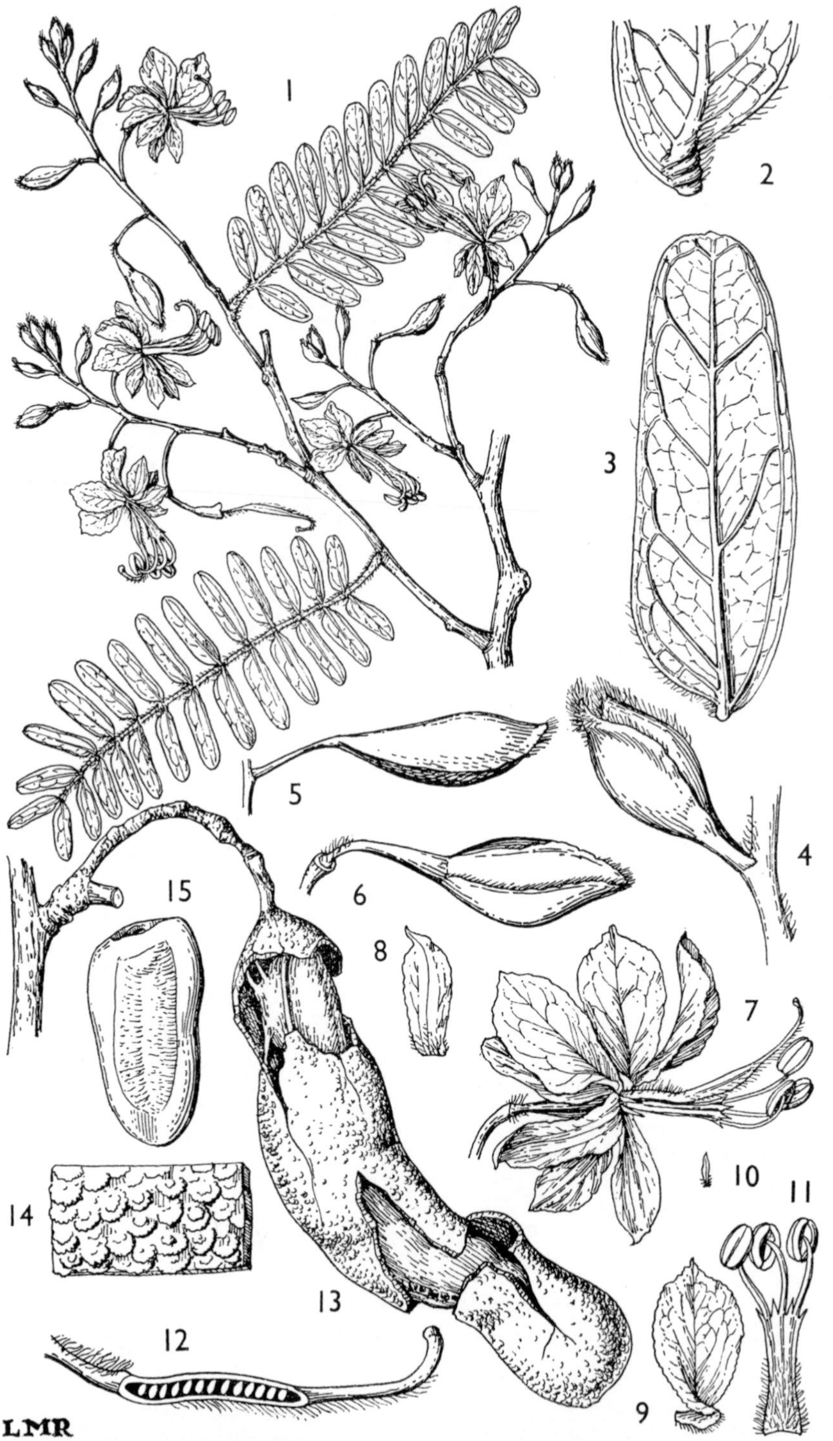

LMR

FIG. 32. *TAMARINDUS INDICA*—**1,** branchlet with racemes and leaves, × $\frac{2}{3}$; **2,** leaflet base, lower surface, × 6; **3,** leaflet, upper surface, showing venation, × 4; **4,** young flower-bud protected by bract, × 4; **5,** young flower-bud protected by bracteoles after fall of bract, × 4; **6,** older flower-bud showing four imbricate sepals, after fall of bract and bracteoles, × 2; **7,** flower, × 2; **8,** sepal, × 2; **9,** one of three upper large petals, × 2; **10,** one of two lower minute petals, × 2; **11,** stamens showing filaments fused below into a band, × 2; **12,** ovary, cut longitudinally, × 4; **13,** mature pod, breaking up, × $\frac{2}{3}$; **14,** part of surface of pod, × 4; **15,** seed showing areole, × 2. 1–12, from *Semsei* in *F.H.* 2867; 13–15 from *Hughes* 5.

A single species in the Old World tropics, but so widely planted that its native range is hard to determine—it is doubtless indigenous in Africa. Allegedly distinct species from north-eastern tropical Africa seem best included within the range of variation of *T. indica.*

T. indica *L.*, Sp. Pl.: 34 (1753); L.T.A.: 702 (1930); T.S.K.: 63 (1936); T.T.C.L.: 106 (1949); U.O.P.Z.: 461, photo, fig. (1949); J. Léon. in F.C.B. 3: 436 (1952); I.T.U., ed. 2: 69, fig. 17 (1952); Torre & Hillcoat in C.F.A. 2: 217 (1956); Roti-Michelozzi in Webbia 13: 134–141, fig. 1 (1957); K.T.S.: 109, fig. 21 (1961); F.F.N.R.: 128 (1962). Type: uncertain

Tree 3–24 m. high; bark rough, grey or grey-black; crown rounded; young branchlets pubescent or puberulous. Leaves: petiole with rhachis 5–12(–16) cm. long, ± pubescent; leaflets in 10–18(–21, *fide* F.C.B.) pairs, narrowly oblong, (0·8–)1·2–3·2 cm. long, 0·3–1·1 cm. wide, rounded to rounded-subtruncate rarely slightly emarginate at apex, rounded and asymmetric at base, glabrous except for a tuft of yellowish hairs at base, sometimes pubescent up midrib and margins, rarely all over both surfaces*; venation ± reticulate-raised on both surfaces. Racemes 1–15(–22) cm. long; axis subglabrous to densely pubescent; pedicels 3–14 mm. long, glabrous to pubescent. Flower-buds red. Hypanthium 3–5 mm. long. Sepals 8–12 mm. long, pale yellow inside, reddish outside. Large petals 10–13 mm. long, elliptic or obovate-elliptic, gold with red veins. Pods curved or sometimes straight, sausage-like, (3–)6·5–14 cm. long, 2–3 cm. in diameter, usually obtuse at base and apex, sometimes irregularly constricted, closely covered outside with brown scurf, 1–10-seeded. Seeds chestnut-brown, ± rhombic to trapeziform, 11–17 mm. long, 10–12 mm. wide. Fig. 32.

UGANDA. Karamoja District: Mt. Moroto, Feb. 1959, *Wilson* 664!; Bunyoro District: Bulesa [? Bulisa], Jan. 1941, *Purseglove* 1105!; Busoga District: Butaleja Camp, July 1926, *Maitland* 1154!

KENYA. Northern Frontier Province: Dandu, 13 June 1952, *Gillett* 13435!; Masai District: NW. of Lake Magadi, 28 Dec. 1958, *Greenway* 9549!; Kilifi District: Malindi, 10 Aug. 1949, *Bogdan* 2556!

TANGANYIKA. Musoma, 1933, *Emson* 365!; Morogoro District: Morogoro–Turiani road, Nov. 1949, *Semsei* in *F.H.* 2867!; Lindi District: Nachingwea, 4 Jan. 1953, *Anderson* 834!

ZANZIBAR. Zanzibar I., Mbweni, 4 Feb. 1929, *Greenway* 1327! & Chwaka, 21 Dec. 1930, *Vaughan* 1736!

DISTR. **U**1–3; **K**1, 2, 4–7; **T**1–8; **Z**; widespread in the tropics of the Old World

HAB. Woodland, wooded grassland, deciduous bushland; near sea-level to 1520 m.

NOTE. This is the well-known Tamarind, the acid pulpy part of whose pods is edible and used for preserves, jams, sweets, etc., and also yields a refreshing drink; the seeds are also edible.

28. ANTHONOTHA

P. Beauv., Fl. Oware & Benin 1: 70, t. 42 (1806); J. Léon. in Mém. 8°, Classe Sci., Acad. Roy. Belg. 30(2): 215 (1957)

[*Macrolobium* sensu auct. afr., pro parte, e.g. J. Léon. in F.C.B. 3: 409 (1952), pro parte, excl. *M. coeruleum, non* Schreb.]

Unarmed evergreen trees. Leaves paripinnate; stipules free, or connate into an intrapetiolar scale; petiolules not twisted; leaflets opposite, in (1–)2–7 pairs, petiolulate, equal-sided or nearly so at base, without translucent gland-dots. Flowers in lateral or terminal panicles, or sometimes racemes; bracteoles 2, well-developed, valvate, completely enclosing the flower-buds, persistent. Hypanthium ± shortly tubular or cupuliform. Sepals 4–5, equal, or with one larger than the other. Petals 2–6, equal, or with 1–3 larger than the rest. Stamens (6–)9(–10), free; usually 3(–5) of them

* Especially to the south of the Flora area.

large and 1–7 ± reduced small and staminodial. Ovary with 2–5 ovules. Pods dehiscing into 2 flattened woody valves with ± transverse nerves. Seeds large, ± compressed, non-areolate, with thin testa and short funicle.

A genus of 27 species, all tropical African, mainly in the forest regions of the centre and west.

A. noldeae (*Rossberg*) *Exell & Hillcoat* in Bol. Soc. Brot., sér. 2, 29: 39 (1955) & in C.F.A. 2: 201 (1956). Type: Angola, Malange, Quela, *Nolde* 213 (B, holo. †, BM, iso.!)

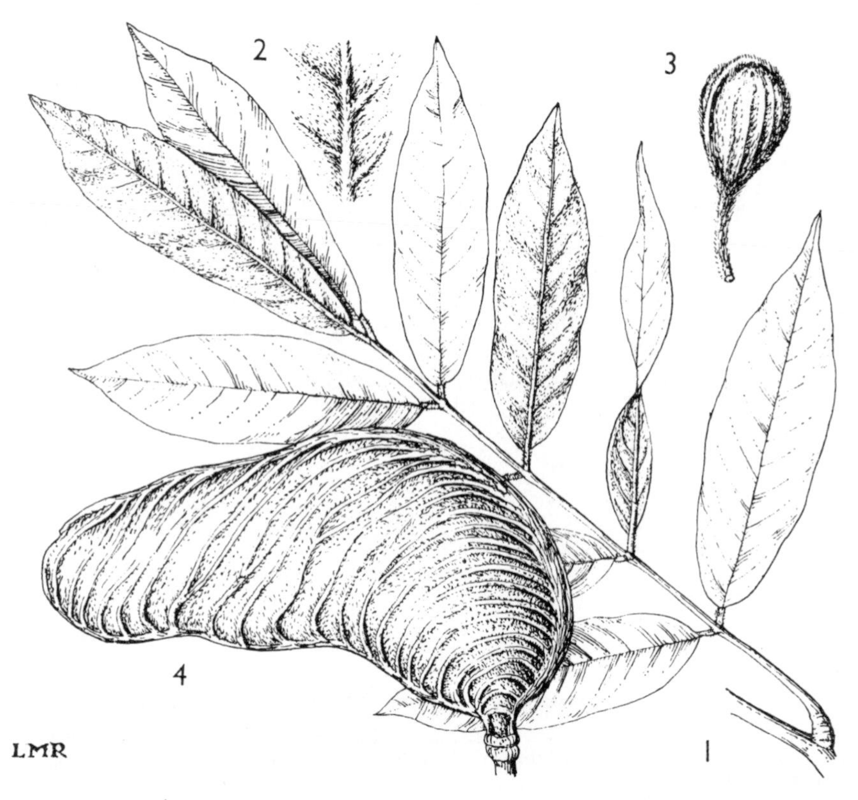

FIG. 33. *ANTHONOTHA NOLDEAE*—**1**, leaf, × $\frac{2}{3}$; **2**, part of under-surface of leaflet, × 3; **3**, bud, × 3; **4**, pod, × $\frac{2}{3}$. 1–3, from *Harley* 9219; **4**, from *Jefford, Juniper & Newbould* 2336.

Tree up to 15–25 m. high, not buttressed, with smooth yellowish-grey bark scaling off in patches; crown dense, round; young branchlets densely brown-pubescent; older twigs glabrous. Leaves: petiole with rhachis 7·5–15(–20) cm. long, brown-pubescent; leaflets in 4–6 pairs, oblong-lanceolate to narrowly oblong-elliptic, 5–12(–17·5) cm. long, 1·2–3(–6·9) cm. wide, acuminate, rounded to cuneate-rounded at base, glabrous above, beneath with a very dense minute appressed indumentum concealing the surface and with a silky copper-coloured sheen; lateral nerves 12–18 on each side of the midrib. Hypanthium glabrous. Petals 5, one of them much larger than the rest and deeply bilobed; 4 very reduced. Stamens (4–)5 large, well-developed; 4 minute, staminodial. Pods elliptic to irregularly oblong, 5–9·5 cm. long, 4–5 cm. wide, densely dark brown-tomentellous; obliquely transverse nerves numerous and raised. Fig. 33.

Tanganyika. Kigoma District: Mahali Mts., Kabesi valley, 31 Aug. 1958, *Newbould & Jefford* 1972! & Ujamba, 5 Sept. 1958, *Newbould & Jefford* 2336! & between Pasagulu and Musenabantu, 8 Aug. 1959, *Harley* 9219!

Distr. **T4**; Congo Republic (Kivu Province, Fizi) and Angola

Hab. Riverine or upland rain-forest; 1520–2130 m.

Syn. *Macrolobium noldeae* Rossberg in F.R. 39: 156 (1936); J. Léon. in F.C.B. 3: 419 (1952)

Note. The remarkable copper-coloured sheen all over the lower surface of the leaflets makes this species outstanding among the East African *Caesalpinioïdeae*. This feature occurs, however, in various other species of *Anthonotha*, among which *A. noldeae* is remarkable in having 4–5 large stamens.

29. PARAMACROLOBIUM

J. Léon. in B.J.B.B. 24: 348 (1954) & in Mém. 8°, Classe Sci., Acad. Roy. Belg. 30(2): 230 (1957)

Unarmed evergreen trees. Leaves paripinnate; stipules connate into a ± elongate persistent intrapetiolar scale which is biapiculate to rounded at apex; petiolules twisted; leaflets opposite, in 2–5 pairs, petiolulate, unequal-sided at base, without translucent gland-dots. Flowers in compact terminal corymbose panicles; bracteoles 2, well-developed, valvate, completely enclosing the flower-buds, persistent. Hypanthium shortly cup-shaped. Sepals 4, unequal, the posticous one larger than the others and 2-toothed at apex. Petals 5, bluish-mauve; the upper one relatively very large, 2 lateral ones much smaller, the 2 lower ones minute. Stamens 9, their filaments unequally connate at base; normally 3 anticous ones large, well-developed, fertile, and 6 posticous ones reduced, staminodial. Ovary long-stipitate, tomentellous; ovules ± 6–8; style elongate, glabrous above; stigma abruptly enlarged, reniform-peltate. Pod dehiscing (initially at least along upper suture only) into 2 flattened woody valves without obvious transverse nerves; upper suture with a narrow wing-like ridge along each side. Seeds large, hard, ± compressed, on short funicles; each side marked with a large areole whose continuous margin is a small but abrupt change in level of the surface of the seed (the areole being slightly sunken in relation to the adjacent surface).

A tropical African genus with a single species.

P. coeruleum (*Taub.*) *J. Léon.* in B.J.B.B. 24: 348 (1954) & in Mém. 8°, Classe Sci., Acad. Roy. Belg. 30(2): 230 (1957); Exell & Hillcoat in C.F.A. 2: 200 (1956); K.T.S.: 105 (1961). Types: Tanganyika, Bagamoyo District, Viansi, *Stuhlmann* 6088 & Dar es Salaam, *Stuhlmann* 7575 (both B, syn. †)

Tree 5–40 m. high (*fide* F.C.B.), usually about 5–15 m. in East Africa; bark grey-brown, fairly smooth, with very fine longitudinal striations; branchlets glabrous, or sometimes (especially in Kenya) ± pubescent when young. Leaves glabrous, or rarely with some pubescence especially on rhachis, petiolules and near leaflet-bases; leaflets ± falcate and lanceolate, oblong-lanceolate, or sometimes ovate, (2·5–)4–10(–15, *fide* F.C.B.) cm. long, (1·4–)1·8–4·3(–6, *fide* F.C.B.) cm. wide, tapering-acuminate at apex, rounded to broadly cuneate at base; network of veins very close and rather prominent on both surfaces. Inflorescence glabrous except for bracts; pedicels 1·5–3 cm. long; bracteoles thick and hard, elliptic or oblong-elliptic, 2–3–3·2(–3·7, *fide* F.C.B.) cm. long, 1–1·8 cm. wide, glabrous outside, densely tomentellous within. Hypanthium glabrous. Largest petal obovate, 3–3·5(–4·7, *fide* F.C.B.) cm. long, 1–2·3 cm. wide. Pods 8·5–18(–20, *fide* F.C.B.) cm. long, 2·5–5(–6, *fide* F.C.B.) cm. wide. Seeds dark brown, ± quadrate, 1·3–2·5 cm. long, 1–2 cm. wide. Fig. 34, p. 156.

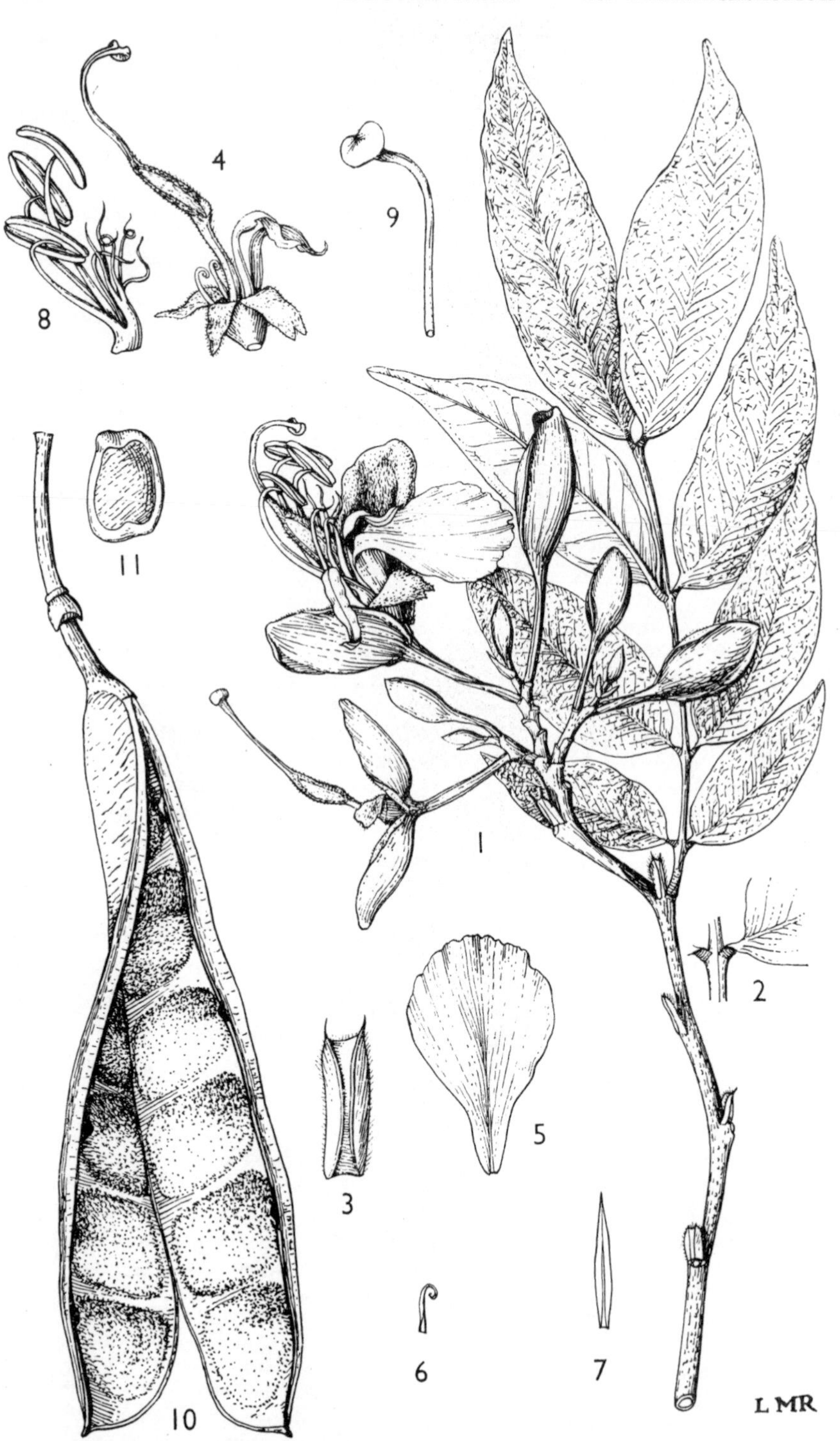

FIG. 34. ***PARAMACROLOBIUM COERULEUM***—**1,** branchlet with leaf and inflorescence, × $\frac{2}{3}$; **2,** petiolule, × 1; **3,** fused stipules, × 2; **4,** flower with bracteoles, one large petal and stamens removed, × 1; **5,** large petal, × 1; **6,** one small lower petal, × 1; **7,** one lateral petal, × 1; **8,** stamens, × 1; **9,** style and stigma, × 2; **10,** dehisced pod ,× 1; **11,** seed, × 1. 1–10, from *Greenway* 9804; 11, from *Deighton* 5539.

Kenya. Kwale District: Shimba Hills, Feb. 1937, *Dale* 3628! & about 32 km. SW. of Jardini towards Mrima Hill, Maseseni [? Mafisini], 28 Jan. 1961, *Greenway* 9804!
Tanganyika. Uzaramo District: Pugu Forest Reserve, June 1954, *Semsei* 1764! & Vikindu Forest Reserve, 29 Jan. 1954, *Omari* 14!
Distr. **K7**; **T6**; Guinée Republic, Sierra Leone, Cameroun Republic, Congo Republic and Central African Republic
Hab. ? Lowland rain-forest; 8–360 m.

Syn. *Vouapa coerulea* Taub. in E.J. 19, Beibl. 47: 31 (1894) & in P.O.A. C: 199, t. 23 (1895)
Macrolobium coeruleoïdes De Wild. in Ann. Mus. Congo, Bot., sér. 5, 2: 137 (1907); L.T.A.: 679 (1930). Types: Congo Republic, District Forestier Central, Lukolela, *Pynaert* 179 & Ibali, *E. & M. Laurent* & Eala, *M. Laurent* 734 & Bas-Congo, Kisantu, *Gillet* 3784 (all BR, syn.)
M. coeruleum (Taub.) Harms in V.E. 3(1): 475 (1915); L.T.A.: 679 (1930); T.S.K.: 62 (1936); T.T.C.L.: 105 (1949); J. Léon. in F.C.B. 3: 411 (1952)

Note. Although the distribution of this species is notably discontinuous, it does not seem correlated with any important geographical variation.

30. BRACHYSTEGIA*

Benth. in G.P. 1: 582 (1865) & Trans. Linn. Soc. 25: 311 (1866); Hoyle in F.C.B. 3: 446 (1952); Hoyle & White in F.F.N.R.: 101 (1962)

Trees (but see note below). Leaves paripinnate, stipulate (see note below); leaflets sessile, very diverse in number, size and shape, normally opposite, in 2–72 pairs, usually furnished with highly variable, often obscure translucent dots. Racemes simple or paniculate, usually terminal, rarely lateral on older branchlets. Flowers ± zygomorphic, completely enclosed in bud by 2 opposite valvate bracteoles which persist during flowering. Tepals 0 or 1–10(–11), much shorter than the bracteoles, free or with 2–3 partly united, imbricate, valvate or open in aestivation; either all sepaloid and grading in size and shape from broad to narrow, or variously differentiated into two whorls, or (in 1, *B. spiciformis*) minute to rudimentary or 0; outer whorl usually 4–6, relatively broad, subequal to very unequal, usually ciliate; inner whorl, when distinguishable, 1–3(–5), narrow, often non-ciliate. Stamens usually 10, all fertile (abnormally 9 or 11), or (in 8, *B. stipulata* only) 13–18 all fertile (or sometimes, with staminodes, totalling ± 20), alternately long and short, free or shortly united, often obscurely diadelphous; filaments or tube continuous externally with the margin of the very short cupular or turbinate hypanthium; mouth of hypanthium with or without a disc formed of obvious or obscure internal glandular swellings. Ovary oblong or naviculiform, stipitate; stipe shorter than or subequal to the ovary, inserted centrally or subcentrally in, and usually closely invested at the extreme base by (but free from) the shorter hypanthium; style long; stigma small, subcapitate; ovules (4–)5–10. Pod flat, woody, soon glabrous, oblong or naviculiform, beaked at apex, dehiscing elastically, the valves becoming spirally twisted; adaxial suture with a flange-like wing on each side, sometimes also a longitudinal nerve near this suture on each valve. Seeds compressed, without areoles, with a hard testa, subsessile.

The genus is confined to tropical Africa except for the type (1, *B. spiciformis*), which ranges to 25° S. in Mozambique. Of about 30 species or major complexes, 15 are recorded from the Flora area, all confined there to Tanganyika except *B. spiciformis* which extends to Kenya. Previous Uganda records were based on a misinterpreted locality.

Except for two or three natural suffrutices and one tree sometimes behaving as a thicket-forming shrub (neither form in the Flora area), all species are normally trees, though often dwarfed or even prostrate under exposure or starvation. Coppice-shoots and root-suckers are, however, very readily produced and sometimes bear flowers and pods near the ground; such growths have been confused with true suffrutices. Moreover, juvenile features shown by coppice- and sucker-shoots and even by trees originating from

* By A. C. Hoyle, Commonwealth Forestry Institute, Oxford.

them can closely simulate the vegetative characters of other species; thus leaflets are often fewer (though often not !) and can develop up to three times normal size, (with or without changes in the position of the midrib) and usually have more acute or acuminate apices but sometimes, by contrast, become more obovate and obtuse though normally ovate. The behaviour of such material is so unpredictable that it cannot be included in keys and descriptions and its identification without comparison is often impossible.

Stipules are usually valuable in diagnosing species but occasionally fail to develop basal auricles when these are normally present, or scarcely meet when normally (Group B) shortly connate, intrapetiolar. They tend to persist longer and are often broader and rather shorter on sterile shoots (especially coppice and suckers) and occasionally persist on vigorous fertile shoots (especially of trees originating from coppice) even when normally caducous.

" Stipels " vary from conspicuous short local expansions of the edges of the channel of the leaf-rhachis, with or without free apices directed forwards or laterally, to narrow wings extending throughout each interval between pairs of leaflets; they are very variable in putative hybrids and in some species with many leaflets and can fail to develop on very vigorous or juvenile material, but are often diagnostic (see figures).

The flush of young foliage in *Brachystegia* is striking (though not unique) in its elegant and often brilliant colours, salmon pink to dark crimson or wine-red, usually brighter in glabrous species and forms because not masked by hairs. The change through maroon, bronze or yellowish shades to pale green and finally the mature leaf-colour usually takes place rapidly in all the leaves simultaneously, before or during flowering, but especially in 6, *B. floribunda* and 7, *B. manga*, some leaves may change from pinkish or crimson through maroon to glaucous more rapidly than others, thus producing a mixture of shades. These transitory colours are more or less characteristic of a species at any one time and place and, after careful observation, can be used in assessing the extent of populations as seen from vantage points. The recorded flushing-colours are here omitted from descriptions unless they are believed to be diagnostic. The mature leaf-colour, present for most of the year, is given when known and reasonably constant, but may often be affected by local nutrient and climatic conditions and may also change, perhaps temporarily, in mid-season (e.g. from glaucous to green in 6, *B. floribunda*) or finally from green or glaucous to yellow in age. Leaf-colours and other useful notes are given in the descriptions of *Brachystegia* species in Appendix I of B. D. Burtt's Field Key, Part II (rev. & ed. Glover & Jackson, 1953), in which the nomenclature is similar to that used here except for *B. longifolia* (q.v.).

Mature bark is often specific but its colour, roughness and behaviour may be profoundly affected by fire. Natural shedding of thin or thick flakes with bevelled smooth edges seems normal in Group A (see below) and sometimes occurs in fairly young trees (notably in 2, *B. bussei* and 5, *B. microphylla*). Long-persistent relatively thick bark, usually with a predominance of vertical furrows, characterizes most species in Group B and, when flakes eventually fall from these, the edges are irregularly broken. These characters need checking on burnt and unburnt trees of all ages, as information is sadly lacking and specimens far too few.

Characters given for pods, especially colour, are those of mature pods unless otherwise stated; young and half-grown pods are seldom diagnostic but in some species develop a characteristic bloom, sometimes lost at maturity or in storage. Poisoning in herbaria often lightens and reddens colours considerably. " Pods up to . . ." means the rarely attained maximum normal size known, with exceptional maxima sometimes added in brackets. Abortion of some or all the seeds (usually due to insect-attack) is so prevalent that much smaller sizes are usual and abnormal shapes are common. Prominent, sometimes rough warts on the valves, usually due to pests, are not to be confused with the finely scurfy surface that develops in 9, *B. allenii*, 10, *B. angustipulata* and 13, *B. boehmii* and in patches in forms of 12, *B.* × *longifolia* (see fig. 39/11). All pods are beaked at apex but the beak is so variable and so easily broken off that it is omitted from specific descriptions. " Sutural wings " are the flange-like wings, always ± prominent on each side of the adaxial (upper) suture of the pod; they are fairly stout and stiffly spreading at right angles in most species or can be suberect, notably in immature pods (or in any position when pressed), but 2, *B. bussei*, 3, *B. utilis* and 4, *B. puberula* have them exceptionally thin and normally revolute.

The basal nerves departing fanwise from the pulvinulus on the proximal side of the midrib of leaflets are described simply as " basal, fanwise nerves ". Their number is sometimes diagnostic.

Many species are very variable in indumentum of nearly all parts and hairs usually persist, but some species are remarkably constant in the distribution and type of hairs or in their absence from some or all external parts. Vegetative bud-scales are usually hairy or at least ciliate, most bracts and tepals are ciliate, and (in East Africa) all ovaries are more or less densely crispate-setose with long, rufous or ferruginous to dark brown hairs, rapidly lost from the young pod; description of a species as glabrous ignores these parts.

All species in East Africa lose their leaves as new growth begins or earlier in drought, except that 15, *B. taxifolia* sometimes retains some old leaves in the second year in moister areas.

Published subdivisions of the genus are confused and cannot be fully unravelled here. Harms in E. & P. Pf., Nachtr. 3: 152 (1908) proposed a section *Neobrachystegia*, based on *B. stipulata* with " 20 " stamens (see Group B1, below); later, in V.E. 3(1): 475–81 (1915), he mentioned this section again and went on to group most of the other recorded species into two " not sharply distinct " sections, 1. *Paucijugae* with 2–8 pairs and 2. *Multijugae* with 9–10 or more pairs of leaflets, bringing together several unrelated species in each and again mentioning *B. stipulata* as exceptional with 7 pairs of leaflets, etc. The East African species cited by Harms appear in the list in T.T.C.L.: 90 (1949) and are used or cited as synonyms in the present account, but detailed references are omitted.

Burtt Davy & Hutchinson, in their comprehensive revision in K.B. 1923: 129–63 (1923), fully cited below, seemed unaware of Harms' work. They distinguished two sections based on the form and relative persistence of stipules (somewhat confused with auricles in their sense); 1. *Stipulatae* included *B. stipulata* with its variants and is therefore antedated by Harms' *Neobrachystegia*; 2. *Caducae* was subdivided on the branched or unbranched inflorescence, bringing nearly all the forms of *B. spiciformis* together in the latter group which, together with most of Harms' *Paucijugae*, should be called sect. *Brachystegia*. The remainder of the *Caducae* comprised those among Harms' *Multijugae* not alienated under *Stipulatae*. The tangled remnants need re-grouping and the present occasion is inappropriate.

The key in L.T.A.: 711–733 (1930) and much of the tentative synonymy, were unfortunate; citation is confined to East African records and desirable corrections.

The number, size and shape of leaflets, used alone, are only reliable as key-characters in extreme cases, as they show continuous variation in the genus as a whole. They are misleading as primary group-characters because they bring together unrelated species and some unfortunate misidentifications have thus occurred. Using the persistence and form of stipules alone produces similar results and also tends to separate mature material from coppice-shoots of the same species. Most other available characters arise from different degrees of reduction (e.g. of tepals) or union of parts (e.g. of stamens) and are therefore almost continuously variable; more definite " spot " characters are usually available only for short periods or are shown only by single species or pairs of species. Moreover, numerous pests quite naturally attack so widely dominant a genus and enormous damage is done by lopping of trees, removal of bark, burning and cultivation, so that normal complete material is often difficult to obtain, notably in sporadic collecting by the roadside where repeated damage is usual. These factors, with the added complication of apparently widespread hybridization, ensure that any character may be confidently expected to fail in a minority of specimens.

In the key to the Groups (p. 165), the species represented in the Flora area are first referred to two main groups A and B, each showing wide diversity in leaves but more constant in the flowers, in the form and behaviour of stipules and in the general type and behaviour of bark. The primary characters, based on the shape of the dormant bud and its relation to the free or connate base of the stipules (although the latter is sometimes obscure), seem to be of basic significance; when properly understood by reference to figs. 35/7, 8 and 38–40 and used in conjunction with other stipule-characters, they will be found remarkably reliable; they also have the merit of being nearly always present even on old, sterile material, which is most often available and also presents most difficulty. The dormant buds and stipule-bases are very small (and sometimes unreliable) in spp. 5, *B. microphylla* and 15, *B. taxifolia*; these species are, however, easily recognized by their leaves (see key to the species, p. 161).

Putative hybrids. There is no experimental proof that species of *Brachystegia* hybridize but much field evidence leaves little doubt that many do and the resulting introgression of characters is thought to be an important cause of variation. Hybrids seem to occur mainly within Group A or B; occasional examples suggest crossing between these groups but this seems unimportant. Circumstantial evidence of crossing is most frequent and convincing among variable species, while those that rarely seem to hybridize show remarkably little variation. At the same time it is noticeable that almost every species represented in East Africa shows essentially similar individuals, or (where known) populations, in two or more isolated areas often separated by vast distances, but varies differently in intermediate areas where it is able to cross with other species. Populations seem remarkably constant in the absence of opportunities for crossing or the modifying effects of extreme environments. In these circumstances the usual hierarchy of subspecific categories seems inappropriate, as tending to create an impression of stability (even in diversity) which is as likely to prove false as the hierarchy would need in some species to be complex and unwieldy. Varieties established in F.C.B. (1952) have already proved as misleading as the " species " on which they were based, because further evidence or study has shown them to be either mere forms repeated in many heterogeneous populations or almost certainly local hybrids.

There is a great need for more collections of developmental and seasonal stages from individual trees of all species and for detailed field records. Above all, comparative studies are needed of the variation of populations in habit and bark as well as characters visible on herbarium specimens.

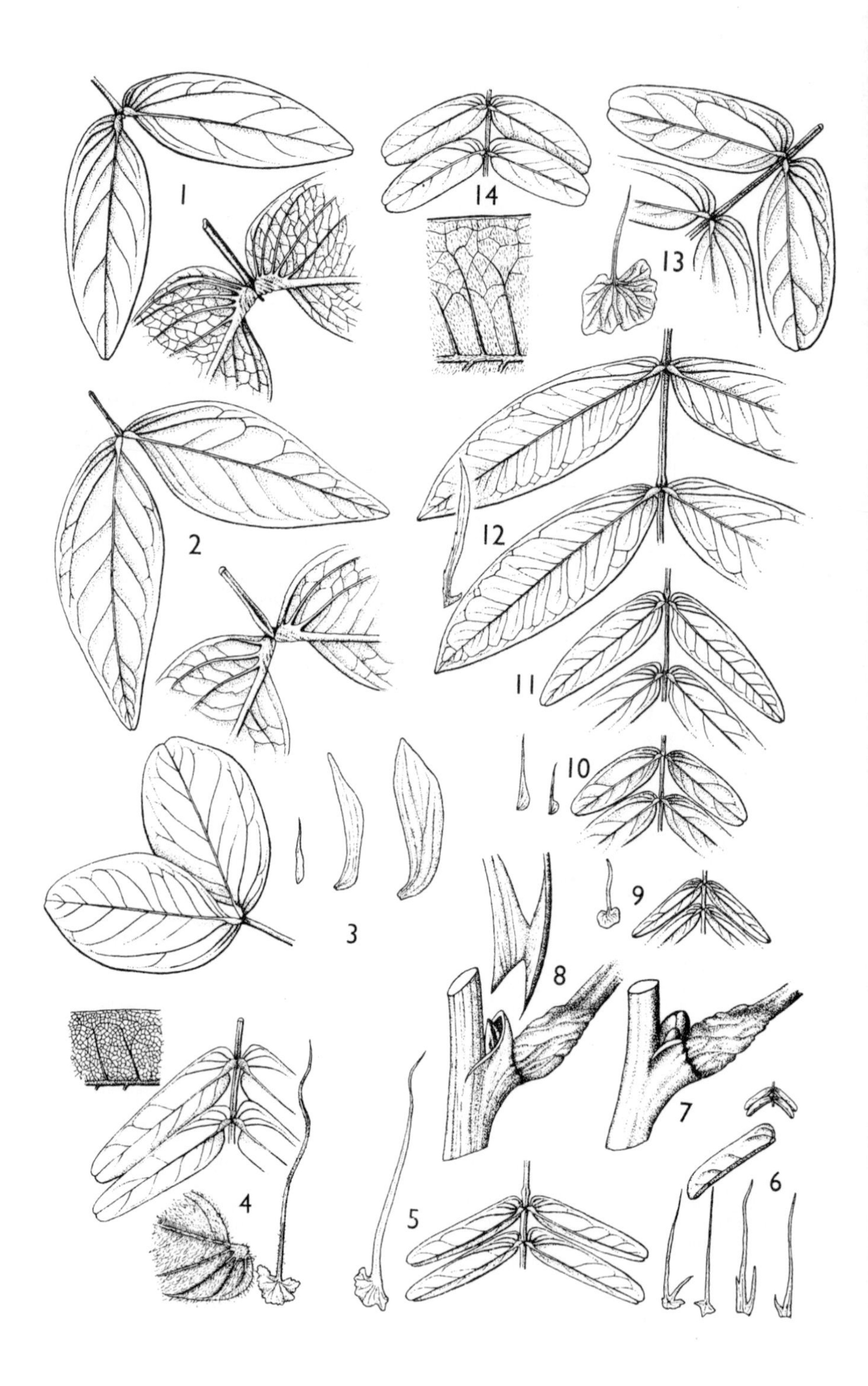
1
2
3
4
5
6
7
8
9
10
11
12
13
14

Following the key to the Groups (p. 165), a numbered list of species summarizes their distribution and variation and also shows where putative hybrids are cited. The nomenclature adopted here follows closely that used in the Forest Flora of Northern Rhodesia (F.F.N.R.), in which will be found a useful complement to the present account, especially in the introduction, plate and figures.

In using keys and descriptions, abnormal extremes of size and number of parts on specimens should be ignored; counts of pairs of leaflets and measurements should be confined to the middle leaves on normal shoots.

1. Leaflets very small and numerous, (3–)5–15(–20) × 1–3(–3·5) mm., in (17–)20–60(–72) pairs (larger and/or fewer on vigorous shoots and in putative hybrids) 2
   Leaflets larger (except in depauperate forms, notably of sp. 14) and usually fewer than above 3
2. Principal nerve of leaflets central (fig. 38/12) . 5. *B. microphylla*
   Principal nerve submarginal (fig. 35/6) . . 15. *B. taxifolia*
3. Small dormant buds* (in the axils of leaves or leaf-scars) not sharply keeled laterally, globose or ovoid, usually clearly visible and not (or scarcely) flattened; stipules usually free from each other, usually caducous leaving lateral scars only and usually without reniform auricles (fig. 35/7) 4
   Small dormant buds sharply keeled laterally and usually much flattened, at least on their inner face; stipules shortly connate or at least meeting in the axil, often persistent and often with reniform auricles; after fall of stipules, their persistent, often prominent, intrapetiolar bases closely subtend and often at first conceal the buds, for which they may easily be mistaken (fig. 35/8) 9
4. Leaves (equally throughout both surfaces of the leaflets and rhachis) and young branchlets and panicles very minutely appressed-puberulous (fig. 35/14); leaflets oblong to elliptic, in (8–)9–18(–21) pairs; flowers as second part of section 5 below . . 4. *B. puberula*
   Leaves etc. variously pubescent or glabrous, not minutely appressed-puberulous throughout; if partly so, then leaflets in fewer than 8 pairs; flowers various 5
5. Racemes normally simple with flowers, even when open, arranged in an obvious close spiral

* See key to the Groups and footnote, p. 165.

FIG. 35. *BRACHYSTEGIA*—some key characters, average examples, mainly of species not shown in other figures; upper surface of leaves unless otherwise stated. **1,** *B. floribunda*, distal pair of leaflets, × ½, and base of same, × 1; **2,** *B. bussei*, distal pair of leaflets, × ½, and base of same, × 1; **3,** *B. manga*, distal pair of leaflets, × ½, and three forms of stipule, × 1 (the latter are similar to those of *B. bussei* and *B. floribunda*); **4,** *B. boehmii*, two middle pairs of leaflets, × ½, base of lower surface of hairy form, × 1, stipule × 1, and reticulation on part of upper surface of glabrous form, × 2; **5,** *B. wangermeeana*, two middle pairs of leaflets, × ½, and one form of stipule, × 1; **6,** *B. taxifolia*, two middle pairs of leaflets, × ½, single leaflet enlarged, × 2, and four forms of stipule, × 1; **7,** basic type of dormant axillary bud of Group A, species 1–7, with lateral stipular scar, much enlarged; **8,** same of Group B, species 8–15, subtended by persistent stipule-base after fall of intrapetiolar connate stipules, similarly enlarged; **9,** *B. gossweileri* (see p. 190, under *B.* × *longifolia*), two middle pairs of leaflets, × ½, and stipule, × 1; **10,** *B. utilis*, two middle pairs of leaflets, × ½, and two stipules, × 1; **11,** *B.* × *longifolia*, one of the commonest forms, × ½; **12,** *B. angustistipulata*, two middle pairs of leaflets, × ½, and stipule from coppice, × 1; **13,** *B. stipulata*, two middle pairs of leaflets, × ½, and stipule, × 1; **14,** *B. puberula*, two middle pairs of leaflets, × ½, and part of surface, × 4. Partly after F.F.N.R., fig. 22.

(latter usually easy to detect up to a late fruiting stage); tepals 0 or 1–2(–4), very small to rudimentary, usually narrow and unequal, inserted too far apart to conceal the usually obvious staminal tube; leaflets in 2–6(–7) pairs, with usually obvious stipels on the rhachis below them (figs. 36, 37) . 1. *B. spiciformis*

Racemes normally paniculate with flowers arranged in an obscure gradual spiral (scars widely spaced after flowers fall); tepals (4–)5–8(–10), at least 4 subequal, imbricate or partly valvate, broad enough to conceal the bases of the free or very shortly united stamens (as in fig. 38) 6

6. Leaf-rhachis channelled above, with short wing-like expansions (sometimes obscure) below each pair of leaflets (fig. 35/2), or very narrowly winged throughout the intervals between the pairs 7

Leaf-rhachis not (or very obscurely) channelled, without expansions or wings (see fig. 35/1, 3) . . . 8

7. Leaflets in 2–4 pairs, usually narrowly ovate, often falcate (fig. 35/2); petiole (1·5–)2–3(–4) cm. long, of which the pulvinus usually forms less than one-fourth; panicles shortly puberulous or sericeous . . . . 2. *B. bussei*

Leaflets in 5–10(–12) pairs, usually oblong to elliptic (fig. 35/10); petiole (0·2–)0·4–0·8(–1) cm. long, of which the pulvinus usually forms more than one-third; panicles pubescent to tomentose externally but tepals sparsely and shortly ciliate to glabrous . 3. *B. utilis*

8. Panicles pubescent to tomentose, usually conspicuous, mainly on older wood below new leafy growth or on leafless branchlets; leaflets ovate to falcate, obtuse to acuminate; basal nerves departing fanwise from the pulvinulus on the proximal side of the midrib (3–)4–5(–6) (fig. 35/1) . . . . . . 6. *B. floribunda*

Panicles finely sericeous (usually on bracteoles and pedicels only), mainly terminal and axillary on new leafy growth; leaflets broadly ovate or subcircular to elliptic or obovate, often irregularly rhombic; apex emarginate or truncate to rounded or rarely obtuse; basal fanwise nerves 2–3(–4) (fig. 35/3) . 7. *B. manga*

9. Bracteoles of open flowers 13–20 mm. long; stamens 13–18(–20), filaments 25–40 mm. long, usually shortly united; leaflets in 5–10 pairs, usually broadly oblong; stipules usually very persistent, shortly linear, 1–2 times as long as their broad reniform auricles (fig. 35/13) . . . . 8. *B. stipulata*

Bracteoles of open flowers 3–10 mm. long; stamens (9–)10(–11), filaments 6–15 mm. long, usually free; leaflets diverse in shape and number; stipules various (figs. 35/4, 5, 11, 12 & 39–41) 10*

10. Leaflets usually 1–2(–2·5) times as long as wide, in 3–5(–6) pairs; apex usually emarginate, more rarely subtruncate to rounded, never (normally) acute or acuminate; stipules usually persistent, shortly linear, 1–2 times as long as their broad reniform auricles; the latter often persistent alone on or near the inflorescence, very rarely undeveloped; mature pod densely scurfy (fig. 39). . 9. *B. allenii*

Leaflets usually 3–7 times as long as wide; apex various, often acute or acuminate (especially when leaflets 12 pairs or fewer); stipules variably persistent, linear or subulate to filiform, usually at least 3 times as long as their auricles (if any); pod various 11

11. Panicles externally glabrous except, rarely, the ciliate margin of the bracts and the apex of some bracteoles; leaves glabrous except for occasional, non-persistent hairs between leaflets of a pair; rhachis not more than 7 times as long as the petiole (but see also 10, *B. angustistipulata*); leaflets in 4–7(–8) usually widely spaced pairs, the middle and distal pairs subequal; upper surfaces very laxly reticulate between the conspicuous nerves; pod smooth (see fig. 40) . . 11. *B. glaberrima*

Panicles puberulous to tomentose (at least on the bracteoles) or, if panicles glabrous, then leaflets 9 pairs or more; leaves puberulous to tomentose (at least on the rhachis), or rarely quite glabrous; leaflets (4–)5–28(–30) pairs, usually not widely spaced (except in 10, *B. angustistipulata* and forms of 12, *B.* × *longifolia*), the middle or lower middle pairs the largest; upper surfaces laxly to very closely reticulate; pod smooth to densely scurfy. . . . 12

12. Principal nerve of leaflets of the middle pairs markedly excentric throughout, closely parallel to the distal margin at a distance of 1–3 mm. and at least twice as far from the other margin at the middle of the leaflet; leaflets often falcate backwards; apex usually very oblique (not merely excentric) and retuse where the principal nerve meets the distal margin (fig. 35/5); pod smooth, ± obscurely pitted-lenticellate . . . . . . 14. *B. wangermeeana*

Principal nerve central or subcentral, at least towards the apex of the leaflet; apex variable,

* It must be emphasized that species 9–14 form a closely-related series with numerous intermediates; many of the latter seem to have arisen by direct and introgressive hybridization among these species and sometimes also with 8, *B. stipulata* and 15, *B. taxifolia*; precise diagnosis is therefore often difficult and in some areas almost impossible. In high exposed places, the usually narrow leaflets of spp. 11–14 are often much broader but, being then in at least 8 pairs, they cannot easily be confused with those of the low-altitude 9, *B. allenii*.

acuminate to emarginate, oblique only in putative hybrids with 14, *B. wangermeeana* or 15, *B. taxifolia*; pod smooth or scurfy, not lenticellate 13

13. Panicles minutely puberulous throughout with very short, erect or subcrispate hairs only, never glabrous; petiole (1·5–)2–5 cm. long; leaflets in (4–)5–7(–9) pairs, often subacute or subacuminate (fig. 35/12); leaves either quite glabrous or appressed-puberulous throughout (without long erect hairs); pod developing a densely scurfy surface very early . . . . 10. *B. angustistipulata*

Panicles, partly or throughout, pubescent to tomentose, with at least some rather long, erect or suberect hairs (with or without shorter hairs) or if panicles glabrous then leaflets in 10 pairs or more (forms of 12, *B.* × *longifolia*); petiole rarely exceeding 1 cm. (except in forms of *B.* × *longifolia*); leaflets in 5–30 pairs, diverse in shape; leaves, at least the rhachis, pubescent to tomentose with some long hairs, or very rarely glabrous; pod smooth to densely scurfy 14

14. Leaflets in (5–)6–14(–17) pairs, mostly narrowly ovate-triangular to narrowly oblong-elliptic, usually distinctly oblique and often falcate; margins mostly convex; apex various, often acute; main nerves above usually discolorous and clearly visible to the unaided eye, rarely obscured by hairs (even when these are rather dense); pulvinus variable but usually forming much less than half of the relatively slender petiole, the whole (7–)10–20 mm. long; stipules falcate to shortly filiform, mostly 15–25 × 2–5 mm., rarely persistent at flowering; auricles reniform to absent; pod smooth or partly (rarely completely) scurfy 12. *B.* × *longifolia**

Leaflets in (13–)14–28(–30) pairs, mostly narrowly oblong or triangular, rarely oblique except at base; margins straight or nearly so (or with the proximal margin somewhat concave and thus leaflets sometimes falcate backwards, notably when midrib rather excentric); apex rarely acute; main nerves slender and obscure (even when surface glabrous); ultimate reticulations above very close and regular, especially when lamina glabrous (see fig. 35/4); pulvinus usually forming more than half (often the whole) of the very stout petiole, the whole (2–)3–7(–10) mm. long; stipules filiform, 25–50 × 1–2 mm., usually persistent at flowering and grading downwards into large conspicuous subpersistent

* Comprising all intermediates between 11, *B. glaberrima* and 13, *B. boehmii*; see fig. 41 and notes under these species.

bud-scales; auricle usually rather large, reniform, often separately persistent, rarely undeveloped; pod densely scurfy throughout (as in 9, *B. allenii*, fig. 39/11) . . . 13. *B. boehmii*

KEY TO THE GROUPS

Small dormant buds* (in the axils of leaves or leaf-scars) globose or ovoid, not or scarcely flattened, not sharply keeled laterally except sometimes the small, outer scales; stipules usually free, usually caducous, leaving lateral scars only, but if (very rarely) slightly connate or partly intrapetiolar, then their bases not closely investing the clearly visible bud (see fig. 35/7); stipular auricles rare, not separately persistent; tepals usually shortly ciliate or much reduced; stamens ± 10, up to ± 15 mm. long; pod rather thinly woody, smooth unless parasitized:

Tepals 0 or 1–2(–4), small to vestigial, inserted too far apart to conceal the usually more obvious staminal tube (see figs. 36, 37); racemes usually simple; stipules with or without auricles; leaflets in 2–6(–7) pairs; rhachis channelled at least below each pair of leaflets and there provided with stipels or short wings, usually conspicuous; pod usually without surface bloom . . . . **A(1)** (sp. 1)

Tepals (4–)5–7(–10); outer usually 5, sepaloid, imbricate or rarely valvate, mostly broad, shortly ciliate or glabrous, at first concealing the base of the ± free stamens (see fig. 38/2–7); inner tepals 0 or 1–5, usually narrow, usually glabrous; racemes paniculate; stipular auricles lateral or very small or 0; leaflets in 2–72 pairs; rhachis channelled or not; stipels or wings (often obscure) present or not; pod with surface bloom (at least when immature):

Rhachis channelled; stipellar expansions (see figs. 35/2 & 38/12, 14) usually present below each pair of leaflets but often obscure, or replaced by long, very narrow wings extending throughout each interval; leaflets in 2–72 pairs, usually not glaucous . . . . **A(2)** (spp. 2–5)

Rhachis not or scarcely channelled, without stipels or wings (see fig. 35/1, 3); leaflets in 2–4(–5) pairs, usually glaucous . . . . **A(3)** (spp. 6, 7)

Small dormant buds much flattened (at least on their inner face) and sharply keeled laterally (at first enclosed in two large keeled scales); stipules shortly connate or at least meeting in the leaf-axil, often persistent and often with basal reniform auricles completely concealing the axil; after stipules fall, their persistent intrapetiolar bases (often becoming prominent, rarely obscure) closely subtend, compress and often at first conceal the

* As distinct from the larger buds (destined to produce main shoots or inflorescences) which often become conspicuous on mature branchlets before flushing (see fig. 36/2). All buds naturally change shape and size as they swell before growing out, so that they cannot be used in diagnosis for some time before and during the early stages of flushing.

buds, for which they may easily be mistaken (see fig. 35/8); outer tepals (at least) densely and usually long-ciliate (see figs. 39–41); stamens ± 10, ± free, or 13–18(–20), ± united; pod often thickly woody, smooth or partly to completely scurfy when mature, often warted through insect-attack, especially when otherwise smooth; leaf-rhachis channelled; stipels variable:

Flowers large, few, conspicuous; bracteoles 13–20 mm. long; tepals 6–8(–11); outer usually 6–8, imbricate, broad to narrow, densely long-ciliate, subequal to the staminal tube; stamens 13–18 (sometimes with staminodes, totalling ± 20), 25–30(–40) mm. long; racemes simple or paniculate; stipules and their large auricles normally persistent; leaflets in 5–8(–10) pairs; pod smooth or nearly so . . . . . **B(1)** (sp. 8)

Flowers medium to small; bracteoles 3–10 mm. long; tepals (4–)5–7(–9); outer usually 4–6, imbricate, densely and usually long-ciliate, at first enclosing the base of the free or very shortly connate stamens; filaments 8–12(–15) mm. long; racemes paniculate (unless depauperate, as often in sp. 15); stipules and auricles various, caducous to very persistent; leaflets in 3–45 pairs, very broad to very narrow:

Principal nerve of leaflets subcentral to very excentric; leaflets usually reticulate above, very broad to narrow, in 3–30 pairs; stipules falcate to filiform, caducous to persistent, often with reniform auricles which can persist alone; pod normally large and woody, smooth to densely scurfy . . . . **B(2)** (spp. 9–14)

Principal nerve of leaflets submarginal; leaflets smooth to transversely nervose or rugose above, very small, in 18–45 pairs; stipules stiffly subulate, often very persistent; auricles tooth-like to digitate, not persistent alone; pod small, smooth . . . . **B(3)** (sp. 15)

Group A(1)

1. *B. spiciformis.* Widely dominant; variation wide and apparently chaotic but vaguely clinal both generally and locally; most populations seem to be a mixture of forms; hybrids evidently frequent with sp. 5, resulting in many intermediate forms, e.g. *B.* × *fischeri* Taub. (pro sp.) and occasional with spp. 2 (and ?7, ?12, ?14). See 2 × 1 under sp. 2, 5 × 1 under sp. 5 and notes under spp. 12 and 14.

Group A(2)

2. *B. bussei.* Widespread, locally dominant; variation slight; most populations seem uniform; hybrids seem occasional with sp. 5, rare with sp. 1 (and ?7). See above and also 2 × 5 under sp. 5.
3. *B. utilis.* Widespread but local, dominant or co-dominant; variation moderate; populations seem fairly uniform; hybrids seem occasional with sp. 5, rare with spp. 1 (and ?7). See 5 × 3 and notes under sp. 3.
4. *B. puberula.* Very rare, little-known.
5. *B. microphylla.* Widespread, locally dominant; variation considerable but most populations fairly uniform; hybrids seem frequent with sp. 1 (q.v.) and occasional with spp. 2, 3, 6 and 7. See 5 × 1, 2 × 5, 6 × 5 and 7 × 5 under sp. 5; also 5 × 3 under sp. 3.

Group A(3)

6. *B. floribunda*. Widespread in S. & W. Tanganyika, often dominant; variation slight; most populations uniform; hybrids seem rare with spp. 5 (and ?7). See 6 × 5 under sp. 5 and notes under sp. 6.
7. *B. manga*. Widespread, locally dominant; variation slight; most populations seem uniform; hybrids seem rare with spp. 5 (and ?6). See 7 × 5 under sp. 5 and notes under sp. 6.

Group B(1)

8. *B. stipulata*. Rare (?) in SW. Tanganyika; variation outside East Africa considerable but populations appear uniform; hybrids seem occasional elsewhere with spp. 9–13 (and ?15) but none recorded in East Africa.

Group B(2)

9. *B. allenii*. Frequent in SW. & SE. Tanganyika, locally dominant; variation moderate but most populations seem uniform; hybrids seem locally frequent with sp. 13 (and/or ? " sp." 12), producing *B.* × *schliebenii* Harms (pro sp.) and other intermediates. See 9 × 13 and notes, under sp. 9.
10. *B. angustistipulata*. Locally dominant in W. Tanganyika; variation moderate; populations usually seem uniform; hybrids seem rare with spp. 13 (and ?11, ?12, ?1). See 10 × 13 under sp. 10.
11. *B. glaberrima*. Frequent, locally dominant in W. & SW. Tanganyika; variation moderate; populations seem uniform; hybrids seem numerous with sp. 13, producing " sp." 12, and frequent with sp. 14. See notes under spp. 11 and 12 and 11 × 14 under sp. 14.
12. *B* × *longifolia*. Widespread, mainly in S. Tanganyika, locally dominant; variation wide and apparently chaotic but some populations superficially uniform; back-crosses with spp. 11 and 13 seem frequent and hybrids with spp. 14 (and ?1) occasional. See notes under spp. 11 and 12, and 12 × 14 under sp. 14.
13. *B. boehmii*. Widespread, locally dominant or co-dominant; variation moderate; most populations uniform; hybrids seem numerous with sp. 11, producing " sp." 12 (q.v.), local with spp. 9, 10 and 14 (q.v.).
14. *B. wangermeeana*. Widespread, locally dominant in W. & SW. Tanganyika; populations uniform to variable; hybrids seem frequent with spp. 11, 12 and 13, perhaps amounting to a complex. See 11 × 14, 12 × 14 and 13 × 14 under sp. 14.

Group B(3)

15. *B. taxifolia*. Rare and local in SW. Tanganyika; hybrids seem occasional outside East Africa with spp. 11–14.

1. **B. spiciformis** *Benth.* in Trans. Linn. Soc. 25: 312 (1865); F.T.A. 2: 306 (1871); Burtt Davy & Hutch. in K.B. 1923: 159 (1923), pro parte, excl. *Welwitsch* 580 (BM, COI & K); L.T.A.: 727 (1930), excl. syn. *B. manga*; C. H. N. Jackson in Journ. S. Afr. Bot. 6: 39 (1940); B. D. Burtt in Journ. Ecol. 30: 74–78, 82–85, 111, 117, 141, 142, t. 2 (1942); T.T.C.L.: 92 (1949); Hoyle in F.C.B. 3: 452, fig. 38, t. 32 (1952); K.T.S.: 97, fig. 19 (1961); Hoyle & White in F.F.N.R.: 107, 117, 118, fig. 23, t. 1/H (1962). Type: Angola, Huila, Mumpulla-Nene, *Welwitsch* 578 (LISU, holo. !, BM, iso. !)

Tree 5–25 m. high to stunted treelet of 1–4 m.; young bark smooth, grey to whitish, becoming rough, reticulately or vertically fissured, grey or

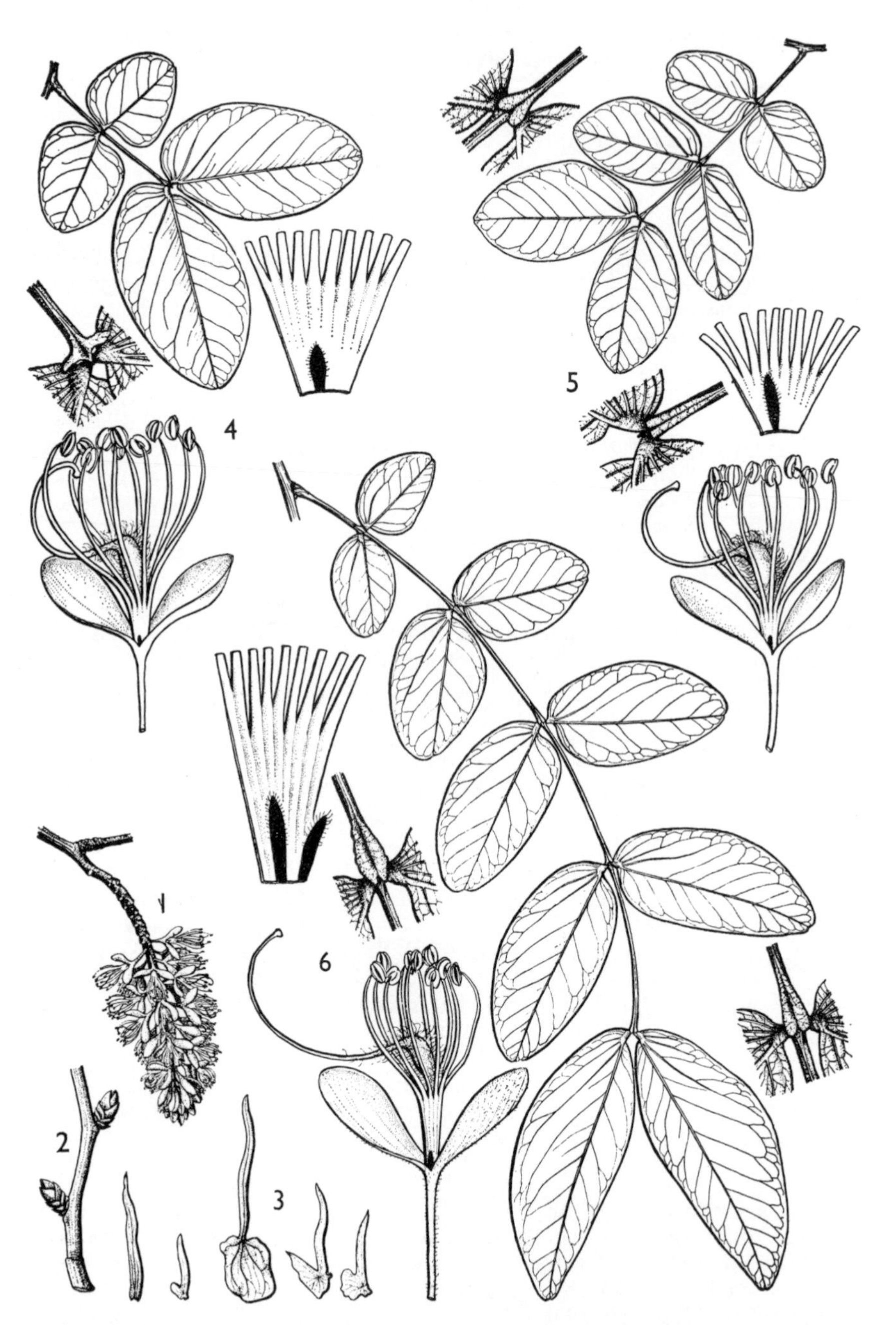

FIG. 36. *BRACHYSTEGIA SPICIFORMIS*—**1,** an average raceme, showing close scars of fallen flowers, × ½; **2,** large fertile or main-shoot buds on leafless mature branchlet, × ½; **3,** five forms of stipule, × 1; **4–6,** three extreme forms of the species each represented by semi-diagrammatic drawings of a leaf, × ⅓, part of upper surface of leaf-rhachis showing stipels and local expansions, × 2, flower with bracteoles, × 2, and exterior view of staminal tube opened out, showing (in black) reduced tepal(s) where usually present, × 4. 4, 5, from *Harley* 9150, 9153 (cf. " *B. hockii* " and " *B. mpalensis* "); 6, from *Wigg* 37 (cf. " *B. appendiculata* "). See also fig. 37 and footnote opposite.

brownish, shedding slowly in thick rectangular or irregular scales; crown rounded to spreading, flatter with age; foliage spreading to pendulous, maturing dark green, shiny when glabrous. Stipules free, filiform to linear, 0·5–3 cm. long, usually caducous; auricle 0, or a lateral tooth only, to irregularly reniform or subfoliaceous. Leaves glabrous to tomentose, with 2–6(–7) pairs of leaflets; petiole (0·5–)1–3(–4) cm. long; rhachis (3–)5–15(–18) cm. long, channelled at least below each pair of leaflets and there provided with stipellar expansions, which are oblong to subfoliaceous, often with free apices (stipels) directed forwards; leaflets (1–)2–8(–10) × (0·5–)1–4(–5) cm., subcircular to narrowly ovate or oblong-elliptic to narrowly obovate, emarginate to acuminate at apex, subsymmetrical to very oblique and cordate to cuneate at base; midrib central to very excentric; basal fanwise nerves (3–)4–5(–6); upper surface maturing reticulate and shining, less with age. Racemes (very rarely branched) terminal, up to 6 × 4·5 cm., glabrous to tomentose. Flowers green with white filaments and red anthers; pedicels usually slender, inserted in a close spiral or subverticillate, their scars persistent on the fruiting axis; bracteoles (4–)5–10(–12) × (3–)4–5(–6) mm. Tepals 0 or 1–2(–4), usually unequal and narrow or rudimentary, inserted well apart. Stamens ± 10; filaments 8–15(–18) mm. long, connate at base; staminal tube (1–)2–7 mm. long, entire or split, usually subequal to or longer than the tepals. Ovary 2–4 × 1–1·5 mm., ± crispate-setose; style 8–15(–20) mm. long, stigma usually prominent. Pod thinly woody, up to 16·5(–20) × 4·5(–5·5) cm., smooth, usually maturing mid-brown to yellowish and ± shiny; sutural wings spreading, each 4–8 mm. broad; apical beak exceptionally long. Figs. 36, 37 (p. 171).*

KENYA. Kwale District: E. of Kwale Boma, fl. 15 Dec. 1956, *Greenway* 9657!; Kilifi District: Sokoke Forest, fl. 27 Feb. 1945, *Jeffery* 102! & W. of Kikuyuni, fr. 19 Sept. 1958, *Moomaw* 940!

TANGANYIKA. Mpanda District: Kwasimba, fl. 9 Aug. 1948, *Semsei* 38 in *F.H.* 2470!; Kondoa District: Bereku ridge, S. scarp, fl. & fr. 1 Oct. 1951, *Hughes* 126!; Rungwe District: Masukulu, fl. 18 Nov. 1912, *Stolz* 1692!

DISTR. **K**7; **T**1–8; Congo Republic, Mozambique, Malawi, Zambia, Rhodesia and Angola

HAB. Deciduous woodland, the most widespread and probably most frequent dominant or co-dominant in " miombo ", from 15 m. altitude on sandy areas near the coast to 2350 m. on ridges and escarpments. Develops best on moist, deep red soils at lower altitudes; usually avoids badly drained or poor shallow soils, but tolerates a wide range of soil, climate and exposure. Associates with most " miombo" woodland species and occupies gaps in coastal forest and thickets.

SYN. *B. appendiculata* Benth. in Trans. Linn. Soc. 25: 313, t. 42 (1865). Type: Malawi, Zomba District, Magamero, *Meller* (K, holo.!, FHO, photo.!)

B. itoliensis Taub. in P.O.A. C: 197 (1895); Burtt Davy & Hutch. in K.B. 1923: 158 (1923); L.T.A.: 726 (1930). Type: Tanganyika, Bukoba District, Itolio, *Stuhlmann* 925 (B, holo.!, K, photo.!)

B. oliveri Taub. in P.O.A. C: 197 (1895); Burtt Davy & Hutch. in K.B. 1923: 159 (1923); L.T.A.: 728 (1930). Type: Kenya, Mombasa, *Wakefield* in *Schweinfurth* (? B, holo. †, K, iso.!)

B. mpalensis Micheli in Compt. Rend. Soc. Bot. Belg. 36: 73 (1897). Type: Congo Republic, Baudouinville District, Mpala [Pala], *Descamps* 27 (BR, lecto.!, K, photo.!)

B. randii Bak. f. in J.B. 37: 433 (1899); Burtt Davy & Hutch. in K.B. 1923: 160, fig. 5 (1923), pro parte, excl. legumina et specim. fruct. *Rand* 611; L.T.A.: 727 (1930). Type: Rhodesia, Salisbury, *Rand* 610 (BM, lecto.!)

B. euryphylla Harms in E.J. 30: 82 (1901); Burtt Davy & Hutch. in K.B. 1923: 162 (1923); L.T.A.: 730 (1930). Type: Tanganyika, Morogoro District, E. Ukami, *Stuhlmann* 8666 (B, holo.!, K, photo.!)

B. hockii De Wild. in F.R. 11: 512 (1913). Type: Congo Republic, Haut Katanga, *Hock* (BR, holo.!, FHO, drawing!, K, photo.!)

* In these figures, hairs have been largely omitted; most forms can be glabrous to pubescent or tomentose, partly or throughout, often within one population. The characters occur associated as shown, or re-combined, in widely separated areas. Some forms attain 6–7 pairs of leaflets; narrower forms can occur; apices can be subacute or acuminate.

B. edulis Burtt Davy & Hutch. in K.B. 1923: 162 (1923); T.S.K.: 50, 109 (1926); L.T.A.: 727 (1930); T.S.K.: 63 (1936). Type: Zambia, Batoka Highlands, *Kirk* (K, holo.!)

B. taubertiana Burtt Davy & Hutch. in K.B. 1923: 150, fig. 8 (1923); L.T.A.: 731 (1930). Type: Tanganyika, Tanga District, Doda, *Holst* 3023 (K, holo.!, ? B, iso. †)

B. venosa Burtt Davy & Hutch. in K.B. 1923: 158, fig. 2 (1923), excl. cit. Uganda (potius NE. Zambia); L.T.A.: 729 (1930), haud tamen syn. *B. lujae*. Type: Zambia/Rhodesia, near Victoria Falls, *Allen* 165 (K, holo.!, PRE, iso.!)

VARIATION. Figures 36 and 37 show part only of the remarkable variation in size, shape and spacing of pairs of leaflets. Glabrous and pubescent forms seem to be mixed in all populations; thus, e.g. *B. D. Burtt* 5294 & 5296 (Tanganyika, Singida–Kondoa, Rift scarp, fl. Oct. 1935) have hairy and glabrous leaves respectively; *Hoyle* 1093 & 1092 (Tanganyika, Njombe–Songea, 40–50 km., fr. 24 July 1949) are a similar pair from adjacent trees. Extreme hairiness is, however, usually associated with high altitude and exposure. Racemes are usually hairy but often combined with glabrous leaves. Early, widely scattered gatherings, showing various combinations of characters, were made the basis of separate species now relegated to synonymy, but known combinations of characters are now becoming so numerous that even the definition of subspecies and varieties seems increasingly futile. On the whole, fewer ovate leaflets predominate in W. and SW. Tanganyika while rather more leaflets, with the distal pair tending to be obovate, occur mainly in NE. Tanganyika and Kenya; in the same direction the length of the staminal tube and of the petiole tend to decrease. There are, however, numerous exceptions and almost identical forms occur hundreds of miles apart. Depauperate, apparently distinct local forms such as " *B. itoliensis* " near Lake Victoria and " *B. oliveri* " in Kenya and Tanga Province are associated with exceptional soil and climate at the limits of range but intergrade with the general population. In many places, some of the variation (especially narrow, pointed leaflets) may well be accounted for by introgressive hybridization, notably with sp. 5, *B. microphylla* (see 5 × 1 under the latter). Naturally enough, hybrids with *B. microphylla* are relatively easy to detect because of wide differences in obvious characters like the number, shape and size of leaflets and the nature of the inflorescence and flowers; this may partly account for their relative frequency among available specimens because they puzzle collectors. Rare putative hybrids with spp. 2, *B. bussei* and 3, *B. utilis* are cited under these species.

Hybrids with species in Group B, although theoretically improbable, are suspected (see notes under spp. 12, *B.* × *longifolia* and 14, *B. wangermeeana*).

2. **B. bussei** *Harms* in E.J. 33: 155 (1902); Burtt Davy & Hutch. in K.B. 1923: 157 (1923); L.T.A.: 732 (1930); C. H. N. Jackson in Journ. S. Afr. Bot. 6: 38 (1940); B. D. Burtt in Journ. Ecol. 30: 76, 85, 86, 141, t. 3 (1942); T.T.C.L.: 91 (1949); Hoyle in F.C.B. 3: 463, fig. 39/A (1952); Hoyle & White in F.F.N.R.: 108, fig. 22/B (1962). Type: Tanganyika, Songea District, Ungoni, *Busse* 729 (B, holo. †, EA, iso.!, K, fragm.!)

Tree 6–20(–?30) m. high, usually reported as slender; bark at first smooth, pale grey, flaking thinly, exposing yellow or buff patches, becoming darker and often rough near base (when burnt?); crown rounded or flattish, rather loose; branches slender; foliage pendulous, maturing bright green. Branchlets usually soon rusty. Stipules free, linear to ovate-oblong or falcate, often subfoliaceous, 0·5–2·5 cm. long, usually caducous; auricles 0. Leaves glabrous or rarely puberulous, with (2–)3–4 widely spaced pairs of leaflets, the distal pairs usually larger; petiole (1·5–)2–3(–4) cm. long, slender; rhachis (2–)4–8(–12) cm. long, slender, channelled; stipellar expansions usually semilunar, sometimes obscure; leaflets narrowly ovate to falcate, (2·5–)4–8(–10) × (1–)1·5–4(–5) cm., obtuse or retuse at apex, often acuminate, ± obliquely subcordate to cuneate at base; midrib subcentral; basal fanwise nerves 2–3(–4). Panicles terminal or terminal and axillary, up to 5 × 5 cm., ± puberulous. Flowers small, greenish-white; bracteoles 4–6(–7) × 3–4(–5) mm., finely sericeous. Tepals usually all sepaloid, imbricate, outer (4–)5(–6), up to 2 × 1·7 mm., very shortly ciliate, inner 0 or 1–4, narrow. Stamens ± 10; filaments free or irregularly connate at base, 7–10 mm. long. Ovary 2–2·5 × 1 mm., subdensely crispate-setose; style 6–9 mm. long, stigma prominent.

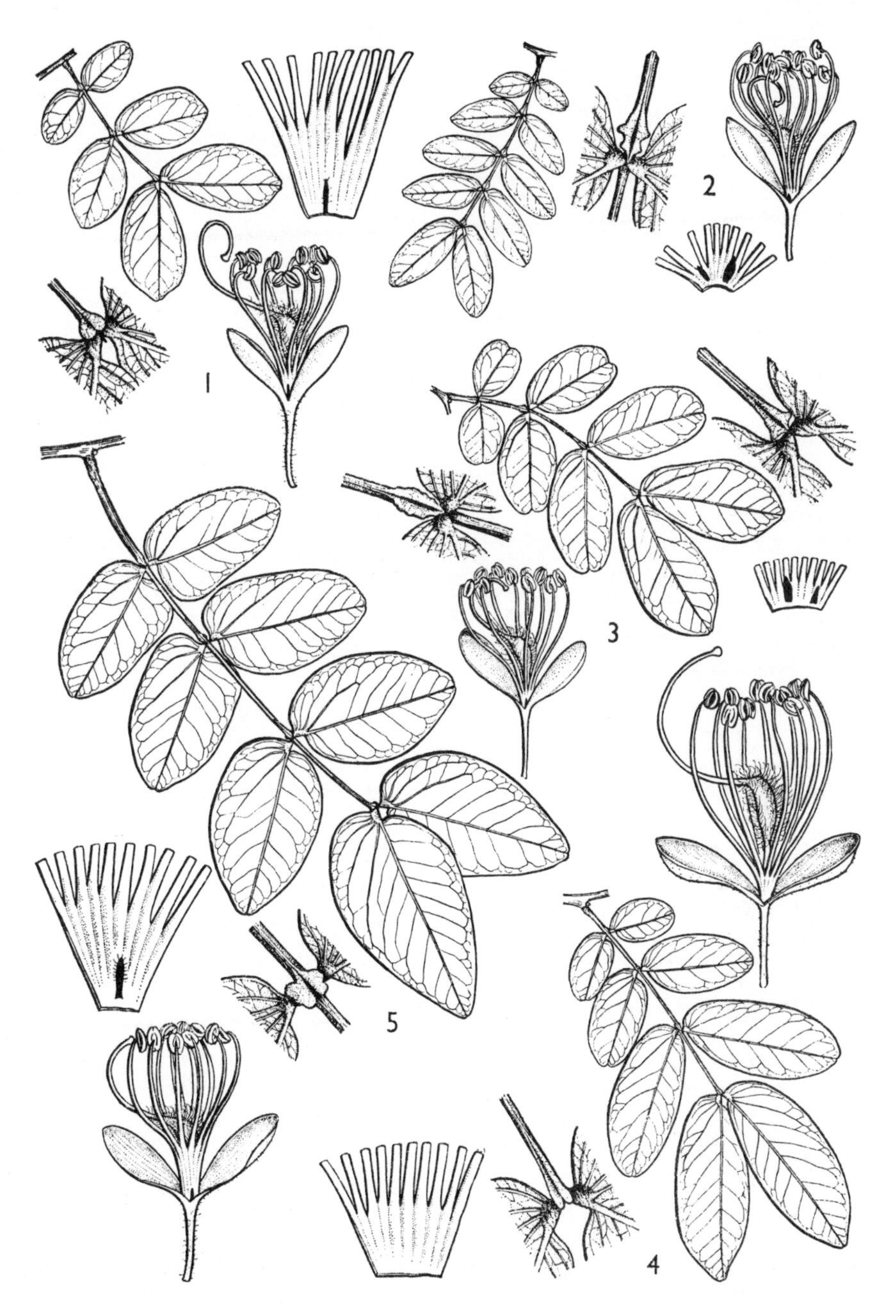

FIG. 37. *BRACHYSTEGIA SPICIFORMIS*—five of the numerous intergrading forms, each represented by semi-diagrammatic drawings of a leaf, × ½, part of upper surface of leaf-rhachis showing stipels and local expansions, × 2, flower with bracteoles, × 2, and exterior view of staminal tube opened out, showing (in black) reduced tepal(s) where usually present, × 4. **1,** a depauperate form, " *B. itoliensis* "; **2, 3,** two variants of " *B. oliveri* "; **4,** a variant of " *B. taubertiana* "; **5,** a common broad form. **1,** from *Stuhlmann* 925; 2, from *Jeffery* 102; 3, from *R. M. Graham* 259; 4, from *Vaughan* 2737; 5, from *Lindeman* 707. See also fig. 36.

Pod thinly woody, up to 15 × 3·5 cm., smooth, pale to mid-brown, ± pruinose; sutural wings thin, revolute, each 6–8 mm. wide. Fig. 35/2, p. 160.

TANGANYIKA. Handeni District: Negero, fl. Dec. 1957, *Semsei* 2727! & Handeni–Turiani, fr. 14 Oct. 1938, *Greenway* 5815!; Lindi District: Rondo [Muera] Plateau, fl. 22 Jan. 1935, *Schlieben* 5898!
DISTR. **T3**–8; Congo Republic, Mozambique, Malawi, Zambia and Rhodesia
HAB. Deciduous woodland, widespread, but local and usually a pure dominant; 240–1700 m. Typically on poor, dry, steep, rocky slopes, notably coarse soils from quartzites, schists and granite, but sometimes on heavy reddish soils; locally common with *B. microphylla* and *B. utilis* or *B. boehmii* and *Julbernardia globiflora*.

VARIATION. *B. bussei* normally shows little variation and is easy to recognize in the field or from fertile material, which is in short supply. Sterile dried material can only be separated from some forms of 1, *B. spiciformis* by comparison and with practice. Hybrids are apparently rare, but crossing seems to occur with 5, *B. microphylla* and more rarely (?) with *B. spiciformis*; the appressed puberulence on the leaves of a minority of otherwise normal specimens may derive from the former (see 5 × 1, under *B. microphylla*); crispate hairs may derive from the latter (see 2 × 1, below).

2 × 1. B. bussei × B. spiciformis

Rare (?) forms with crispate pubescence on the leaf-rhachis, appressed puberulence on some or all leaflets and with more obvious and variable stipellar expansions than usual in *B. bussei* (see details in note below).

TANGANYIKA. Handeni District: S. of Handeni, Feb. 1962, *Procter* 2000!; Kigoma District: 24 km. on Kasulu road, Machazo scarp, Nov. 1956, *Procter* 551!; Mbeya District: Njele, May 1937, *Ross* 18 in *F.H.* 1199!
DISTR. **T3, 4, 7**; Mozambique, Malawi and Zambia
HAB. Where both putative parents occur

NOTE. *Procter* 2000, collected as *B. bussei* which he knows well, bears the axis of a simple raceme (with flowers fallen) similar to that of the local form of *B. spiciformis* for which the petiole is much too long; some of the 5 pairs of leaflets are narrow, so that *B. microphylla* may be involved here, as well as the local *B. spiciformis* with up to 6 pairs. The other cited specimens have only 2–4 pairs but the leaves vary in pubescence and in the development of stipels.

Procter also collected a specimen (Tanganyika, near Iringa golf course, on granite kopje, Feb. 1961, *Procter* 1765) which he considers unusual in the very copious panicles; the population was noted as variable in shape of leaflets, both between trees and on the same tree; the 3–4 pairs of leaflets on the specimen have very prominent stipels on some leaves and are also variable in pubescence. The variation described and observed could be caused by introgression from the local *B. spiciformis* or *B. microphylla*, the panicles in particular recalling those of the latter; both these species are frequent locally and probably hybridize. *Procter* 1835 (Tanganyika, 35 km. from Mbeya, by waterfall, Mar. 1961), sterile, is similarly indeterminate.

3. **B. utilis** *Burtt Davy & Hutch.* in K.B. 1923: 155 (1923); C. H. N. Jackson in Journ. S. Afr. Bot. 6: 39 (1940); B. D. Burtt in Journ. Ecol. 30: 83, 141 (1942); T.T.C.L.: 93 (1949); Hoyle in F.C.B. 3: 468, fig. 40/C (1952); Hoyle & White in F.F.N.R.: 114, 117, 118, fig. 22/I (1962). Type: Malawi, Mlanje, *Purves* 193 (K, holo.!)

Tree 6–15(–20) m. high; bark narrowly fissured and rather finely reticulate, grey at first, becoming rough and shedding slowly in rather thick scales, often brown to black in age; crown rounded or flattish, dense, bushy; mature foliage rather dark green. Stipules free, subulate from a broader oblique base or falcate-subfoliaceous, 0·5–1·5 cm. long, usually caducous; auricle 0. Leaves with 5—10(–12) subequal pairs of leaflets, the pairs contiguous or slightly spaced; petiole (2–)4–8(–10) mm. long; rhachis (4–)5–9(–12) cm. long, channelled, very narrowly winged, usually pubescent to tomentose with spreading hairs; stipels consisting of short semilunar expansions, closely below each pair of leaflets, or 0; leaflets mostly oblong to elliptic and rounded at both ends, (1–)2–4(–5) × 0·5–1·5(–1·8) cm., often retuse, sometimes mucronate, rarely narrowed at apex, asymmetric but rarely very oblique at base; midrib subcentral; basal fanwise nerves 2–3(–4); surfaces usually mat and concolorous, closely reticulate, densely pubescent to glabrous. Panicles terminal or terminal and axillary, up to 5 × 5 cm., pubescent to tomentose (more so than the leaves) with spreading brown or rusty hairs. Flowers small, greenish; bracteoles 4–5(–6) × 3–4 mm. Tepals usually all sepaloid and broad, imbricate; outer 5(–6) subequal, up to 3 × 2 mm., free or with 2 or 3 partly connate, thin, margin usually translucent, glabrous or shortly and sparsely ciliate; inner 0 or 1, narrow or rudimentary. Stamens ± 10; filaments free, 6–8 mm. long. Ovary 2–3 × 1 mm., very densely crispate-setose; style 6–7 mm. long; stigma small. Pod thinly woody, up to 12 × 3·5 cm. (usually seen much smaller), smooth, maturing yellowish or pinkish-brown, at first ± pruinose; sutural wings thin, normally spreading, each up to 4 mm. wide. Fig. 35/10, p. 160.

TANGANYIKA. Ufipa District: Kasanga–Sumbawanga, near Kasoti, fl. 27 Oct. 1933, *Michelmore* 699 !; Singida or Kondoa District: W. scarp of Rift wall, fl. 25 Oct. 1927, *B. D. Burtt* 661 in *Game Dept.* 5661 !; Lindi District: Ruponda, fr. 28 Aug. 1947, *Semsei* in *F.H.* 2190 !

DISTR. **T**4, 5, 7, 8; Congo Republic, Mozambique, Malawi, Zambia, Rhodesia and Angola

HAB. Deciduous woodland, locally dominant or co-dominant; 300–1830 m., mainly above 1000 m. (lower in **T**8 on sandy valley soils). Typically in zones or groups on ridges, scarps and slopes, notably in shallow, stony or gritty soils over granite, etc., below *B. microphylla* (q.v.); often with *Julbernardia globiflora* or in mixed woodland.

VARIATION. Stipules sometimes persist on the inflorescence on apparently vigorous shoots (perhaps of coppice-trees only); broad subfoliaceous ones up to 2·5 × 0·8 cm. on coppice shoots can sometimes lack their usual long apices and resemble auricles. Otherwise, *B. utilis* varies little except in indumentum of leaves and size of leaflets. Forms with more than 10 pairs of small leaflets (especially if narrowed to the apex) may be hybrids with 5, *B. microphylla* (q.v.). Exceptionally large distal leaflets, or very few large leaflets, especially if subacute, e.g. *Anderson* 1292 (Tanganyika, Songea District, Suluti, 5 Dec. 1960), can indicate young trees, coppice or perhaps hybrids with 1, *B. spiciformis*. There is a shortage of good flowering and fruiting material and descriptive notes of the species from Tanganyika.

5 × 3. **B. microphylla** × **B. utilis**; F.F.N.R.: 118 (1962)

Sundry forms are available, notably two series showing a gradation of number, shape, venation and indumentum of leaflets, in the presence of both putative parents. Details are given in notes below.

TANGANYIKA. Tabora District: about 6 km. S. of Tabora, Uruma Forest Reserve, 26 June 1949, *Hoyle & Greenway* 1043 ! & 1043/A !; Chunya District: Kipembawe, Kisambya Hill, *Hoyle* 1068 ! & 1069 !

DISTR. **T**4, 7, ? 8; Mozambique, Malawi and Zambia

HAB. In localities with both putative parents, which frequently occupy adjacent zones. The steepness and relative inaccessibility of the habitat may explain the rarity of material although my own observations suggest that it is often present.

NOTE. *Hoyle* 1068–1070 occurred together; No. 1068 has 10–12 pairs of leaflets, many tapering, the tree tending to have the habit of No. 1069; the latter tree resembled *B. microphylla* but had only (19–)20–24 pairs of small but rather broad obtuse leaflets, many pubescent on the midrib beneath and some like those of *B. utilis* in miniature; No. 1070, with 8–9 pairs of leaflets, was from a coppiced tree of *B. utilis* becoming normal above. An almost normal example of *B. microphylla*, *Hoyle* 1072, with (14–)20–24 pairs of leaflets, still rather broad and variable, came from the next hill (Ngulwe Hill). All the above specimens show the peculiar yellowish colour usually associated with old leaves of *B. utilis*; *Groome* in *F.H.* 2237 (Tanganyika, Chunya District, N. Lupa, Tikwe [or ? Kikwe] Hill, fl. Nov. 1947) with 26–29 pairs of small narrow leaflets, seems to represent the nearest to normal *B. microphylla* in the district. *Hoyle* 1043 & 1043/A also occurred together, in the interzone between normal *B. microphylla* above and normal *B. utilis* below, the latter represented by *Wigg* in *F.H.* 2762 collected at the same time. No. 1043, from a small tree tending to have the bark of *B. utilis* and essentially its crown-habit, has (8–)12–13 pairs of leaflets, mostly oblong or tapering, with pubescent midrib and ciliate margin, many retuse; No. 1043/A, a fragment, has 12–13 pairs of small tapering leaflets obviously approaching those of *B. microphylla*; *Wigg* 2762 has (5–)6–7 pairs of leaflets, large for *B. utilis*, especially the distal pair; this last fact and also the presence, on one leaf of *Hoyle* 1043, of some stipels like those of the local 1, *B. spiciformis*, suggests that the latter species may also be involved in the complex. All three specimens have the yellowish colour mentioned above.

Four other aberrant specimens from Tanganyika seem to belong here but may be merely extreme variants of *B. utilis*. *Michelmore* 707 (Ufipa District, Mbisi Mts., Lake Rukwa escarpment, fl.-buds 1 Dec. 1933) has 7–9 pairs of glabrous, glaucous leaflets, many narrow and tapering; *Semsei* in *F.H.* 2201 (Masasi District, Kilimarondo, 24 Aug. 1947) has 7–8 pairs, similar but not glaucous; *Anderson* 747 (Lindi District, Nachingwea, 21 Feb. 1952) has 10–13 pairs of leaflets of variable intermediate shape, in the presence of both putative parents; the colour and texture (especially at this date) suggest that sp. 7, *B. manga*, also locally frequent, may be involved; the pod mounted with this specimen in the East African Herbarium is from *Julbernardia magnistipulata*. *Smith* 410 (Kahama District, Ushirombo, fl.) has (8–)10–12 pairs of glabrous leaflets, many strongly tapering and mucronate; sp. 5 is frequent in the district.

4. **B. puberula** *Burtt Davy & Hutch.* in K.B. 1923: 156 (1923); L.T.A.: 721 (1930), nec tamen aff. *B. bequaertii* nec *B. woodianae*; Hoyle in C.F.A. 2: 231, t. 46, figs. 23, 32 (1956); Hoyle & White in F.F.N.R.: 114, t. 1/M (1962). Type: Angola, Bié, between R. Cusaba and R. Cunene, *Gossweiler* 2867 (BM pro parte, holo. !, COI, K, LISJC, iso. !)

Tree 6(–?12) m. high; young parts evenly clothed with minute appressed hairs; bark blackish; crown spreading (see note below). Stipules free, linear to falcate, 1–2 cm. long, usually caducous; auricle 0. Leaves with (8–)9–18(–21) contiguous or slightly spaced pairs of leaflets, the middle pairs usually the largest; petiole 7–10(–12) mm. long; rhachis 7–14(–16) cm. long, channelled, each interval with narrow revolute wings throughout, sometimes locally expanded; leaflets narrowly oblong or elliptic to narrowly ovate, rarely falcate, (1–)2–4(–4·5) × (0·4–)0·8–1·5(–2) cm., rounded to emarginate at apex, subobliquely rounded-cuneate at base; midrib subcentral, slender, obscure above; basal fanwise nerves obscure, (2–)3–4(–5); lateral nerves and veins more obscure, almost concolorous with both very similar surfaces. Panicles terminal or terminal and axillary, 3–7 × 3–5 cm., slender-branched, rusty-sericeous. Flowers small, crowded; bracteoles 5–6(–7) × 3·5–5 mm. Tepals 5(–6), free, outer imbricate, up to 2 × 1·5 mm., shortly ciliate to glabrous, inner 0 or 1, narrow. Stamens ± 10, free, 7–9 mm. long. Ovary 2–3 × 1 mm., ± rufous-setose. Pod thinly woody, pale reddish-brown, ± pruinose; sutural wings thin, venose, each 8–10 mm. broad. Fig. 35/14, p. 160.

TANGANYIKA. Kigoma District: E. of Uvinza, at 1133·6 km. [on railway?], 5 Feb. 1926, *Peter* 36267!

DISTR. **T4**; Angola, Congo Republic and Zambia

HAB. Deciduous woodland, evidently rare, local and (in Zambia) gregarious; known in the Flora area from the cited gathering only, at 990 m. Elsewhere up to 2000 m. altitude, on poor Kalahari sand soils and schistose or siliceous rocky ridges.

NOTE. More good material and field notes on this interesting species are much needed. It combines some of the characters of species 2, *B. bussei*, 3, *B. utilis* and 7, *B. manga* so as to suggest a relict species allied to their origin. The scanty field notes and most of the description necessarily come from Angolan, Congo and Zambian specimens and the 12 m. height record (from the type) is suspect as evidently confused with *B. boehmii*. The sole East African collection is apparently coppice but bears the note " grosser Baum, blüht nicht." Although it is sterile there is no reasonable doubt of its identity.

5. **B. microphylla** *Harms* in E.J. 28: 397 (1900); Burtt Davy & Hutch. in K.B. 1923: 153 (1923); L.T.A.: 716 (1930); C. H. N. Jackson in Journ. S. Afr. Bot. 6: 35, 39 (1940); B. D. Burtt in Journ. Ecol. 30: 74–76, 78, 82–85, 109, 116, 141, t. 2, 5, 7 (1942); T.T.C.L.: 91 (1949), haud tamen qua syn.; Hoyle in F.C.B. 3: 477 (1952); Hoyle & White in F.F.N.R.: 115, 118, fig. 22/L (1962). Type: Tanganyika, Iringa District, Ndegere, *Goetze* 603 (B, holo. †, K, fragm. !)

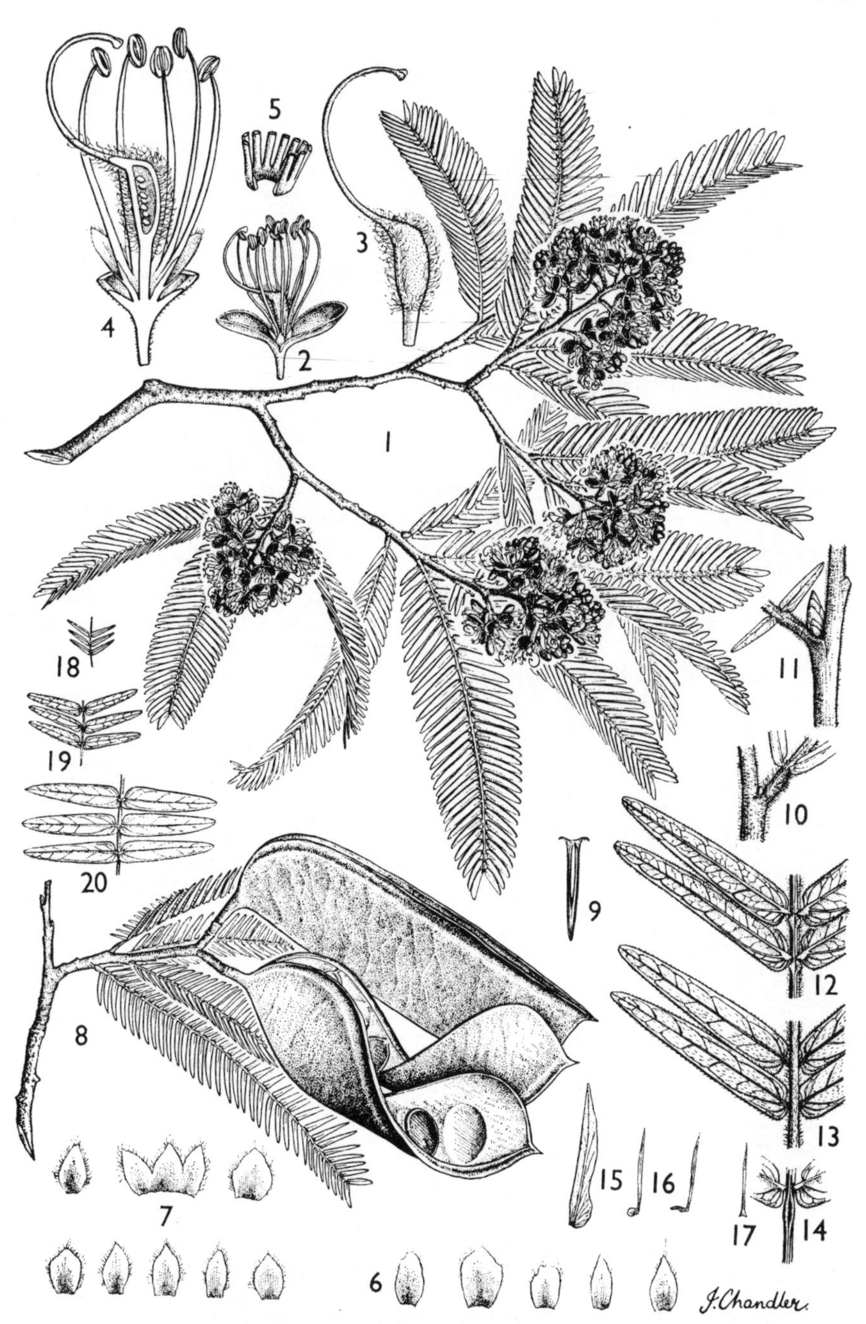

FIG. 38. *BRACHYSTEGIA MICROPHYLLA*—**1,** flowering branchlet, × ½; **2,** flower, × 2; **3,** gynoecium, × 4; **4,** longitudinal section of flower, × 4; **5,** hypanthium and base of filaments, usual form, × 4; **6,** tepals × 4; **7,** tepals, two other variants, × 4; **8,** fruiting branchlet, × ½; **9,** transverse section of pod, × ½; **10,** small ovoid dormant bud subtended by base of leaf and (laterally) by scar of separate stipule × 2; **11,** large inflorescence- or main shoot-bud of characteristic shape, × 2; **12,** leaflets of middle pairs, upper surface, showing reticulation and a frequent form of adnate stipels on rhachis, × 2; **13,** same, lower surface, × 2; **14,** stipellar wings of longer form, sometimes decurrent throughout the interval between pairs of leaflets, × 2; **15,** stipule from vigorous coppice, × 1; **16,** two stipules from young normal sterile branchlet, × 1; **17,** stipule from flowering branchlet, × 1; **18, 19,** two extreme sizes of normal leaflets, × ½; **20,** leaflets from coppice, × ½. 1, from *Brenan* 8266, *Groome* 16; 2–6, from *Eggeling* 6734; 7, 17, from *B. D. Burtt* 663; 8, 9, 11, from *Hoyle* 1060; 10, from *Groome* 16; 12, 13, 15, from *Brenan* 8266; 14, 16, from *Lewis* 99; 18, from *B. D. Burtt* 5748; 19, from *B. D. Burtt* 5798; 20, from *B. D. Burtt* 6449.

Tree 4–20(–27) m. high; young bark smooth, pale silver-grey, soon flaking off thinly on bole and main branches, exposing yellowish or buff-coloured patches; older bark becoming rough, darker grey or brownish, flaking in thicker scales exposing large lenticellate patches; crown usually spreading, flat-topped; upper branches divaricate; branchlets ± horizontal with delicate fern-like foliage. Young branchlets usually pubescent to tomentose, brown or rusty, soon grey with small circular lenticels. Stipules fugacious (rarely seen), free, linear to filiform, 0·5–1 cm. long, with or without a small lateral lobe at the base. Leaves with (17–)25–60(–72) closely pectinate pairs of leaflets, the middle pairs the largest; petiole 1–4 mm. long; rhachis very slender, channelled, with ± conspicuous wings and often local stipellar expansions, usually obscured by spreading, often rusty pubescence; leaflets linear to very narrowly triangular, often falcate, (3–)5–15(–20) × 1–2(–3·5) mm., acute to rounded, often minutely apiculate, obliquely truncate to subcordate at base; midrib subcentral, very slender; surfaces usually glabrous, more rarely appressed-puberulous, but usually ciliolate. Panicles terminal or terminal and axillary, 2–4 × 2–4 cm., brown-, rusty- or tawny-pubescent to -tomentose. Flowers small, greenish-white or yellowish; bracteoles 4–6 × 3–4·5 mm., puberulous to tomentose. Tepals apparently (4–)5 only, free or 2–3 partly connate, always (?) all sepaloid, broad and imbricate, up to 2 × 1·5 mm., shortly and sparsely to subdensely and (for their size) rather long-ciliate, otherwise glabrous. Stamens ± 10, usually free, 8–10 mm. long. Ovary 2·5–3 × 0·7–1 mm. Pod thinly woody, smooth, up to 12 × 3 cm. (usually much smaller), immature drying pale or mid-brown, mature blackish-purple and usually pruinose, with numerous but minute and obscure pale lenticels; sutural wings thin, spreading, each up to 4 mm. wide. Fig. 38, p. 175.

TANGANYIKA. Handeni District: Kideleko road, fl. 14 Jan., fr. 7 April 1954, *Faulkner* 1325! & 1401!; Mpanda District: 11 km. on Uruwira–Nyonga road, fr. 3 July 1949, *Hoyle & Greenway* 1060!; SE. Tabora District: Itulu Hill, fl. Sept. 1951, *Groome* 16!; Newala escarpment, fl. Nov. 1953, *Eggeling* 6734!

DISTR. **T**1–8; Congo Republic, Mozambique, Malawi, Zambia (and ? Rhodesia)

HAB. Deciduous woodland, widespread but mainly local; 300–2200 m., mainly in highlands. Typically as pure dominant on crests and summits of rocky hills and escarpments, especially on granite; often as a zone above 3, *B. utilis*; very frequent on Iringa Highlands, often mixed with 1, *B. spiciformis*; less often on level ground and on sandy loams but then sometimes very large; sometimes mixed with 2, *B. bussei* and 3, *B. utilis*, notably in **T**8.

SYN. [*B. tamarindoïdes* sensu T.T.C.L.: 90, 91 (1949), pro parte, et in herb. mult., *non* Benth.]

VARIATION. In T.T.C.L.: 91 (1949), *B. microphylla* was cited as a " form " of the Angolan *B. tamarindoïdes*. The latter, although belonging to the same group, is a complex of uncertain status which now seems to be very closely related to *B. utilis*.

Although a manifestly distinct species, *B. microphylla* varies considerably in number of leaflets on normal specimens, without significant change in other characters, in each area from which several gatherings are available. There is also a vague, clinal variation in maximum number of leaflets per specimen, from 40–72 pairs in E. and NE. Tanganyika to 25–45 pairs in the west and south-west. Occasional specimens, otherwise normal but (for the locality) with exceptionally broad and/or few leaflets, interrupt this cline at several points. Some of these are locally associated with highly aberrant material showing various sizes and numbers of leaflets intermediate in shape, venation and indumentum between local forms of *B. microphylla* and 1, *B. spiciformis*; the flowering specimens sometimes have simple racemes with intermediate flowers showing various abnormalities usually associated with hybridity. One of these was described by Taubert as a species which is now more appropriately called *B.* × *fischeri* Taub. (pro sp.) and enlarged to include other collections of similar apparent origin. These are dealt with in some detail below under 5 × 1. Other hybrids, very similar in appearance at first sight, seem to occur less often with 2, *B. bussei*, 6, *B. floribunda* and 7, *B. manga* (see below under 2 × 5, 6 × 5 and 7 × 5 respectively). For putative hybrids with 3, *B. utilis* see under the latter.

5 × 1. **B. microphylla × B. spiciformis**; T.T.C.L.: 91 (1949) in obs. sub "*B. tamarindoïdes*"; F.F.N.R.: 118 (1962)

Rather frequent forms showing among them every combination of characters of the putative parents and often also displaying unusual variability (with or without obvious abnormality) from one branchlet, leaflet, inflorescence or flower to another. The most obvious examples are, naturally enough, centrally intermediate specimens with numbers of leaflets in the range 8–18 pairs, especially when flowering and showing simple racemes and/or leaflets more pubescent on the midrib beneath than on the lamina (the reverse condition being usual in pubescent leaflets of *B. microphylla*). Fruiting and sterile specimens with glabrous leaflets can usually be diagnosed by comparison only, as the putative parent species do not differ definitely nor reliably enough in the leaf-rhachis or the pods. Details and further individual examples are given below the citation of the most obvious, complete specimens.

TANGANYIKA. Mwanza, fr. 1 May 1945, *Greenway* 7389 A!; Iringa District: Great North Road, half-way between Mbeya and Iringa, fl. 18 Oct. 1936, *B. D. Burtt* 5786! & 5799! & Image Mt., fl. Nov. 1959, *Procter* 1559!

DISTR. T1, 3–5, 7, 8; Congo Republic, Mozambique, Malawi, Zambia and Rhodesia

HAB. Rather frequent (?) where both putative parents are recorded

SYN. *B.* × *fischeri* Taub. (pro sp.) in P.O.A. C: 197 (1895); Burtt Davy & Hutch. in K.B. 1923: 154 (1923); L.T.A.: 726 (1930). Type: Tanganyika, Singida District, Usure [Usuri], *Fischer* 148 (B, holo.!, K, fragm.!)

NOTE. *B. D. Burtt's* specimens form part of an associated series; No. 5786, pubescent throughout, has 12–14 pairs of medium to small leaflets of obviously intermediate shape and venation, on a rhachis with variable stipels unmistakably like those of 1, *B. spiciformis*; the tomentose mostly simple racemes, also recalling *B. spiciformis*, have flowers mostly with 4 variable often narrow tepals, sometimes inserted on a short staminal tube; No. 5799 has 8–11 pairs of leaflets, smaller but otherwise similar, glabrous or sparsely ciliate at base, on a subglabrous rhachis otherwise like that of No. 5786; the tomentose depauperate racemes are in young bud; both these trees were reported as showing a flaming claret-red flush of young leaves instead of the pale salmon-pink and green flush of the normal population of 5, *B. microphylla* in which they were scattered; with these occurred No. 5897, typical *B. microphylla* with up to 46 pairs of small leaflets, and No. 5798 with 30–35 pairs of larger leaflets of which some have irregular bases, probably indicating a back-cross, (sp. 5 × sp. 1) × sp. 5; No. 5773, "resembling *B. spiciformis* but with unusually small foliage", could easily be a back-cross (sp. 5 × sp. 1) × sp. 1. *Procter* 1559, otherwise similar in all respects to *B. D. Burtt* 5786, has 10–14 pairs of somewhat larger leaflets. *Greenway* 7389A (originally mixed with 7389B under 7389 but recorded as a different form) has 7–9 pairs of small broad leaflets and is manifestly intermediate in all characters between the associated No. 7387 (*B. spiciformis* with 2–4 pairs) and No. 7389B with 17–20 pairs of narrow leaflets; the latter has fewer and larger leaflets than more normal *B. microphylla* from the same district (*B. D. Burtt* 6452 attaining 31 pairs) and is therefore probably another back-cross, (sp. 5 × sp. 1) × sp. 5. *Anderson* 933 (Newala escarpment, 23 Sept. 1953) with 11–12 pairs of leaflets, is a rare glabrous intermediate between *B. microphylla* (*Eggeling* 6734, same locality) and *B. spiciformis* (observed here by Hoyle and always collected with glabrous leaves in this area). Other more or less obvious intermediates are: *B. D. Burtt* 1894 (in BM!, = 1296 in K, partly!) & 1388 & 660 (in *Game Dept.* 5660); 132 km. Iringa–Mbeya, *Procter* 2070; Lindi District, Nachingwea, *Anderson* 1053; Handeni District, Swagilo, *Semsei* in *F.H.* 2909; 88 km. Njombe–Songea, *Boaler* 552.

It is important to distinguish the evident manifestations of hybridity from the results of pathological or other growth-stimuli. The former usually show most obviously in differences between leaflets or flowers on the same leaf or inflorescence, or between leaves or inflorescences on different branchlets diverging at the same level; the latter usually show in sudden changes, in the number and size only, of leaflets on regrowth continuing the same branchlet. It is noticeable that many putative hybrid specimens show a surprising constancy in the number of leaflets per leaf throughout one season's growth, even more so than their associated putative parents.

2 × 5. **B. bussei × B. microphylla**

Apparently less frequent than forms of *B. microphylla* × *B. spiciformis*, discussed above, differing from them usually in being less pubescent (often quite glabrous or with only the vegetative buds pubescent); the leaf-rhachis is usually less winged and the stipellar expansions (if any) rather obscure and semilunate like those of *B. bussei*; the leaflets have more open and less prominent reticulation, fewer basal fanwise nerves and a surface and texture essentially recalling *B. bussei*. Details and further examples are given below the citation of the most obvious intermediate specimens; all available from East Africa are sterile (or with detached pods) except those close to either putative parent, which they then resemble in the pod.

TANGANYIKA. Morogoro District: Morogoro–Dar es Salaam road, Kiroka Pass, 19 June 1949, *Hoyle* 1029 ! & 1031 !; Rufiji District: 3 km. S. of Utete, 14 Nov. 1954, *Anderson* 1012 !; Songea District: 106 km. on Songea–Njombe road, Nyungo (?) R. bank, 22 July 1949, *Hoyle & Greenway* 1088 ! & 1089 !

DISTR. T3 (or ? 6), ? 5, 6, 8; Mozambique, Malawi; probably also in Zambia

HAB. In a few areas where the putative parents are associated

NOTE. *Hoyle* 1029 (FHO) is a vigorous glabrous branchlet from a small tree (probably of coppice-origin) bearing leaves with 9–11 pairs of large to medium-sized narrow tapering leaflets and, in the axils of some of the leaves, secondary branchlets with leaves bearing 9–11 pairs of leaflets like the smallest on the main shoot; all the leaves are intermediate in varying degree between the adjacent tree, *Hoyle* 1028 (*B. bussei*) and local examples of *B. microphylla*; *Hoyle* 1031, from a sapling or sucker with puberulous branchlets and leaf-rhachis, has 10–12 pairs of more uniform, often falcate leaflets similar to the smaller ones of *Hoyle* 1029. *Anderson* 1012 has (5–)7–9 pairs of leaflets on a pubescent rhachis, exactly intermediate between his associated No. 1010 (*B. bussei*) and *B. microphylla* (recorded north, west and south of, but not in this thinly-collected area). *Hoyle & Greenway* 1088 & 1089, both glabrous except the vegetative buds, have leaves with (5–)6–8 and (7–)9–10 pairs respectively of both medium-sized and small leaflets, mingled at various levels on the secondary divisions of a branch-system; No. 1087, with 26–33 pairs of very small leaflets, approaches nearer to the local *B. microphylla* which can attain 43 pairs; trees of *B. bussei* were observed on the adjacent slope. *Milne-Redhead & Taylor* 10679 (Tunduru District, 7 June 1956), with (17–)19–23 pairs of small leaflets, is intermediate between *Hoyle & Greenway* 1089 and 1087. *Anderson* 1054 (Lindi District, Nachingwea, 24 April 1955) has 8–10 pairs of medium-sized leaflets on a pubescent rhachis and pods (from ground) identical with those of *B. bussei*; it occurred " in small numbers " among *B. microphylla* in areas next to *B. bussei* (No. 1041 with some puberulence and No. 1042, adjacent, glabrous). *Greenway* 5817 (Turiani–Handeni, fr. 14 Oct. 1938), approaching *B. bussei* but with leaves persistently puberulous in parts, has some very narrow leaflets and a smaller pod than the associated No. 5815 (*B. bussei*), with No. 5816 (*B. microphylla*) also present; *Boaler* 240/2 (129 km. on Morogoro–Korogwe road, 18 July 1962, i.e. same area as the last) has 6–7 pairs of small often falcate leaflets like those of *Hoyle* 1031 (cited above) and occurs with No. 240/3 (*B. bussei*).

6 × 5. **B. floribunda × B. microphylla**

Forms showing intermediate characters in the leaf-rhachis and in the number, spacing, size, shape and venation of leaflets; those with fewer than 12 pairs of leaflets have completely glabrous branchlets and leaves with the rhachis not channelled, as in *B. floribunda*; one attaining 18 pairs of glabrous leaflets has pubescent branchlets and leaf-rhachis, the latter channelled as in *B. microphylla* (for details see note below citation).

TANGANYIKA. Ufipa District: Ufipa plateau, Sumbawanga–Malonje path, 19 Dec. 1934, *Michelmore* 1073 !; Mbeya District: Chimala scarp, Feb. 1961, *Procter* 1775 ! & 1776 ! & 1778 ! & 1779 !

DISTR. T4, 7; Zambia

HAB. Rare, where both putative parents are recorded

NOTE. *Procter* collected his (sterile) specimens at two altitudes. At 2070 m., normal *B. floribunda* (Nos. 1773 & 1774, with non-flaking bark and 3–4 pairs of leaflets) was common on top of the scarp but one aberrant tree (No. 1775) had flaking bark and 4–5 pairs of narrow, very acute leaflets of a shape and venation unprecedented in this species. At 1935 m., a group of trees included normal *B. floribunda* (No. 1777 with 3 pairs of leaflets) and three apparent hybrids; No. 1778 with 6–8 pairs of leaflets narrower and smaller than those of *B. floribunda*; No. 1776 with 6–8 pairs still smaller but relatively broader; No. 1779 with 15–18 pairs of small oblong leaflets much broader than in normal *B. microphylla*, of which several small populations were noted on the slope of the scarp. *Michelmore* 1073 (Ufipa), when collected at 1980 m., was thought to be a hybrid or form of *B. microphylla* [which occurs on the plateau], " one tree only, in company with Muputu [*B. spiciformis*] and Musompa [*B. floribunda*] "; glabrous except for vegetative buds and the apparent remnants of panicles, it has leaves with 9–11 pairs of very narrow, acute leaflets on a non-channelled rhachis.

In addition to the above, *Procter* 1849 (Chimala scarp at 1525 m., fr. May 1961), with " bark smooth and grey like *B. microphylla* ", pubescent vegetative buds like *B. spiciformis* but otherwise glabrous, and 7 pairs of medium-sized leaflets on a narrowly channelled rhachis with obscure stipellar wings, suggests by its characters a triple hybrid, (sp. 5 × sp. 1) × sp. 6; all three putative parents were present.

7 × 5. **B. manga** × **B. microphylla**

Very rare (?) forms showing intermediate characters in the leaf-rhachis and in the number, spacing, size, shape and venation of glabrous leaflets (see details in note below citation).

TANGANYIKA. Iringa District: 6·4 km. on Iringa–Dodoma road, Kihesa, Feb. 1961, *Procter* 1761 ! & 1762 ! & Iringa Highlands " at head of north-east valley ", 31 Jan. 1932, *Lynes* I.h.52 !

DISTR. **T7**; not recorded elsewhere

HAB. Rare, where the putative parents are associated

NOTE. *Procter* 1761, nearer *B. manga*, has 4–6 pairs of rather broad glabrous leaflets, very variable in size, on a non-channelled rhachis with occasional hairs between the leaflets of some pairs on a few leaves; No. 1762, nearer *B. microphylla*, has 7–9 pairs of smaller narrower leaflets on a channelled slightly winged rhachis with tufts of hairs between the leaflets of most pairs on most leaves; both specimens have the characteristic purplish-pruinose surface shown by No. 1760 (*B. manga*), occurring close by, with 2–3 pairs of broad leaflets on a glabrous non-channelled rhachis. *Lynes* I.h.52 has 9–11 pairs of small leaflets variable in shape; his I.h.210 (Iringa, Mt. Tarik, 12 Mar. 1932), with 15–29 pairs of small oblong leaflets, seems also hybrid but much nearer sp. 5; a specimen of *B. manga* (fr. 15 Mar. 1932) is also numbered I.h.210. Both putative parents are frequent around Iringa.

6. **B. floribunda** *Benth.* in Hook., Ic. Pl. 14: 43 (1881); Burtt Davy & Hutch. in K.B. 1923: 157 (1923); C. H. N. Jackson in Journ. S. Afr. Bot. 6: 38 (1940); B. D. Burtt in Journ. Ecol. 30: 78, 141, t. 4 (1942); T.T.C.L.: 92 (1949); Hoyle in F.C.B. 3: 465, fig. 39/B (1952); Hoyle & White in F.F.N.R.: 107, fig. 22/A, t. 1/J (1962). Type: Malawi, Shire Highlands, Zomba, *Buchanan* 10 (K, lecto. !, E, isolecto. !)

Tree (4–)6–12(–15) m. high; bark at first smooth, whitish or silvery-grey, becoming rather rough and darker (when burnt?), slowly shedding in irregular scales or completely* to expose a pale grey finely muricate surface; crown at first narrow, erect-branched, finally spreading and irregularly rounded; mature foliage usually glaucous*, usually bunched; leaves pendulous, fluttering in the slightest breeze. Branchlets usually soon rusty. Stipules free, linear to falcate-subfoliaceous, 0·5–2 cm. long, usually fugacious; auricle 0. Leaves glabrous, with (2–)3–4 widely spaced pairs of leaflets, the distal pair usually the largest; petiole (1·5–)2–5(–7) cm. long, slender; rhachis (2·5–)3–8(–12) cm. long, slender, not (or very obscurely) channelled, without stipels or expansions; leaflets ovate or rhombic to falcate, (2–)3–8(–10) × (1–)2–3·5(–5) cm., obtuse to acute or acuminate, usually very oblique and rounded- or subcordate-cuneate at base; midrib subcentral; basal fanwise nerves (3–)4–5(–6); surfaces mat, concolorous, both equally reticulate with age. Panicles usually conspicuous, often clustered, mainly

* See notes on leaf-colour and bark below.

on older wood or leafless branchlets (more rarely also terminal and then smaller), up to 10 × 8 cm., pubescent to tomentose with brown or rusty hairs. Flowers greenish-white; bracteoles 4–6(–8) × 2·5–4(–5) mm. Tepals 5(–7), usually all sepaloid but usually narrow and then scarcely imbricate; outer 5(–7), up to 3 × 1 mm., free, shortly ciliate; inner 0 or 1–2, narrow. Stamens ± 10, usually free, filaments 8–10 mm. long. Ovary 2·5–3 × 1 mm., ± densely crispate-setose; style 7–8 mm. long, stigma small. Pod thinly woody, up to 12·5 × 4 cm., pendulous, smooth, blue-black to brownish-purple, ± pruinose over a mat, minutely papillose surface; sutural wings spreading, each 3–4 mm. wide. Fig. 35/1, p. 160.

TANGANYIKA. Kigoma District: Mahali Mts., Kabesi valley, fl. 31 Aug. 1958, *Jefford, Juniper & Newbould* 1973!; Mbeya District; Mbozi plateau, near Vawa, fl. 28 Sept. 1936, *B. D. Burtt* 5756! & Mbeya Mt., Mbeya–Chunya road, fr. 8 July 1949, *Hoyle & Greenway* 1077!

DISTR. T4, 7, 8; Congo Republic, Mozambique, Malawi, Zambia and Angola

HAB. Deciduous woodland, locally dominant or co-dominant between 700 and 2075 m., often over large areas on high plateaux, mainly above 1200 m. in SW. Tanganyika. Typically on rather fertile orange and red soils on plateaux, rocky hill-slopes and scarps, sometimes next below mountain forest or stunted on edges of mountain grassland. Occurs almost pure or in mixed woodland with *Uapaca* or *B. spiciformis*, *Julbernardia paniculata*, etc., less often in drier types.

SYN. *B. polyantha* Harms in E.J. 30: 319 (1901); Burtt Davy & Hutch. in K.B. 1923: 157 (1923); L.T.A.: 731 (1930). Type: Tanganyika, Rungwe District, Kiwira valley, *Goetze* 1478 (B, holo. !, E, iso. !, K, photo. !)

VARIATION. Normal fertile material varies little except in width and size of leaflets, but those on sterile branchlets are often gradually acuminate and conspicuously falcate. Confusion with *B. bussei* is then probable, especially because the usually glaucous leaflets of *B. floribunda* can evidently turn green for a while at maturity, or even fail to become glaucous in moister areas. Immature sterile material of *B. floribunda* and the related *B. manga* is often very difficult to distinguish, notably if it comes from young vigorous trees or even older ones derived from coppice. Such an origin, rather than hybridity, may well explain intermediate forms from Mpanda and Kigoma Districts of Tanganyika.

Descriptions of the bark of *B. floribunda* vary widely. There is evidence that intense fire may cause the outer bark to be sloughed off over the whole bole, but that where fire is absent or slight, the outer bark may persist smooth and pale much longer. Moderate fire and slow shedding of bark can probably be regarded as " normal ".

Hybrids of *B. floribunda* seem very rare. There are only two examples available (both outside the Flora area) of apparent crossing with 1, *B. spiciformis*, although the species are widely associated. In two localities in southern Tanganyika, hybrids with 5, *B. microphylla* seem obvious and a triple hybrid is also suspected (see 6 × 5 below *B. microphylla*).

7. **B. manga** *De Wild.*, Contrib. Fl. Katanga, Suppl. 2: 48 (1929); L.T.A.: 728 (1930), haud tamen aff. *B. spiciformi*; Hoyle in F.C.B. 3: 464 (1952); Hoyle & White in F.F.N.R.: 110, fig. 22/C, t. 1/I (1962). Type: Congo Republic, Katanga, Kapiri valley, *Homblé* 1245 (BR, lecto. !)

Tree (4–)7–12(–20) m. high; bark at first smooth, silver-grey, becoming roughly reticulate-fissured in age, persistent (?), grey, darker when burnt; crown rounded; foliage usually bunched, mature bluish-glaucous, old pale, yellowish grey-green. Branchlets usually soon rusty. Stipules free, linear to falcate-subfoliaceous, (0·7–)1–1·5(–2·5) cm. long, usually fugacious; auricles 0. Leaves glabrous (see note), with 2–4(–5) pairs of leaflets, the middle or distal pairs the largest; petiole (1·5–)2–3(–3·5) cm. long; rhachis (3–)5–10(–15) cm. long, not (or very obscurely) channelled, without stipels or expansions; leaflets very variable in shape but usually broad, subcircular or ovate to obovate, often ± irregularly rhombic, (1·5–)2–8(–10) × (1–)1·5–4·5(–6) cm., emarginate to obtuse, normally not acuminate, ± obliquely cuneate to cordate at base; midrib subcentral; basal fanwise nerves 2–3 (very rarely 4, even when base very broad); surfaces mat, ± concolorous, both equally but obscurely reticulate, glabrous or very rarely puberulous (see note). Panicles terminal and often also axillary (rarely lateral on older wood), slender, up to 9 × 5 cm.; axis and branches usually glabrous, remainder finely sericeous. Flowers very small; bracteoles 3–5(–6) × 3–4 mm., yellowish green. Tepals (4–)5(–6), usually all sepaloid and subequally broad but very small, up to 2 × 1 mm., pale green, shortly and sparsely ciliate or glabrous. Stamens ± 10, usually free, filaments 6–8 mm. long. Ovary 2–2·5 × 1 mm., subdensely setose with very dark brown hairs; style 6–8 mm. long; stigma rather prominent. Pod thinly woody, up to 10 × 3·5 cm., pendulous, smooth, pinkish- or purplish-brown, ± pruinose over a mat, minutely papillose surface; sutural wings thin, spreading, each 4–6 mm. wide. Fig. 35/3, p. 160.

TANGANYIKA. Dodoma District: Manyoni, fl. 21 Nov. 1931, *B. D. Burtt* 3429 !; Chunya District: Kipembawe, Ngulwe Hill, fr. 6 July 1949, *Hoyle & Greenway* 1073 !; Mbeya District: foothills of Poroto Mts., W. of Chimala, fl. 16 Nov. 1949, *Hoyle* 1374 !

DISTR. T4–8; Congo Republic, Mozambique, Malawi, Zambia and Rhodesia

HAB. Deciduous woodland, locally dominant, mainly between 1100 and 1830 m. (down to 300 m. in T8), often over wide areas on dry aspects of escarpments. Typically on over-drained rocky hills and slopes, notably Rift walls, more rarely on sandy loams of valley slopes. Usually occurs pure, less often with *B. spiciformis*, *B. allenii*, *B. microphylla* or *Julbernardia globiflora*.

SYN. *B. burttii* C. H. N. Jackson in Journ. S. Afr. Bot. 6: 37 (1940), anglice tantum; B. D. Burtt in Journ. Ecol. 30: 78, 141 (1942); T.T.C.L.: 90 (1949)

VARIATION. The usually very broad leaflets often vary characteristically from leaf to leaf and are often irregular in shape, partly due to very frequent insect-attack during youth. The appearance of the bunched foliage at flowering-time, usually with a mixture of colours from the claret-red or maroon flush to bluish plum-colour, is very characteristic, although sometimes similar in *B. floribunda*. The habit of a pure stand bears a remarkable resemblance to an orchard, especially when the old leaves turn a pale, greyish apple-green.

Forms with 5 or more pairs of leaflets (or fewer when puberulous) are suspected of being hybrids with 1, *B. spiciformis* or 5, *B. microphylla*. See notes below 6, *B. floribunda* and also 7 × 5 below *B. microphylla*.

8. **B. stipulata** *De Wild.* in Ann. Mus. Congo, Bot. sér. 4: 44, t. 12 (1902), in t. sub nom. *B. appendiculata* De Wild. *non* Benth.; Burtt Davy & Hutch. in K.B. 1923: 150 (1923); C. H. N. Jackson in Journ. S. Afr. Bot. 6: 39 (1940); B. D. Burtt in Journ. Ecol. 30: 141 (1942); T.T.C.L.: 92 (1949); Hoyle in F.C.B. 3: 457 (1952); Hoyle & White in F.F.N.R.: 106, fig. 22/F (1962). Type: Congo Republic, Katanga, Lukafu, *Verdick* 18 (BR, holo. !, K, photo. !)

Tree or treelet 1–7(–9) m. high; mature bark rough, reticulate to fissured, persistent, grey; crown umbrella-shaped or flat; foliage maturing green, often glaucous beneath. Stipules shortly connate, linear to falcate, 1–2 cm. long, usually persistent; auricle usually very persistent and conspicuous, 1–2 cm. long, broadly reniform; intrapetiolar stipule-bases persistent, subtending flattened and laterally keeled dormant buds. Leaves with (5–)6–8(–10) pairs of leaflets, the middle pairs usually the largest; petiole 1–2 cm. long; rhachis (6–)8–15(–20) cm. long, channelled; stipellar expansions variable, often obscure or 0; leaflets oblong-elliptic or more rarely narrowed to the retuse or rounded apex, (2–)3–6(–8) × (1–)1·5–2(–4) cm., ± obliquely rounded to cordate at base; midrib subcentral; basal fanwise nerves (3–)4–5(–6). Racemes simple or few-branched, terminal or terminal and axillary. Flowers few, large; bracteoles 13–20 × 6–10 mm. Tepals 6–8(–11), imbricate, usually all sepaloid and grading from broad to narrow, 4–8 mm. long, densely long-ciliate, or obscurely in 2 whorls with the inner spathulate to vestigial. Stamens 13–18 (sometimes with staminodes, totalling ± 20); filaments 25–30(–40) mm. long, ± connate, the tube usually subequal to the tepals. Ovary 8–10 × 1·5–3 mm., ± densely crispate-setose, tapering gradually into the 20–30 mm. long erect style with small stigma. Pod woody, up to 15(–22) × 5 cm., smooth but often transversely nervose, rarely scurfy in patches, maturing pinkish- (usually pale) brown; sutural wings stiffly spreading, each 4–8 mm. wide. Fig. 35/13.

TANGANYIKA. Ufipa plateau, Rwassi Forest Reserve, fl. 16 Oct. 1928, *Wigg* in *F.H.* 387!
DISTR. T4; Congo Republic, Mozambique, Malawi and Zambia
HAB. Deciduous woodland, known in Tanganyika only from the cited gathering, at 1460 m.; elsewhere mainly between 900 and 1700 m. and locally frequent. Typically gregarious on edges of ill-drained dambos on poor or shallow soils, especially over schists or ironstone; less often on quartzite ridges; commonly acts as understorey to *B. boehmii* or *Julbernardia globiflora*.

SYN. *B. velutina* De Wild. in F.R. 11: 512 (1913). Type: Congo Republic, Katanga, Elisabethville, *Hock* (BR, holo. !, K, photo. !)
See also additional synonymy in F.C.B. 3: 457–60 and following note.

VARIATION. The species, in a broad sense, is easily recognized by its unique large flowers and persistent, short stipules with large auricles. Variation is, however, considerable in hairiness and in the shape, size and number of leaflets; extreme forms, when sterile, are easily confused with *B. allenii* and its hybrids or with forms of the *B.* × *longifolia* complex. The sole record from the Flora area represents one of several forms with spreading fulvous hairs on leaves and inflorescence and is indistinguishable from *B. velutina* var. *quarrei* De Wild., the flowering stage of *B. velutina* De Wild. This specific epithet was used (F.C.B. 3: 459) under *B. stipulata* in a varietal combination which is now seen to be illegitimate and should not be used. It has become evident, not only that various more or less hairy forms occur widely both inside and especially outside the Congo, but that occasional forms, glabrous like the type (but without apparent connection with it), appear sporadically here and there within the total range. Hairy inflorescences may also be combined with glabrous leaves. Moreover a hairy form can apparently hybridize with *B. boehmii* to produce *B.* × *bequaertii* De Wild. (pro sp.), and a glabrous form with *B. glaberrima*, so that the position is potentially complex. In these circumstances varietal names seem more misleading than useful.

9. **B. allenii** *Burtt Davy & Hutch.* in K.B. 1923: 156 (1923); L.T.A.: 731 (1930); C. H. N. Jackson in Journ. S. Afr. Bot. 6: 37 (1940); B. D. Burtt in Journ. Ecol. 30: 104, 141, t. 3 (1942); T.T.C.L.: 92 (1949); Hoyle in F.C.B. 3: 467 (1952); Hoyle & White in F.F.N.R.: 108, fig. 22/D (1962). Type: Mozambique, Niassa, Msalu R., *Allen* 93 (K, holo. !, FHO, photo. !)

Tree 3–15(–20) m. high, glabrous; bark rough, persistent, usually deeply fissured and transversely cracked, pale grey; crown rounded; foliage bushy (see notes below), maturing blue-grey, especially beneath. Branchlets often rusty-yellow where pale grey epidermis peels. Stipules shortly connate (or ± free but partly intrapetiolar), linear to falcate, 0·5–1(–2) cm. long, usually subpersistent; auricle often more persistent, broadly reniform, 0·3–1(–2) cm. long, rarely undeveloped; intrapetiolar stipule-bases persistent but often very short and obscure, subtending flattened and laterally keeled dormant buds. Leaves with (3–)4–5(–6) subequal pairs of leaflets or the middle pairs the largest; petiole (1–)1·5–2·5 cm. long; rhachis (4–)6–15(–18) cm. long, ± channelled, with or without spreading stipels or short narrow wings; leaflets subcircular or broadly oblong to obovate, (1·5–)2–6 × (1–)1·5–4 cm., emarginate or subtruncate to rounded at apex, cordate to truncate and subsymmetrical at base; midrib subcentral; basal fanwise nerves (3–)4–5(–6); nerves conspicuous; veins very closely reticulate. Panicles terminal or terminal and axillary, slender, 4–7 × 3–7 cm.; branches rather long, usually subtended at anthesis by pairs of stipular auricles or complete stipules, with or without reduced 1–3-jugate leaves. Flowers cream-yellow or with rose-pink bracteoles and white tepals; bracteoles 5–8 × 3–4 mm. Tepals (4–)5(–8), all sepaloid or more rarely (4–)5 + 1–3, usually all free; outer (4–)5, ovate to spathulate, usually unequal, ± imbricate, 2–4 × 1–2 mm., ± densely long-ciliate, sometimes grading into the (0–)1–3 inner, or the latter

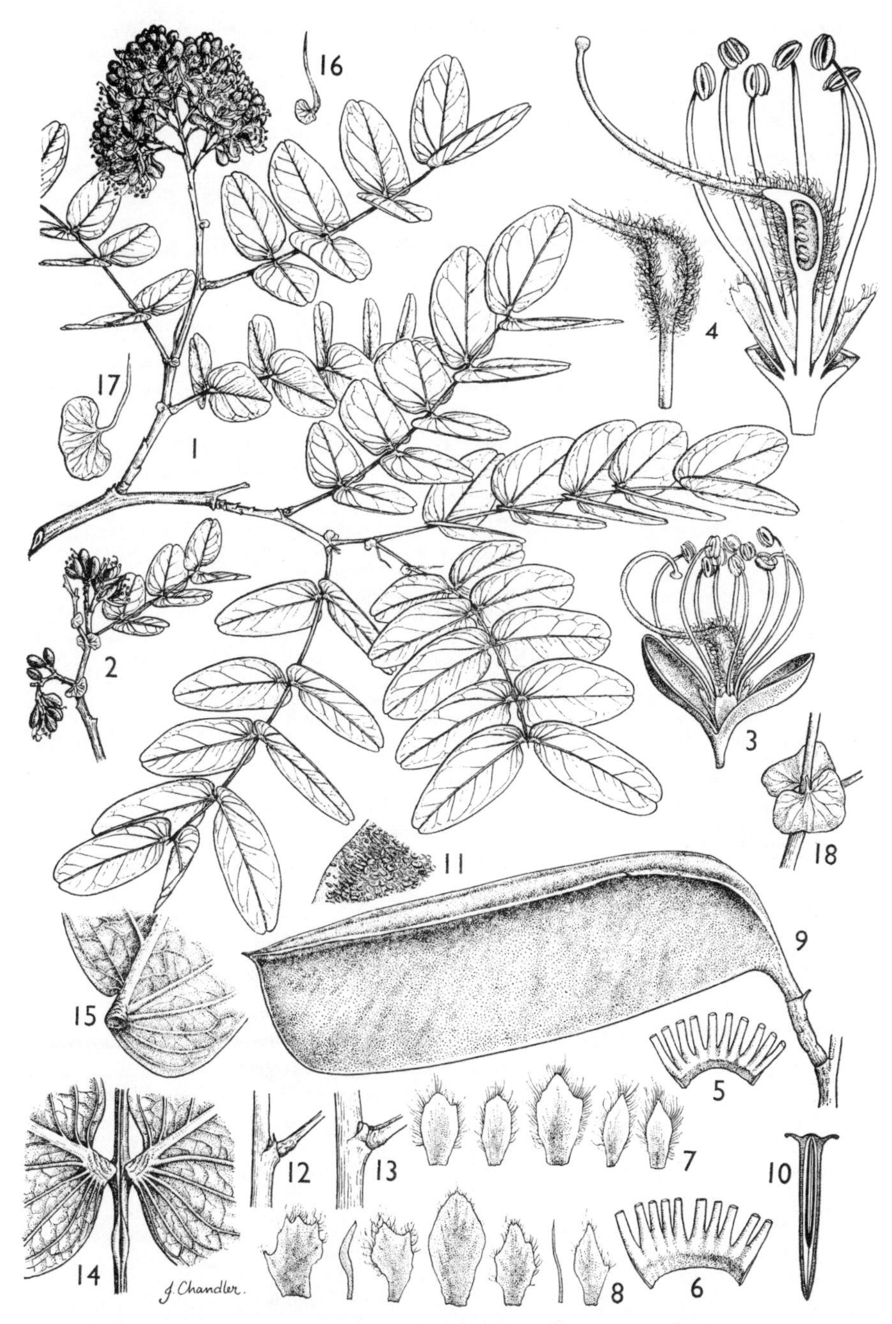

Fig. 39. *BRACHYSTEGIA ALLENII*—**1,** flowering branchlet, × ½; **2,** part of another inflorescence showing reduced leaf and characteristic stipular auricles subtending branches, × ½; **3,** flower, × 2; **4,** gynoecium and longitudinal section of flower, × 4; **5, 6,** hypanthium and base of filaments, two variants, × 4; **7, 8,** tepals, two variants, × 4; **9,** pod, × ½; **10,** transverse section of pod, × ½; **11,** scurfy surface of pod, × 5; **12,** dormant bud (on young branchlet) subtended by pulvinus and persistent intrapetiolar base of connate stipules, × 1; **13,** same on old branchlet, × 1; **14,** base of leaflets of a middle pair, upper surface, showing one form of the variable (sometimes absent) adnate stipels, × 2; **15,** base of leaflet, lower surface, × 2; **16, 17,** stipules, two forms, × 1; **18,** stipular auricles of average size in characteristic position, × 1. 1, 3, 4, 6, 8, 12, 14–16, from *Allen* 93; 2, from *R. G. Miller* 244; 5, 7, from *Gillman* 1088; 9–11, 13 from *Hoyle* 1080; 17, from *Lindeman* 793; 18, from *H. Gamwell* 246.

linear to vestigial, usually glabrous. Stamens ± 10, ± free, 8–12 mm. long. Ovary 2–3 mm. long, densely setose mainly on the margins. Pod woody, up to 16 × 4·5 cm., usually held conspicuously like a flag above crown of tree, smooth and drying blue-black and ± pruinose when immature, slowly becoming pinkish-brown and finely scurfy throughout when mature; sutural wings spreading, each 4–8 mm. wide. Fig. 39, p. 183.

TANGANYIKA. Ufipa District: Kipili–Kirando, Lake Tanganyika coast, fr. 16 Apr. 1936, *B. D. Burtt* 5703!; Masasi, fl. 12 Dec. 1942, *Gillman* 1088! & 16 km. Masasi–Newala, fr. 18 July 1949, *Hoyle & Greenway* 1080 (with photographs of habit and bark, FHO, K)!

DISTR. **T4**, 7, 8; Congo Republic, Mozambique, Malawi, Zambia and (?) Rhodesia (see note below)

HAB. Deciduous woodland, locally dominant on well-drained sites, mostly in dry, hot country at lower altitudes, from sea-level or lake-level to 1000 m. Typical of rocky escarpments around the southern half of Lake Tanganyika and of sandy slopes to the Ruvuma valley, otherwise rather rare; usually in pure stands below *B. bussei* and *B. boehmii*, less often in mixture with these and other species.

SYN. *B. pruinosa* De Wild., Contr. Fl. Kat., Suppl. 2: 53 (1929); L.T.A.: 730 (1930), haud tamen syn. *B. bakerana*. Type: Congo Republic, Katanga, Kiambi, *Delevoy* 537 (BR, holo.!, FHO, photo!)

VARIATION. In the genuine form of *B. allenii* (as understood), the leaflets are not complanate when growing but each lies in a plane more or less at right-angles to the leaf-rhachis (see fig. 39/1). This certainly applies to the glabrous, Lake Tanganyika form, with (3–)4–5(–6) pairs of broad leaflets and even to one in Mozambique with 5–7 pairs. Sundry specimens differ slightly from type, e.g. *Gillman* 1088 & 1186 (Masasi, 12 Dec. 1942) with rather narrow leaflets and no stipules remaining in the late flowering stage; also *Wigg* 388 (Ufipa District, Kirando, fl. 30 Oct. 1928), similar but with stipular auricles persistent, showing occasional hairs on bracteoles and leaf-rhachis. These are insignificant details unless accompanied by other aberrations.

In several areas, more aberrant forms occur, some with hairy panicles and leaves and with stipules longer relative to their auricles. These forms seem to be direct or introgressive hybrids with 13, *B. boehmii*, the only species in the same group known to be in contact in the areas concerned. The most complete and obvious series of hybrids comes from Rhodesia, where the occurrence of " pure " *B. allenii* seems doubtful; further examples come from localities scattered through northern Mozambique to southern Tanganyika (see below).

Sterile specimens of *B. allenii* and its putative hybrids are very easily confused with 8, *B. stipulata*.

9 × 13. **B. allenii** × **B. boehmii**; F.F.N.R.: 117 (1962)

The putative hybrid specimens available from the Flora area are mostly poor or sterile and approach either *B. allenii* or *B. boehmii* in number and shape of leaflets, centrally intermediate forms in this respect not being represented. Variations mainly concern the apparent introgression of pubescence, relatively longer stipules and more and/or narrower leaflets into *B. allenii* on the one hand and the opposites of these characters into *B. boehmii*. Examples of the former are cited, with details and further examples, in the notes below; examples of the latter are detailed in the notes below 13, *B. boehmii*.

TANGANYIKA. Ufipa District: Kasanga, 9 May 1941, *Lindeman* 793!; Masasi District: 8 km. W. of Masasi, 19 July 1949, *Hoyle & Greenway* 1081!; Tunduru/Masasi District: at R. Ruvuma, 12 Feb. 1901, *Busse* 1020!; Lindi District: Nachingwea, 14 Jan. 1955, *Anderson* 1019!

DISTR. **T4**, 8; Congo Republic, Mozambique, Malawi, Zambia and Rhodesia

HAB. In areas where both putative parents are recorded

SYN. *B.* × *schliebenii* Harms (pro sp.) e descript. in N.B.G.B. 12: 508 (1935); T.T.C.L.: 92 (1949). Type: Tanganyika, ? Newala District, Lukuledi valley 85 km. from Lindi, fl. 23 Oct. 1934, *Schlieben* 5520 (B, holo. †) (vide notam)

NOTE. *Hoyle & Greenway* 1081, a glabrous tree with 5–7 pairs of narrowly ovate to oblong leaflets, grew on the edge of a uniform population of normal *B. allenii* but differed in its bark and drooping foliage; it was found during a search for *B. schliebenii* Harms (type not seen, believed destroyed), of which it seems to represent the fruiting stage; *Milne-Redhead & Taylor* 7663 (16 km. E. of Masasi, nearer the locality of *Schlieben* 5520) is similar but with 5–6 pairs of leaflets; *Semsei* in *F.H.* 2113 (Lindi District, Ruponda), also glabrous, has 5–7 pairs of elliptic to oblong leaflets. *Anderson* 1019 (growing with his No. 1018, glabrous *B. allenii* with 4–5 pairs of broad leaflets, and only 0·8 km. from his No. 1021, pubescent *B. boehmii*) is pubescent with hairs like those of *B. boehmii*, and has 5–6 pairs of leaflets, long stipules and remnants of large persistent bud-scales. *Busse* 1020 (with pruinose immature pod), pubescent throughout, has long stipules and a single leaf, originally with 7 pairs of leaflets, of which the two remaining are long and narrow. *Lindeman* 793 (EA ! & K, partly !) has hairy leaves with 4–6 pairs of leaflets (only 3–5 pairs, glabrous, being evidently usual for *B. allenii* in the area); *Sciwale* 4 (80 km. on Mpanda–Sumbawanga road) has 4–5 pairs of similar but glabrous leaflets on a pubescent rhachis.

10. **B. angustistipulata** *De Wild.* in F.R. 11 : 511 (1913) & in Ann. Mus. Congo, Bot. sér. 4, 2: 43 (1913); Burtt Davy & Hutch. in K.B. 1923: 157 (1923); Hoyle in F.C.B. 3: 470 (1952). Type: Congo Republic, Katanga, Sankishia, *Bequaert* 165 (BR, holo. !, K, photo !)

Tree 2·5–6 m. high; habit gnarled; bark finely to coarsely reticulate-fissured, grey, brownish or black; crown few-branched, flat-topped; foliage pendulous in loose bunches. Stipules shortly connate, linear, ± falcate, 1·5–2 cm. long, caducous; auricle lateral, small, or 0; intrapetiolar stipule-bases persistent, usually prominent, subtending flattened and keeled dormant buds. Leaves glabrous or shortly appressed-pilose throughout, with (4–)5–7(–9) widely-spaced pairs of leaflets, the middle pairs usually the largest; petiole (1·5–)2–5 cm. long, pulvinus prominent, dark; rhachis (13–)15–20(–25) cm. long, channelled; stipellar expansions usually obscure or absent; leaflets oblong to narrowly ovate, rarely falcate, (2–)3–11 × (1–)2–3(–4) cm., obtuse or acute to acuminate, obliquely truncate-subcordate to rounded-cuneate at base; midrib subcentral; basal fanwise nerves (2–)3–4; lateral nerves slender, rather close; surfaces closely but often obscurely reticulate. Panicles terminal and axillary, 5–10 × 4–7 cm., ± densely puberulous. Flowers rather large; bracteoles (6–)7–9(–11) × 5–6 mm. Tepals 4–6(–8), free or some connate; outer broadly imbricate, 1·5–3 × 1–3 mm., densely and usually long-ciliate; inner 0 or 1–2, linear. Stamens ± 10, free, filaments 9–11 mm. long; disc-glands usually prominent. Ovary 3·5–5 × 1–1·5 mm., margins crispate-setose. Pod woody, up to 16 × 4 cm., maturing pale brown, thin, becoming densely scurfy very early; sutural wings thin, each 6–8 mm. wide. Fig. 35/12, p. 160.

TANGANYIKA. Kigoma District: Mahali Mts., Itemba, fl. 1 Oct. 1958, *Jefford, Juniper & Newbould* 2801 ! & 2804 !; Mpanda District: 16 km. S. of Mpanda, fr. 21 Mar. 1961, *Boaler* 242 ! & fl. Sept. 1961, *Procter* 1939 ! & 1943 !

DISTR. **T4**; Congo Republic

HAB. Deciduous woodland; 980–1525 m. Gregarious and local, evidently rare; seems to favour ill-drained soils below hills or wet sites close to drainage-lines with *Acacia, Combretum* or *Terminalia* and *Isoberlinia angolensis*, or in patches in *Julbernardia globiflora-Brachystegia boehmii* woodland. Apparently confined, in the Flora area, to extreme west Tanganyika.

VARIATION. Rather inadequately known from few, scattered localities but obviously distinct and regularly showing a unique character-combination. Many of the flowers seen are galled and this may explain the rarity of what seems to be an ancient relic with suggestive resemblances to high-forest species. It varies little except in width and apex of leaflets but can evidently have leaves glabrous or appressed-puberulous in one population (thus *Procter* 571, Uvinza–Kasulu, young fruit Nov. 1956, is mixed in FHO). The usually prominent disc-glands are not shown by *Rounce* in *EA* 8/31/1 (Buha District, Kasulu), an exceptionally vigorous and otherwise rather aberrant specimen with a tendency to a staminal tube.

Jackson 108 (Kasulu, Aug. 1935) is a mixture (in BM ! & K !) of flowering *B. angustistipulata*, with apparent coppice and also (in K !) sterile, old leaves of 11, *B. glaberrima*. The characters of the latter species and of 12, *B.* × *longifolia* in the region (e.g. *Semsei* in *Herring* 52 in *F.H.* 2484, Mpanda, Kabungu, fr. 25 July 1949), suggest probable introgression. Occasional hybrids with sp. 13, *B. boehmii*, seem to occur (see below).

10 × 13. **B. angustistipulata** × **B. boehmii**

A single gathering, with panicles but most flowers fallen, showing marked variation among the three available branchlets in the number (8, or 12–14 pairs) and shape of leaflets and in length of petiole (2·5 or 1 cm.); most leaflets are oblong and obtuse or retuse but some acute to acuminate and a few clearly abnormal; the long hairs on the leaves obviously come from the associated *B. boehmii*, while the panicles and notably the flowers (some galled) have mixed indumentum but otherwise resemble those of the associated *B. angustistipulata* (see also note below).

TANGANYIKA. Kigoma District: Mahali Mts., Itemba, fl. 1 Oct. 1958, *Jefford, Juniper & Newbould* 2799!
DISTR. T4; not known elsewhere
HAB. In dry, deciduous woodland of *Julbernardia globiflora*, with both putative parents

NOTE. Collected with Nos. 2800, 2801, 2803 and 2804, only slightly variable forms of *B. angustistipulata*, having glabrous leaves with (5–)6–7 pairs of leaflets; these were from a " hybrid swarm? " of about 50 trees, in the same woodland as No. 2808 with (13–)17–18 pairs of leaflets, representing " a few trees " of typical *B. boehmii*. The obvious hybrid No. 2799 has, on one sheet in Kew Herbarium, lateral panicles and one abnormal leaf with leaflet-scars at the apex of the pulvinus and one reduced leaflet of a pair after a long interval, followed by 7 pairs of variable leaflets like those of *B. angustistipulata* or intermediate; the leaves on the other sheet, with 12–14 pairs of leaflets, larger but also variable, have the proximal pairs of small leaflets very diverse in size and shape above a short, very stout petiole; the remnants of panicles are axillary. The FHO " duplicate " has three leaves, each with 8 pairs of large leaflets of variable, intermediate shape above a longer petiole.

11. **B. glaberrima** *R. E. Fries* in Wiss. Ergebn. Schwed. Rhod.-Kongo-Exped. 1: 66 (1914); Burtt Davy & Hutch. in K.B. 1923: 157 (1923); Hoyle & White in F.F.N.R.: 110, 117, fig. 22/G, t. 1/A, C, E (1962). Type: Zambia, Mporokoso, *Fries* 1177 (UPS, lecto.!, K, photo.!)

Tree (2–)5–15(–18) m. high, glabrous; bole often with conspicuous rounded bosses; bark rough, with few long deep furrows, grey; crown ± ovoid, becoming flattish only in age; main branches suberect; branchlets many, slender; foliage evenly spreading in all directions, not pendulous in tufts, maturing green above, glaucous beneath. Stipules shortly connate, linear to falcate, 1–2 cm. long, caducous; auricles 0 or small, usually lateral; intrapetiolar stipule-bases persistent, usually prominent, subtending flattened and keeled dormant buds. Leaves glabrous, with (4–)5–7(–8) widely-spaced pairs of leaflets, the middle and distal pairs subequal, spreading and conspicuously ladder-like in silhouette on the tree; petiole 1·5–3(–4) cm. long, usually subequal to the average of the intervals on the rhachis; pulvinus prominent and (though long) always forming less than one third of the petiole; rhachis 10–18(–20) cm. long (not more than 7 times as long as the petiole), channelled; stipels or local expansions usually obvious but variable; leaflets ± obliquely and narrowly triangular or subtrullate to ovate-elliptic or (rarely) oblong-elliptic, sometimes falcate, 3–10(–12) × (0·7–)1–3(–4) cm., obtuse or acute to rounded or retuse at apex, ± obliquely cuneate to subcordate at base; midrib subcentral; basal fanwise nerves (3–)4–5(–6), prominent and conspicuous above, the inner strongly ascending; main nerves and major reticulation above much more conspicuous than the final venation, both obscure beneath. Panicles terminal, up to 10 × 8 cm., glabrous; peduncle slender, terete, usually glaucous. Flowers green and cream; bracteoles 6–8 × 5–6 mm. Tepals 5–6(–7), usually 5–6 all sepaloid, imbricate, densely long-ciliate; rarely with 1–2 inner, linear or vestigial, usually glabrous. Stamens ± 10, ± free, filaments 9–12 mm. long. Ovary 2–3 × 1 mm., crispate-setose on the margins; style 8–10 mm. long; stigma rather large. Pod woody, up to 15 × 5 cm., smooth and usually shiny but often with warts (due to

FIG. 40. *BRACHYSTEGIA GLABERRIMA*—**1**, flowering branch, × ½; **2**, flower, × 2; **3**, gynoecium, × 4; **4**, longitudinal section of flower, × 4; **5**, hypanthium and base of filaments, × 4; **6–8**, tepals, three variants, × 4; **9**, pod in plan and transverse section, × ½; **10**, dormant bud, subtended by pulvinus and persistent intrapetiolar base of connate stipules, × 1; **11**, base of leaflets of distal pair, showing a frequent form of adnate stipels, × 2; **12**, **13**, two variants of stipels from middle of leaf, × 2; **14**, base of typical leaflet, lower surface, × 2; **15**, stipule from vigorous flowering branchlet, × 1; **16**, stipule from coppice, × 1; **17–19**, leaflets, three variants, × ½. 1–6, 11–14, from *Fries* 177; 7, from *R. G. Miller* 234; 8, 19, from *R. G. Miller* 225a; 9, from *D. Stevenson* 256; 10, from *Angus* 751; 15, from *Wigg* 66 in *F.H.* 441; 16, from *Delevoy* 965; 17, from *Barbosa & Carvalho* 3613; 18, from *R. G. Miller* 233.

insect attack), never finely scurfy; sutural wings spreading, each 4–7 mm. wide. Fig. 40, p. 187.

TANGANYIKA. Kigoma District: near Mkuti R., fl. Sept. 1956, *Procter* 516!; Tabora District: near Kakoma, fl. Aug. 1935, *C. H. N. Jackson* 110! & fr. 25 June 1949, *Hoyle* 1040! & SE. Tabora, 22 km. S. of Lyela sawmill, fl. Sept. 1951, *Groome* 13!

DISTR. **T**4, 7; Congo Republic and Zambia, rare (?) in Angola, Malawi and Mozambique

HAB. Deciduous woodland; 840–1500 m. Widespread in NW. Tanganyika but local and rather gregarious; typically dominant on the first slopes above an " mbuga " or drainage channel; mainly on red " plateau " soils, alone or with *B. boehmii*, *B.* × *longifolia*, *B. spiciformis*, *Julbernardia globiflora* or *Pterocarpus angolensis*; not, so far as known, on shallow soils.

SYN. [*B. longifolia* sensu auct., e.g. C. H. N. Jackson in Journ. S. Afr. Bot. 6: 38 (1940), pro parte (" glabrous forms "); T.T.C.L.: 93 (1949), pro parte (vide notam), *non* Benth.]

NOTE. Jackson (l.c.) and also Hoyle & Brenan in T.T.C.L. did not maintain *B. glaberrima* as a species distinct from what is here regarded as a hybrid complex, 12, *B.* × *longifolia* (apparently derived mainly from *B. glaberrima* × *B. boehmii*). Although very strict, the definition of *B. glaberrima* cannot be regarded as wholly satisfactory and difficulties are bound to arise in the identification of borderline specimens. Introgression of characters seems to have taken place singly as well as in groups; expression of characters may be unaltered or mingled, so that it may be possible to identify glabrous flowering material with all appropriate other characters as *B. glaberrima* and to name fruiting material from the same individual tree " *B.* × *longifolia* " because the pods are scurfy. As reliably correlated flowering and fruiting material from the same individual tree is almost unknown in the group, this unfortunate nomenclatural situation can only be avoided by resourceful collecting. By the same means it is possible to avoid the confusion that also arises between genuine examples of *B. glaberrima* from reasonably uniform populations and similar glabrous segregates of the *B.* × *longifolia* complex in heterogeneous populations. In naming individual specimens, less error is likely if doubtful ones are referred to *B.* × *longifolia* (see notes under this).

The mature bark of *B. glaberrima*, with few long deep furrows as if clawed by a lion, is normally characteristic and remarkably constant in botanically uniform populations. It is, however, shown by individuals in heterogeneous populations of *B.* × *longifolia*, even those approaching *B. boehmii* in many characters.

Hybrids seem to occur with 14, *B. wangermeeana*, notably in NW. Tanganyika (see 11 × 14 and notes under the latter).

12. **B.** × **longifolia** *Benth.* (pro sp.) in Hook., Ic. Pl. 14, t. 1359 (1881); Burtt Davy & Hutch. in K.B. 1923: 154 (1923); L.T.A.: 723 (1930), excl. syn. (*B. holtzii* tantum excepta); C. H. N. Jackson in Journ. S. Afr. Bot. 6: 38 (1940); B. D. Burtt in Journ. Ecol. 30: 76–78, 141, 142 (1942), pro parte majore; T.T.C.L.: 93 (1949), pro parte, excl. syn. *B. glaberrima*; Hoyle in F.C.B. 3: 470, fig. 40/D, E (1952), pro parte; Hoyle & White in F.F.N.R.: 111, 118, fig. 22/H (1962). Type: Malawi, Shire Highlands, *Buchanan* 22 (K, holo. !, E, iso. !)

Tree (2–)4–15(–25) m. high, often stunted by exposure, very variable in all parts especially habit and indumentum, displaying every intermediate expression and most combinations of the characters described for *B. boehmii* and *B. glaberrima*; most forms fall within the range shown in fig. 41. Bole often with conspicuous rounded bosses (even in forms otherwise close to *B. boehmii*); bark deeply furrowed to coarsely reticulate, grey or brownish; crown rounded to obconical or flat; main branching suberect to spreading; branchlets spreading in bunched systems with foliage usually rather pendulous in loose tufts (unless leaflets few). Stipules filiform to falcate, 1–3 cm. long, caducous to subpersistent, usually with reniform auricles. Leaves glabrous to tomentose, with (5–)6–12(–17) pairs of leaflets, the pairs very variable in spacing, the middle pairs usually the largest; rhachis usually 10–15 times as long as the petiole, channelled, usually with obvious but variable stipels or local expansions; leaflets oblong to narrowly triangular, (3–)4–6(–8) × (0·5–)1–2(–3) cm., very acute to broadly rounded or emarginate

FIG. 41. *BRACHYSTEGIA* × *LONGIFOLIA*—four of the numerous forms of the putative hybrid complex *B. boehmii* × *B. glaberrima* (etc.?). Each form represented by a leaf and inflorescence, × ½, flower with bracteoles, × 3, part of upper surface of leaf-rhachis showing stipellar expansions, × 1, and part of lower surface of a leaflet, × 1. Three of these forms have been described as species.—**1,** " *B. holtzii* ", subglabrous, with shorter petiole than *B. glaberrima* and stipules persistent, narrow and auriculate; **2,** a widespread form similar to *B. longifolia* Benth. sensu stricto; **3,** a very hairy form with few leaflets but with pubescent tepals; **4,** " *B. goetzei* ", a form closely similar to *B boehmii* but with shorter stipules and fewer leaflets than the known local form of the latter species. 1, from *Holtz* 31; 2, from *B. D. Burtt* 5759; 3, from *Rea* 76; 4, from *Goetze* 1423.

at apex, obliquely rounded or truncate to cordate at base; main nerves and larger veins usually more conspicuous than the open to rather close final reticulation. Panicles terminal (rarely also axillary), glabrous or fulvous- to dark brown-tomentose; peduncle usually slender and terete, rarely sulcate. Flowers greenish with white or cream stamen-filaments; bracteoles very variable in thickness and shape. Tepals (4–)5(–9); outer 4–6 imbricate, densely long-ciliate, often with 1 or more pubescent outside; inner 0 or 1–4, variable. Pod up to 15 × 5 cm., smooth or partly to completely scurfy, often warted when smooth or nearly so. Figs 35/11, p. 160, & 41, p. 189.

TANGANYIKA. Chunya District: N. Lupa Forest Reserve, fl. 22 Sept. 1962, *Boaler*, Lupa transect I, No. 6!; Mbeya District: Mbozi, fr. 29 July 1949, *Hoyle & Greenway* 1099!; Iringa District: near Iheme, fl. 19 Oct. 1936, *B. D. Burtt* 5759!; Songea District: 3 km. N. of Songea, fr. 15 June 1956, *Milne-Redhead & Taylor* 10824!

DISTR. T4, 6–8; Congo Republic, Mozambique, Malawi, Zambia and Angola

HAB. Deciduous woodland; 275–2000 m., mainly above 1000 m. Widely dominant or co-dominant in the southern half of Tanganyika, especially the highlands; typically on lower hill-slopes and rolling plateau country bordering seasonally wet open plains, " mbugas " or " dambos "; also on and below granite hills; occurs on acid, sandy to deep red soils, alone or with other *Brachystegia* species.

SYN. *B. goetzei* Harms in E.J. 30: 318, t. 13 (1901); Burtt Davy & Hutch. in K.B. 1923: 150 (1923); L.T.A.: 721 (1930); Topham in K.B. 1930: 355 (1930), pro parte, quoad cit. Tanganyika tantum. Type: Tanganyika, Mbeya District, Unyiha [Unyika], near Pisaki, *Goetze* 1423 (B, holo. †, E, iso.!, K, photo.!)

B. holtzii Harms in E.J. 33: 154 (1902); V.E. 1(1): 241 (1910); Burtt Davy & Hutch. in K.B. 1923: 150 (1923); L.T.A.: 724 (1930). Type: Tanganyika, Uzaramo District, Mogo Forest Reserve, *Holtz* 31 (B, holo. †, EA, iso.!, K, photo.!)

NOTE. Recent authors have included *B. glaberrima* among glabrous forms of *B. longifolia*, usually following my earlier opinion. I have now given more careful thought to the question of how to deal with the mass of forms in this group to which further collection continues to add new combinations of the characters of *B. glaberrima* and *B. boehmii*, sometimes with added complications apparently due to introgression from spp. 8–10, 14 and 15. The admitted scrap-heap grouped as putative hybrids under *B.* × *longifolia* seems to have originated through repeated crossing and back-crossing among the local forms of the two species mainly concerned, sometimes evidently to the point of swamping *B. glaberrima* sensu stricto in much of the area of overlap. Swarms of diverse forms occupy intermediate habitats and also seem to have invaded areas made available by destruction of forest, notably on the southern highlands of Tanganyika. Here B. D. Burtt (Iringa and Iringa–Mbeya road, Oct. 1936) collected a series of flowering specimens; Nos. 5762, 5759, 5811 and 5741, all with hairy panicles, range (in that order) from glabrous leaves with 7–10 pairs of leaflets to hairy leaves with 9–12 pairs.

Boaler's valuable series (Chunya District, Lupa Forest Reserve, transect I, fl. 22 Sept. 1962) includes Nos. 6, 8, 1, 4, and 5, ranging (in that order) from glabrous throughout with 6–7 pairs of leaflets to densely pubescent throughout with 8–10 pairs; No. 8 has a clearly defined tomentellous zone on the upper and outer part of each bracteole, but an otherwise glabrous panicle; this peculiar distribution of pubescence, so far unique in the genus, is correlated with tepals more like those of *B. boehmii* than in any other specimen in this series, while the leaves are intermediate between No. 6 and the associated *B. spiciformis* (Nos. 3 and 7)! The possibility of rare crossing between the latter species and *B.* × *longifolia* is also suggested by *Procter* 1319 (Iringa District, Image Mt., July 1959) a remarkable, hairy specimen with 7–9 pairs of ovate leaflets and obscure, scarcely axillary stipule-bases, occurring where the putative parents occupy adjacent zones.

The total range of forms included is largely covered by fig. 41, but *Lindeman* 712 (6 km. S. of Tabora, fl. 14 Sept. 1938), chosen as illustrative specimen of *B. longifolia* in T.T.C.L.: 93 (1949), is a good example of a borderline case, approaching *B. glaberrima* more closely than any of the forms illustrated.

B. gossweileri Burtt Davy & Hutch. was believed to occur in Tanganyika on the basis of a small-leaved sterile specimen (Chunya District, 13 km. E. of Chunya, 22 Dec. 1961, *Boaler* 374) and the species was accordingly included when fig. 35 was prepared. More recent gatherings from the same locality are now available (20 May 1962, *Boaler*, Chunya transect I, Nos. 1–16, again sterile but comprising normal and coppice material in variety), as well as a specimen in young fruit (4 Dec. 1962, *Boaler* 759), stated to be from the same site but this time described as " 10 miles [16 km.] east of Chunya ". These specimens show conclusively that the population forms part of the *B.* × *longifolia* complex as represented in the area generally and covers the leaf-variation of "*B. goetzei*". Whether *B. gossweileri*, as understood elsewhere, can continue to be maintained as a species independent of the complex, is a problem still awaiting a satisfactory solution.

13. **B. boehmii** *Taub.* in P.O.A. C: 197 (1895); Burtt Davy & Hutch. in K.B. 1923: 151 (1923); L.T.A.: 721 (1930); C. H. N. Jackson in Journ. S. Afr. Bot. 6: 37 (1940); B. D. Burtt in Journ. Ecol. 30: 75–78, 85, 86, 141 (1942); T.T.C.L.: 93 (1949); Hoyle in F.C.B. 3: 474, t. 33 (1952); Hoyle & White in F.F.N.R.: 111, 117, fig. 24, t. 1/B–D, F, G (1962). Type: Tanganyika, Tabora District, Igonda, *Boehm* 159a (B, holo. †, K, fragm. !)

Tree (2·5–)5–15(–21) m. high; pubescent to tomentose or more rarely the leaflets glabrous; bole without rounded bosses (so far as known); bark rough, long-persistent, ± coarsely reticulate, with many narrow fissures and transverse cracks, grey to brown (black when burnt only?); crown flat-topped, with few, heavy branches and stout branchlets; foliage long, pendulous in tufts; flush pink to brick-red, turning through buff or yellow to pale green, maturing darker; fallen dead leaves dull reddish. Stipules shortly connate, filiform to linear-filiform or subspathulate, (2–)3–4(–5) cm. long, subpersistent, usually grading downwards into large brown and crimson bud-scales at flushing and upwards into the often bifurcate lower bracts at flowering-time; auricle usually persistent, reniform, (0·5–)1–1·5(–2) cm. long, rarely caducous or undeveloped; stipule-bases persistent, usually prominent, subtending flattened and laterally keeled dormant buds. Leaves with (13–)14–28(–30) pairs of leaflets, the middle or lower middle pairs the largest; petiole stout, 0·3–1(–1·2) cm. long, the very stout pulvinus usually forming half to the whole; rhachis (8–)10–30(–35) cm. long, 20–50 times as long as the petiole, ± channelled; stipels or wings variable or 0, often obscured by coarse pubescence; leaflets narrowly oblong to narrowly triangular, (2·5–)3–6 × (0·7–)1–1·5 cm., rounded or obtuse to emarginate and then often asymmetric at apex, obliquely rounded or subtruncate to cordate at base; midrib subcentral or somewhat excentric; main nerves above (except the 4–6 fanwise basal nerves) scarcely more obvious (in mature leaflets) than the very closely reticulate veins (all areolations usually complete and equally prominent, 10–20 per square mm. at maturity, especially close in glabrous forms, otherwise often obscured by hair); lower surface typically pubescent to floccose with long yellowish crispate hairs throughout or with much denser rusty to dark brown spreading hairs on the midrib, but often glabrous or nearly so. Panicles ± erect above the foliage, terminal or terminal and axillary, up to 10 × 8 cm., much-branched, dense-flowered, always pubescent to tomentose with yellow to rusty and often dark brown hairs; peduncle short or 0, typically stout, sulcate when dry. Flowers yellowish-green with white filaments; bracteoles 6–10 × 4–7 mm., rather thick, pubescent to tomentose, with longer darker hairs on the usually prominent keel. Tepals (4–)5(–8), usually free; outer (4–)5 sepaloid, broad, imbricate, up to 4·5 × 3·5 mm., densely ciliate, often 1 or more pubescent outside; inner 0 or 1–2(–3), linear to spathulate, 4–6 mm. long, glabrous or sparsely ciliate. Stamens ± 10, ± free, 10–12 mm. long. Ovary 4 × 2 mm., densely setose. Pod thickly and rigidly woody, up to 16(–20) × 5 cm., held conspicuously like a flag above crown of tree, maturing pale to yellowish-brown or pinkish, surface becoming finely scurfy throughout, usually at an early age, never (?) prominently warted through insect-attack (cf. 11, *B. glaberrima* and 12, *B.* × *longifolia*); sutural wings stiffly spreading, 5–9 mm. wide. Fig. 35/4, p. 160.

TANGANYIKA. Buha District: Kasulu–Kigoma road, fl. Oct. 1954, *Procter* 262 !; Rungwe District: fl. Dec. 1912, *Stolz* 1717 ! & 1757 !; Kilwa District: Liwale, fr., *Gillman* 1044 !
DISTR. T1, 3–8; Congo Republic, Mozambique, Malawi, Zambia, Rhodesia, Botswana and Angola

HAB. Deciduous woodland, widespread, locally dominant, mainly 150–1500 m. Typically on paler, poorer soils; notably on poorly drained, ± flat or shallow sites, especially over schists, quartzites and ironstone; often forms zones on lower parts of escarpments, valley-slopes and stony hills; usually seen in pure stands but commonly associated with *Julbernardia globiflora* and often with most other species; exceptionally resistant to elimination by fire and cultivation.

SYN. *B. flagristipulata* Taub. in P.O.A. C: 198 (1895); Burtt Davy & Hutch. in K.B. 1923: 152 (1923); L.T.A.: 722 (1930); Topham in K.B. 1930: 355 (1930), pro parte, excl. *Hendry* 525. Type: Tanganyika, Uzaramo District, *Stuhlmann* 6400 (B, holo. †, K, fragm. & photo. !)
B. filiformis Burtt Davy & Hutch. in K.B. 1923: 150, t. 2, fig. 2 (1923); L.T.A.: 722 (1930); Topham in K.B. 1930: 356 (1930), quoad cit. Tanganyika. Type: Zambia, Broken Hill, *F. A. Rogers* 8605 (K, holo. !, BM, iso. !)
[*B. woodiana* sensu auct. quoad cit. Tanganyika, e.g. K.B. 1923: 151 (1923) & L.T.A.: 720 (1930) & K.B. 1930: 356 (1930), *non* Harms]
Further synonymy in F.C.B. 3: 474 (1952)

VARIATION. Apart from a rather wide range in size and number of leaflets and in their relative width, variation is greatest in the quantity and distribution of pubescence on the leaves. In this respect there are two main variants, connected almost everywhere by numerous intermediates:

1. A very hairy (" typical ") form, very widespread but apparently most common in western Tanganyika. This was described by Taubert as having leaflets decreasing in size from apex towards base of the rhachis; I have not seen his holotype but find this statement difficult to believe.

2. A less hairy and usually more slender form, often with quite glabrous leaflets and very rarely glabrous rhachis, is also widespread but found mainly (in extreme forms at least) in eastern Tanganyika; the leaflets are usually more oblong with more excentric midrib and often with ± emarginate asymmetric apex (cf. 14, *B. wangermeeana*, but less extreme). Such forms occurring in the west may well have acquired some characters by introgression from *B. wangermeeana* and there is suggestive evidence of clinal variation across north-central Tanganyika.

Putative hybrids with 14, *B. wangermeeana* are cited below the latter. Those with 11, *B. glaberrima*, are here treated as a widely dominant hybrid complex, 12, *B.* × *longifolia* (see note under this).

The comparatively high proportion of *B. boehmii* with glabrous leaflets in **T8** may be due to introgression from *B. allenii* (see 9 × 13 and notes under the latter). *Gillman* 1228 and *Tanner* 162 (both from Masasi District, Mkwera), attain only 17 pairs of glabrous leaflets as compared with the local maximum per specimen, in obvious *B. boehmii*, of 21–28 pairs; *Schlieben* 5575 (Lindi District, Lake Lutamba) attains 21 pairs but has a glabrous inflorescence-axis, unique if it is *B. boehmii.*

14. **B. wangermeeana** *De Wild.* in F.R. 11: 513 (1913): Burtt Davy & Hutch. in K.B. 1923: 152, fig. 1 (1923); L.T.A. 3: 718 (1930), haud tamen syn. *B. tamarindoïdes*; C. H. N. Jackson in Journ. S. Afr. Bot. 6: 40 (1940); B. D. Burtt in Journ. Ecol. 30: 76, 141 (1942); T.T.C.L.: 90 (1949); Hoyle in F.C.B. 3: 478, fig. 40/A (1952); Hoyle & White in F.F.N.R.: 116, fig. 22/M (1962). Type: Congo Republic, Katanga, Elisabethville, *Hock* (BR, holo. !, K, fragm. & photo. !)

Tree (1·5–)4·5–10(–14) m. high with pubescent to tomentose inflorescences and subglabrous to pubescent leaves; bark ± smooth to rather finely fissured and transversely cracked in young to mature trees, very hard, long-persistent, rather dark grey; old bark rough, black (? due to fire) and much cracked; crown flat or umbrella-shaped; foliage long, in tufts, vertically pendulous when young and exceptionally sensitive, leaflets folding early in the evening or when plucked; flush soon green with pink tips, turning to pale bluish-green, maturing green or greyish; fallen dead leaves orange. Stipules usually caducous before flowering but often at first conspicuous, shortly connate, filiform to subspathulate, (1·5–)2–4(–5) cm. long, usually grading downwards into large (pinkish ?) bud-scales at flushing, the latter caducous before flowers open; auricles ± irregularly reniform or very small or 0, not separately persistent; stipule-bases persistent, often prominent but sometimes very short, subtending flattened and keeled dormant buds. Leaves with (12–)13–22 pairs of leaflets (but see note on variation below), the middle or lower middle pairs the largest; petiole stout, 0·5–0·8(–1·3) cm. long, the stouter pulvinus usually forming one-third to half; rhachis 7–24 cm. long (15–30 times as long as the petiole), channelled, usually with short often conspicuous spreading expansions, always ± pubescent (at least above) with long, spreading hairs; leaflets obliquely and narrowly triangular to narrowly oblong, often falcate backwards in some middle pairs, (1–)1·5–5(–6) × (0·3–)0·5–1·2(–1·7) cm., rounded to obtuse, usually oblique and ± retuse where the very excentric midrib meets the distal margin, obliquely truncate or rounded to subcordate at base; midrib ± parallel to the distal margin at a distance of 1–3 mm. and (2–)3–4 times as far from the other margin at the middle of the leaflet; basal fanwise nerves (4–)5–6(–7); main nerves and major reticulation above slender but more obvious than the ultimate close reticulation; surfaces glabrous or more rarely pubescent but typically with forwardly-directed hairs on the midrib beneath. Panicles terminal, up to 6 × 6 cm., always pubescent to tomentose with yellowish to dark brown hairs. Flowers (apparently) greenish; bracteoles 6–8 × 4–5 mm., pubescent to tomentose with yellowish to dark brown hairs, especially on the keel. Tepals (4–)5(–7), usually all sepaloid, mostly broad, imbricate, up to 3 × 2·5 mm., ± densely ciliate with long dark hairs; (?) very rarely 1 inner tepal linear to ± spathulate, or vestigial, ± ciliate or not. Stamens ± 10, ± free, 9–12 mm. long. Ovary 2–3 × 1 mm., densely setose on the sutures. Pod rather thinly woody, up to 12(–?15) × 3·5(–?5) cm., smooth and ± obscurely pitted-lenticellate mostly near the margins, maturing brown to pale greyish-brown; sutural wings spreading, each 3(–4) mm. wide. Fig. 35/5, p. 160.

TANGANYIKA. Tabora District: Simbo Forest Reserve, Plot 2, fr. 27 June 1949, *Hoyle & Greenway* 1045 (with photographs of habit & bark, FHO, K)!; Ufipa District: Chapota, fl. 3 Dec. 1949, *Bullock* 1991!; Iringa District: Malangali, fl. 21 Jan. 1941, *Lindeman* 1009!

DISTR. T4, 7; Congo Republic, Zambia and Angola

HAB. Deciduous woodland, locally abundant and often dominant, mainly 1100–2000 m. Typically on rocky hills and upper parts of escarpments; also frequent and co-dominant around seasonally flooded areas in Tabora District (see note below).

SYN. [*B. flagristipulata* sensu Topham in K.B. 1923: 355 (1923) pro parte quoad *Hendry* 525, *non* Taub.]
See also synonymy in F.C.B. 3: 478 (1952)

VARIATION. The maximum number of leaflets recorded in or near Tanganyika is 22 pairs but these increase westwards to over 30, with submarginal " midrib ". Most of the material available from Tanganyika and adjacent Burundi is atypical; more than half of it is unusually variable in the position of the midrib, in the shape and apex of leaflets and in their number, width and pubescence. The degree of aberration, like the density of gatherings, seems to be greatest in the more accessible areas of western Tanganyika. It is here, judging from my own observation and other records, that *B. wangermeeana* is most often closely associated with spp. 11, 12 and 13, belonging to the same group, and that intermediate forms are most frequent. Throughout the area from Kasulu to Mbeya, the majority of available material shows a maximum per specimen of 12–17 pairs of leaflets, many uncertain in shape or with a subsymmetrical rounded apex and often with a tendency to a repand margin; these last three characters are most often shown in the north and in subglabrous leaves with the least, or most variable, number of leaflets, as if instability had resulted from a conflict between the different basic shapes and numbers of leaflets in *B. glaberrima* and *B. wangermeeana*. Specimens with hairy leaves and/or more leaflets are less irregular, as might be expected if they were variants of *B. wangermeeana* or hybrids with the predominantly hairy *B. boehmii* or *B. × longifolia* with leaflets less different in shape and number. From Mbeya southwards, where *B. boehmii* seems often to have glabrous leaflets, the relation between pubescence and stability of shape does not seem to apply. Examples of the three groups of putative hybrids are detailed below under 11 × 14, 12 × 14 & 13 × 14.

In addition to these, Procter collected an extraordinary series of forms at various altitudes on Chimala scarp, Mbeya District, in Feb. 1961 and Jan. 1962; all have glabrous leaflets but ± densely pubescent rhachides. At 1310 m. (No. 1782), *B. boehmii* with 20 (or more ?) pairs of leaflets with rather excentric midrib; at 1460 m. (No. 1784), an intermediate with up to 20 pairs of leaflets, very closely reticulate but with variable (± triangular) shape, variable apex and position of midrib; at 1768 m. (No. 1788), a form nearer *B. wangermeeana*, with up to 17 pairs of triangular leaflets showing a moderately excentric midrib; at 1830 m. (No. 1993), a similar form attaining only 13 pairs of leaflets. Further, at 1525 m. (No. 1848) and at 2073 m. (No. 1772), two extraordinary forms with up to 15 pairs of very broad leaflets with excentric midrib and very irregular apices; No. 1772 especially strongly suggests by the shape and venation of leaflets the probability of hybridity with *B. spiciformis*, which is frequent at the higher altitudes (*Procter* 1780 and 1786 and his reports). This remarkable putative hybrid has been suspected in other areas (e.g. Tabora). Altitude and exposure do not seem sufficient to explain such extreme variation.

11 × 14. **B. glaberrima** × **B. wangermeeana**; F.F.N.R.: 117 (1962)

Frequent (?) forms with intermediate numbers of glabrous leaflets on a sparsely pubescent rhachis; shape of leaflets varying from narrowly triangular to oblong-elliptic or narrowly ovate, often on the same leaf and often with irregularly repand margin; midrib varying from subcentral (at least towards the apex) to distinctly excentric throughout, and apex from rounded to obliquely retuse; inflorescence, where known, pubescent to tomentose, variable in details; pod little-known, smooth or doubtfully scurfy in patches.

TANGANYIKA. Biharamulo District: 24 km. N. of Biharamulo, fl. Sept. 1957, *Procter* 705!; Buha District: Kasulu, Oct. 1930, *Rounce* B2!; Kigoma District: Kasulu–Uvinza road, fl. Oct. 1956, *Procter* 531!

DISTR. **T**1, 4, 7; Congo Republic, Burundi and Zambia

HAB. In areas where both putative parents are recorded

VARIATION. *Rounce* B2, cited above, is exceptionally variable. Both sheets in the East African Herbarium have a mixture of flowering and young fruiting material, probably from at least two individual trees; the main flowering shoot on Sheet 1 has several leaves with (9–)10–11 pairs of leaflets, rather variable in shape, apex and position of midrib but approaching normal *B. wangermeeana*; it also bears a depauperate lateral branch whose leaves approach those of *B. glaberrima*, having only (4–)6–9 pairs with a higher proportion of subsymmetrical apices; the other three shoots have (10–)11–16(–17) pairs of leaflets, less variable and much nearer those of *B. wangermeeana*. Procter, on Kasulu Boma hill, in Oct. 1956, collected five examples from 1·6 m. high coppice, all fertile, Nos. 542, 544, 549 & 550 in flower, No. 548 in fruit; these are probably from the same population as *Rounce* B2, and No. 542 is similarly variable. In this collection the maximum number of leaflets per specimen varies from 14–17 pairs and the shape, apex and position of midrib vary widely; the variation is believed not to be due solely to coppicing, because these characters are not modified by it to anything like the same extent in other areas. *Procter* 554 (Kigoma–Kasulu road, Nov. 1956), an isolated tree in woodland of *B. floribunda*, contrasts with the specimens cited above in having up to 19 pairs of glabrous leaflets, scarcely variable and in all essentials typical of *B. wangermeeana* as represented in the district. It may be noted that *B. glaberrima* is well represented from Kasulu and N. Kigoma and neighbouring areas but 12, *B.* × *longifolia*, especially hairy forms, seems absent or scarce north of Tabora and Mpanda.

There are a few examples of putative *B. glaberrima* × *B. wangermeeana* hybrids from Kahama District (e.g. *B. D. Burtt* 4505), Tabora District (e.g. *C. H. N. Jackson* 19 & 21, *Lindeman* 370, *Peter* 35026 & 35074) and Mbeya District (e.g. *Anderson* 1186). It will be realized, however, that the same leaf-characters could be produced by a cross between *B. wangermeeana* and a subglabrous form of 12, *B.* × *longifolia*, which predominates in the last two districts.

12 × 14. **B.** × **longifolia** × **B. wangermeeana**; F.F.N.R.: 118 (1962)

Rare (?) forms combining hairy leaves having (10–)11–17 pairs of leaflets, variable in shape and apex and in the position of the midrib, with smooth pods (some quite large) resembling those of *B.* × *longifolia* or 11, *B. glaberrima* rather than those of *B. wangermeeana*; the smooth pods and few leaflets suggest that 13, *B. boehmii* is not directly involved (see 13 × 14 below). Alternatively, specimens combine a rather long petiole and glabrous leaflets having a moderately excentric midrib, with a panicle showing characters of *B. boehmii*.

TANGANYIKA. Kigoma District: 18·5 km. on Uvinza–Mpanda road, fr. 30 June 1949, *Hoyle & Greenway* 1048! & same road, July 1956, *Procter* 476!; E. Tabora District: near Rubugwa, fl. 25 Oct. 1934, *C. H. N. Jackson* 20!; Mpanda District: 25 km. on Uruwira–Nyonga road, fr. 3 July 1949, *Hoyle & Greenway* 1061!
DISTR. T4, 7; Congo Republic and Zambia
HAB. In areas where both putative parents are recorded

NOTE. *Hoyle* 1048 and *Procter* 476 both attain 16 pairs of yellowish-pubescent leaflets and are otherwise so similar that they could have come from the same tree. *Hoyle* 1061 is of similar form and pubescence, with up to 14 more widely spaced pairs of leaflets and smooth lenticellate pods. *Groome* in *F.H.* 2234 (Chunya District, N. Lupa, Nov. 1947) closely resembles *Hoyle* 1061 but the pairs of leaflets are less spaced and more pubescent. *C. H. N. Jackson* 20 combines an inflorescence approaching that of *B. boehmii*, with up to 17 pairs of variable glabrous leaflets.

13 × 14. **B. boehmii × B. wangermeeana**; F.F.N.R.: 117 (1962)

Rare (?) forms combining a more or less excentric midrib with scurfy pods and/or other characters of the local form of *B. boehmii*, e.g. very hairy or more oblong leaflets with (for *B. wangermeeana*) exceptionally close reticulation, etc.

TANGANYIKA. Kigoma District: Mahali Mts., 30 Sept. 1958, *Jefford, Juniper & Newbould* 2775!; S. Tabora District: Kakoma, 25 June 1949, *Hoyle & Greenway* 1041!; Mpanda District: 13 km. N. of Mpanda, Kabungu, 29 June & 1 July 1961, *Boaler* 285! & 286!
DISTR. T4, 7; Congo Republic and Zambia
HAB. In areas where both putative parents are recorded

NOTE. *Hoyle* 1041 attains 17 pairs of stiffly pubescent oblong leaflets with subcentral to excentric midrib and variable apex; the pods are scurfy though not yet mature. Close by, on the same day, *Hoyle* 1042 (in old flower) was collected from typical *B. boehmii* for comparison; this has up to 21 pairs of leaflets with central midrib and symmetrical apex, floccose-tomentose beneath; both species are common in the district. *Jefford, Juniper & Newbould* 2775 attains 19 pairs of narrowly oblong glabrous leaflets, very closely reticulate and with variable midrib and apex; the pods are partly scurfy, partly lenticellate. *Boaler* 285 attains 14 pairs of hairy triangular leaflets with variable, mostly excentric midrib and apex; the immature pod is partly scurfy. *Boaler* 286 has larger variable leaflets, with pods scurfy and lenticellate in patches; both species are common locally and also 11, *B. glaberrima* and 12, *B.* × *longifolia*. *Ross* 15 in *F.H.* 1196, & 16 in *F.H.* 1197 (Mbeya District, Njele, May 1937) are instructive though sterile; No. 15 is mostly 13, *B. boehmii*, attaining 24 pairs of glabrous oblong leaflets variable in apex and position of midrib, probably indicating introgression from *B. wangermeeana*; No. 16 attains only 14 pairs of more triangular leaflets with much more excentric midrib but variable apex and may belong to the category sp. 11 × sp. 13 × sp. 14.

15. **B. taxifolia** *Harms* in E.J. 33: 155 (1902); Burtt Davy & Hutch. in K.B. 1923: 153 (1923); L.T.A.: 717 (1930); C. H. N. Jackson in Journ. S. Afr. Bot. 6: 39 (1940); B. D. Burtt in Journ. Ecol. 30, t. 4 (1942); Hoyle in F.C.B. 3: 480 (1952); Hoyle & White in F.F.N.R.: 116–118, fig. 22/N (1962). Type: Tanganyika, SW. Iringa District, Ngominyi[Gominyi]-Bueni, *von Prittwitz & Gaffron* 54 (B, holo. †, K, fragm.!)

Tree, treelet or bush, (1–)2–6(–16) m. high, with bole up to 1 m. in diameter; all young parts (except the leaflets) rather coarsely pubescent with fulvous to dark brown hairs; bark at first grey to whitish, becoming deeply and irregularly fissured, dark grey to black below (? only when burnt); crown flat, obconic or umbrella-shaped, dense, the dark foliage semi-evergreen; branchlets short, stiff. Stipules free or very shortly connate, intrapetiolar, subulate, stiff, 1–2 cm. long, usually very persistent; auricles variable, tooth-like to digitate or 0. Leaves with (20–)25–35(–45) pairs of very small leaflets, the middle pairs the largest; petiole 1–2 mm. long; rhachis channelled, (3–)4–8(–10) cm. long, intervals narrowly winged; leaflets narrowly oblong to falcate, 5–10(–15) × 1–2(–3) mm., acute or rounded at apex, very obliquely rounded or truncate to subcordate at base; principal nerve submarginal; surfaces very discolorous, glabrous or puberulous, upper dark green, rugose with transverse nerves (or the nerves very obscure or invisible), lower glaucous or whitish. Panicles terminal, small, dense, reduced to racemes when (often) depauperate. Flowers greenish-white; bracteoles 6–9(–10) × 4·5–6(–7) mm. Tepals 5–7(–8), usually all sepaloid, broadly imbricate, 2–4 × 0·5–3 mm., grading from very broad to narrow, densely ciliate with long brown hairs. Stamens ± 10, free; filaments 9–15 mm. long. Ovary 3–4 × 1–2 mm., ± densely crispate-setose; style 10–12 mm. long; stigma rather prominent. Pod up to 12 × 4 cm., thinly woody, smooth but often transversely nervose, maturing pale to mid-brown, often lenticellate; sutural wings spreading, each 4–5 mm. wide. Fig. 35/6, p. 160.

TANGANYIKA. Chunya District: near Kipembawe, fl. 31 Aug. 1936, *Pitt* 508!; Chunya or Mbeya District: Kipembawe–Madibira, *Orr* in *Procter*!; Iringa District (near Mbeya District boundary): Ngominyi–Bueni, fl. Aug. 1901, *von Prittwitz & Gaffron* 54

DISTR. **T**7; Congo Republic, Malawi and Zambia

HAB. Scarcely known in the Flora area but reported by Procter as scattered in patches along the lower western slopes of the Southern Highlands of Tanganyika. Elsewhere local in woodland and thicket, mainly at high altitudes (1200–1800 m.) and in areas of rather high rainfall close to forest conditions; typical of white leached sands and poor stony ridges; often forming dense pure stands as trees or dwarf thicket, especially on edges of wet " dambos ".

NOTE. Very distinctive, evidently rare in Tanganyika and so far represented only by the specimens cited above, of which one seems now to be largely destroyed and the others depauperate (*Pitt* 508 in FHO) or fragmentary (*Orr* in *herb. Procter*).

Although not strictly evergreen, the dark green old leaves often persist until new growth is well advanced, at least in moister areas, thus increasing the resemblance of the tree to *Taxus*.

31. MONOPETALANTHUS

Harms in E. & P. Pf., Nachtr. zu 3(3): 195 (1897) & in E.J. 26: 265 (1899); Pellegr. in Bull. Soc. Bot. France 89: 118 (1942); J. Léon. in Mém. 8°, Classe Sci., Acad. Roy. Belg. 30(2): 255 (1957)

Unarmed evergreen trees. Leaves paripinnate, with leaflets in one to many pairs; stipules persistent or quickly falling off, united and intrapetiolar, sometimes auriculate at base*; leaflets opposite, very asymmetric at base, with midrib marginal or sometimes ± central; translucent gland-dots absent. Flowers in axillary dense racemes which are strobiliform (with imbricate bracts) when young; racemes sometimes aggregated into lateral or terminal panicles; bracteoles 2, well-developed, valvate, completely enclosing the flower buds, persistent. Hypanthium very short indeed, hardly present. Sepals very small or absent, 0–5. Petals: 1 relatively large, well-developed; the other 4 absent or rudimentary. Stamens (8–)9–10, fertile; filaments very shortly connate at base, or one of them free. Ovary densely pubescent, very shortly stipitate; stipe free; ovules 2–3(–6, *fide* Pellegrin); style elongate, with an abruptly and peltately enlarged stigma. Pods compressed, dehiscent, 2-valved; each valve with 1 (rarely 2) strong longitudinal nerve running from stipe to style. Seeds compressed, apparently without areoles, borne on short funicles.

About 14 species, all tropical African, nearly all in evergreen forest, and reaching their eastern limit in Tanganyika and Zambia.

M. richardsiae *J. Léon.* in Exell & Mendonça, C.F.A. 2: 204 (1956) & in Mém. 8°, Classe Sci., Acad. Roy. Belg. 30(2): 257 (1957); F.F.N.R.: 126

* A specimen from Nigeria (*Ejiofor* FHI.21893), without flowers or fruits but apparently a *Monopetalanthus*, shows free stipules. Whether this is more than a chance abnormality remains to be seen.

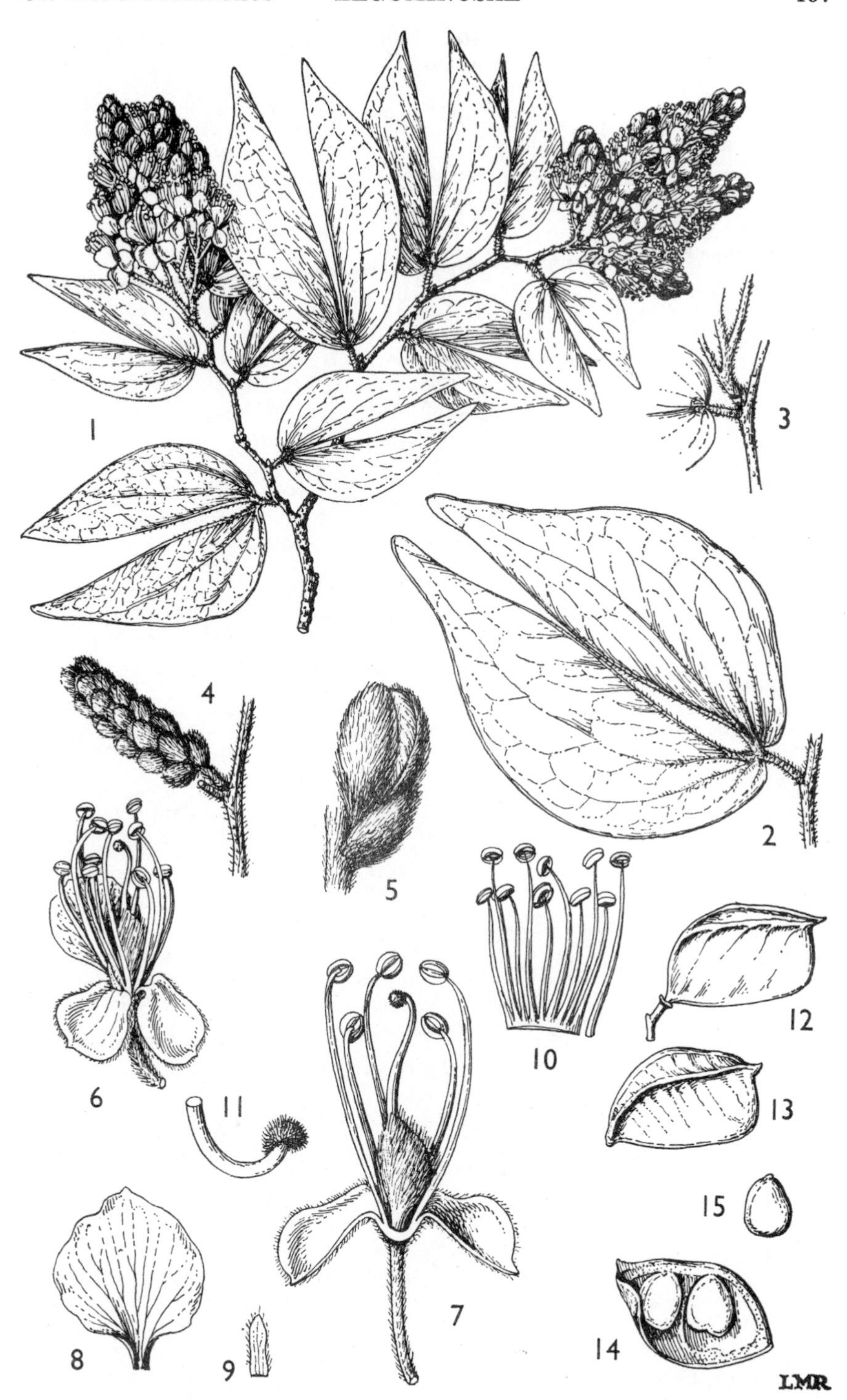

FIG. 42. *MONOPETALANTHUS RICHARDSIAE*—**1,** branchlet with leaves and inflorescences, × $\frac{2}{3}$; **2,** leaf, lower surface, from juvenile shoot, × $\frac{2}{3}$; **3,** basal part of leaf and inflorescence, showing pubescence, × 2; **4,** young inflorescence, × $\frac{2}{3}$; **5,** flower-bud enclosed by two bracteoles and with a basal bract, × 4; **6,** flower, × 2; **7,** flower, cut longitudinally, × 3; **8,** large petal, × 3; **9,** sepal, × 3; **10,** stamens, nine with filaments connate at base, the tenth (opposite large petal) free, × 2; **11,** stigma, × 8; **12,** undehisced pod, × $\frac{2}{3}$; **13,** one valve of dehisced pod, × $\frac{2}{3}$; **14,** inside of one valve, showing seeds, × $\frac{2}{3}$; **15,** seed, × $\frac{2}{3}$. 1, 3, 5–11, from *Richards* 4483; 2, from *Richards* 10203; 4, from *Lawton* 150; 12, from *Glover* 1; 13–15, from *B. D. Burtt* 5995.

(1962). Type: Zambia, near Abercorn, Inono stream close to Mpulungu road, *Richards* 4483 (K, holo. !, BM, EA, iso. !)

Tree 3–12(–25, *fide* Devred & Bamps) m. high; young branchlets glabrous to shortly pubescent. Leaves with stipules quickly falling off; petiole 1–6 mm. long; leaflets in one pair, sessile, coriaceous, glabrous or almost so, semi-elliptic or semi-ovate-elliptic, 1·7–7·5(–10) cm. long, 0·6–3·2(–4·5) cm. wide (to 9 × 4·5 cm. on juvenile shoots), acutely or sometimes obtusely ± acuminate at apex, rounded on outer side at base; vein-network prominent on both surfaces. Racemes 1–3·5 cm. long, brown-pubescent. Bracteoles subcircular, 4·5–7 mm. long and up to 6 mm. wide, brown-pubescent outside. Sepals 0, or 1–2 of them up to 3·5 × 2·5 mm. and 1–2 minute. Large petal white, 5–8 mm. long, 4–6 mm. wide, shortly clawed, with subcircular lamina; other petals 0, or 1–2 minute. Stamens 9–10; filaments glabrous. Pods asymmetric, ± oblong-elliptic or obovate-elliptic, 2·5–3·7 cm. long, 1·5–2·3 cm. wide, glabrescent, with 1(–2) prominent longitudinal nerves on each valve, 1–2-seeded. Seeds ± 1·2 × 0·9–1 cm., apparently dark brown. Fig. 42, p. 197.

TANGANYIKA. Kigoma District: Uvinza–Mpanda road, on bank of Niamanzi [Nyamanzi] R., 30 June 1949, *Hoyle* 1052!; Mpanda District: 40 km. N. of Mpanda, July 1956, *Procter* 479! & Sept. 1961, *Procter* 1948!
DISTR. T4; Congo Republic and Zambia
HAB. Riverine forest; altitude range uncertain

SYN. *M. leonardii* Devred & Bamps in B.J.B.B. 30: 111, fig. 18 (1960). Type: Congo Republic, Kivu Province, Bunyakiri, *A. Léonard* 3331 (BR, holo., K, iso. !)

NOTE. *M. richardsiae* will probably be found in the SW. part of Tanganyika adjacent to Abercorn in Zambia.

M. leonardii was separated from *M. richardsiae* by having only one petal (not two), and by having a " submedian " not " submarginal " longitudinal nerve on each valve of the pod. *Fanshawe* 4343 (Zambia, Kawambwa) has only one petal per flower, and is from the area of *M. richardsiae*, with which it is clearly conspecific. The shape of the pods and the position of the longitudinal nerve of *Léonard* 2227 (cited under *M. leonardii*) differ in no significant way from those shown by the abundant series of pods of *B. D. Burtt* 5995, from Zambia, Abercorn. These were, however, scarcely accessible at Kew when *M. leonardii* was described. I do not consider, therefore, that *M. leonardii* can be maintained as a distinct species.

32. CRYPTOSEPALUM*

Benth. in Benth. & Hook. f., Gen. Pl. 1: 584 (1865); J. Léon. in Mém. 8°, Classe Sci., Acad. Roy. Belg. 30(2): 270 (1957); Duvign. & Brenan in K.B. 20: 1–23 (1966)

Unarmed subshrubs, shrubs or trees. Leaves paripinnate, with leaflets in one to many pairs, or (but not in Flora area) unifoliolate; stipules linear, free, quickly falling off; leaflets sessile or subsessile, opposite, without translucent gland-dots. Flowers arranged in more than 2 ranks in terminal or axillary many-flowered racemes; bracteoles 2, well-developed, petaloid (pink or white), opposite, valvate, completely enclosing the flower-buds, persistent. Hypanthium short, cup-shaped. Sepals usually small, 0–6. Petals (0–)1(–3), well-developed and elliptic when present. Stamens 3–6(–8) (in East African species normally 3); staminodes sometimes present. Ovary stipitate; stipe adnate to the side of the hypanthium; ovules 1–5; stigma terminal, abruptly enlarged. Pods compressed, woody, dehiscent, glabrous; valves twisted after dehiscence, without longitudinal nerves, but upper suture with a longitudinal wing-like ridge on either side projecting in a plane at right-angles to the valve. Seeds compressed, elliptic, without areoles.

About 11 species, all in tropical Africa.

* I am greatly indebted to Professor P. Duvigneaud, of the Université Libre, Brussels, for much help and advice over this difficult genus.

Shrub or small tree, sometimes suffruticose but then stems branched above and bearing several inflorescences, or at least with clustered lateral inflorescences; stems apparently persisting for at least two seasons . . 1. *C. exfoliatum*
Suffruticose, with a thickened rootstock and erect simple annual stems arising at or near ground level and each ending in a single terminal inflorescence . . . 2. *C. maraviense*

In a very large number of specimens I have only seen two which are possibly exceptions to the distinction used in the key above: *Harley* 9406, from Tanganyika, Mpanda District, S. end of Kungwe-Mahali peninsula, Kalya, 23 Aug. 1959 and *Hoyle, Greenway et al.* 1050, from Kigoma District, Uvinza–Mpanda road, 30 June 1949. These specimens have thinly woody stems about 20–100 cm. high, which may be simple (though perhaps only in the first season) or with up to two orders of branching. The stems appear to persist for at least two growing seasons, although the primary stem may be leafless after the first. The ultimate leafy branches bearing inflorescences may arise even from the upper nodes. The stems, both sides of the leaflets and the inflorescence-axes are pubescent with spreading hairs.

Although by the key these specimens clearly go under *C. exfoliatum*, yet the lateral branches are more similar to those of *C. maraviense*, particularly in their length (7–20 cm. or more, including inflorescences—those of *C. exfoliatum* being at most up to 10 cm. long). It is difficult to make a final decision with only scanty material and further study of this plant in the field would certainly be valuable.

1. **C. exfoliatum** *De Wild.* in Ann. Mus. Congo, Bot., sér. 4, 1: 41, t. 3/1–8 (1902); L.T.A.: 743 (1930); J. Léon. in F.C.B. 3: 493 (1952); F.F.N.R.: 121 (1962); Duvign. & Brenan in K.B. 20: 8 (1966). Type: Congo, Katanga, 1900, *Verdick* (BR, holo.!)

Shrub or tree 0·2–18(–30) m. high, sometimes suffruticose, but then branched above or at least with several often clustered lateral inflorescences, and with stems apparently persisting for at least 2 seasons; bark grey to brown. Leaves with rhachis 1·5–9·5 cm. long; leaflets in 2–11 pairs, ± asymmetrically oblong to elliptic, 0·5–7·2 cm. long, 0·15–3·3 cm. wide, emarginate or rounded to subacute at apex, asymmetric at base, glabrous or subglabrous to ± pubescent beneath or even somewhat so above. Racemes terminal or lateral, solitary or clustered up to 3 together, 2–10 cm. long, glabrous or subglabrous or pubescent. Flowers white or pale pink, sweetly scented. Bracteoles elliptic, ± 5–8 mm. long and 3–6 mm. wide. Sepals 1–4, triangular, up to ± 2 mm. long. Petals 1(–2), 5–10 mm. long. Stamens 3. Pods 4·5–6 cm. long, 2–2·5 cm. wide. Seeds ± 1·3–1·5 cm. long and 0·8–1 cm. wide.

subsp. **exfoliatum**; Duvign. & Brenan in K.B. 20: 8 (1966)

Shrub or small tree 1·5–7·5 m. high. Branchlets glabrous to densely pubescent. Leaflets large, up to 25–75 mm. long and 10–33 mm. wide.

var. **fruticosum** (*Hutch.*) *Duvign.* & *Brenan* in K.B. 20: 8 (1966). Type: Zambia, Abercorn District, 29 km. NW. of Abercorn, *Hutchinson & Gillett* 3947 (K, holo.!, BM, BR, iso.!)

Branchlets glabrous. Leaflets glabrous, or pubescent only at extreme base on lower side. Inflorescence-axes and pedicels glabrous.

TANGANYIKA. Ufipa District: Kashota, Rwasi Reserve, 16 Apr. 1928, *Wigg* in *F.H.* 451! & Kalambo R., Aug. 1952, *Groome* 30!
DISTR. **T4**; also in Zambia
HAB. " Fairly common on rocky ground " (*Groome* 30)

SYN. *C. fruticosum* Hutch. in K.B. 1931: 238 (fig.), 250 (1931); T.T.C.L.: 100 (1949); J. Léon. in F.C.B. 3: 494 (1952)

NOTE. *Michelmore* 696, from Tanganyika, Ufipa District, above Kasanga (EA, K), has pubescent stems, 4–5 pairs of leaflets similar to those of var. *fruticosum* in size and indumentum, and sparsely puberulous inflorescence-axes. In fact it is intermediate between the two varieties.

var. **pubescens** *Duvign. & Brenan* in K.B. 20: 9 (1966). Type: Zambia, Kalambo Falls, *Bullock* 3002 (K, holo. !, BM, iso. !)

Branchlets with dense spreading pubescence. Leaflets rather densely pubescent on lower side, and ± pubescent above. Inflorescence-axes and pedicels densely spreading-pubescent.

TANGANYIKA. Ufipa District: Ufipa plateau, between Abercorn and Sumbawanga, July 1947, *Bredo* 6104! & 6120! & Ufipa District/Zambia: Kalambo Falls, 6 July 1949, *Hornby* 3051!
DISTR. T4; also in Zambia
HAB. " In *Brachystegia-Pterocarpus* woodland " (*Bullock* 3002)

DISTR. (of species as a whole). T4; Congo Republic, Malawi, Zambia and Angola

OTE. *C. exfoliatum* De Wild. appears to be the earliest name for a large variable and widespread complex, in which *C. pseudotaxus* Bak. f. and *C. fruticosum* Hutch., as well as other alleged species, should be included. The complicated taxonomy is explained more fully by Duvigneaud & Brenan in K.B. 20: 2–12 (1966). In the Flora area only subsp. *exfoliatum* occurs, where it is represented by two varieties, though not by typical var. *exfoliatum*, which is restricted to Katanga.

Although in the Flora area *C. exfoliatum* is apparently always a shrub or small tree, two of its subspecies with a constantly suffruticose habit occur in the Northern Province of Zambia, and may perhaps be found in the south-western part of Tanganyika.

2. **C. maraviense** *Oliv.*, F.T.A. 2: 304 (1871); L.T.A.: 744 (1930); T.T.C.L.: 100 (1949); J. Léon. in F.C.B. 3: 487, fig. 42/A, t. 34 (1952); F.F.N.R.: 122, fig. 25 (1962); Duvign. & Brenan in K.B. 20: 12 (1966). Type: Mozambique, Maravi country W. of Lake Nyasa, *Kirk* (K, holo. !, BR, iso.(fragm.) !)

Suffrutex with a thickened woody rhizomatous rootstock, whence (or from basal part of previous year's stems) arise often tufted erect annual stems, each ± 4–40 cm. high, simple and with a single terminal inflorescence. Leaves with rhachis 3–14 cm. long; leaflets in 3–16(–18) pairs, ± asymmetrically oblong-lanceolate, oblong-elliptic or oblong, the upper ones sometimes with an obovate tendency, 0·6–8 cm. long, 0·2–2·7 cm. wide, rounded to subacute or occasionally acute at apex, asymmetric at base, usually glabrous or subglabrous, sometimes ± pubescent. Racemes terminal, single, 2–12(–16) cm. long, glabrous to ± pubescent. Bracteoles elliptic, 5–15 mm. long, 2·5–8 mm. wide. Sepals 1–6, small or very small. Petal 1, 7–9 mm. long, sometimes in addition with 1(–2, *fide* F.C.B.) smaller ones 3·5–5 mm. long and 1–1·5 mm. wide. Stamens usually 3, rarely with up to 2(–3) smaller fertile ones also; very rarely all subequal. Pods mostly 1–2-seeded, 2·5–5 cm. long, 1·5–2·7 cm. wide. Seeds 1·2–1·3 cm. long, 0·7–1·0 cm. wide. Figs. 43 & 44, p. 202.

TANGANYIKA. Ufipa District: Sumbawanga, new road, Aug. 1952, *Groome* 38!; Rungwe District: Kyimbila, *Stolz* 1679!; Songea rest-camp, 12 Sept. 1956, *Semsei* 2458!
DISTR. T4, 6–8; Congo Republic, Mozambique, Malawi, Zambia and Rhodesia
HAB. Deciduous woodland (*Brachystegia-Julbernardia*); 300–1600 m.

SYN. *C. dasycladum* Harms in E.J. 30: 319, fig. (1901); L.T.A.: 744 (1930); T.T.C.L.: 99 (1949). Types: Tanganyika, Mbeya District, Mbozi Hill, *Goetze* 1384 (B, syn. †, BR, isosyn. !) & Unyika, Kananda village, *Goetze* 1438 (B, syn. †)
C. pulchellum Harms in E.J. 30: 321 (1901); L.T.A.: 745 (1930); T.T.C.L.: 100 (1949). Type: Tanganyika, Rungwe District, Kiwira [Kivira] valley, Untali, *Goetze* 1472 (B, holo. †, BR, P, iso. !)
C. boehmii Harms in E.J. 33: 156 (1902); L.T.A.: 743 (1930); T.T.C.L.: 99 (1949). Type: Tanganyika, ? Mpanda District, between Pa-kakombue [? pa-Kabombue] and Mdani [? Ndani], *Boehm* 17a (B, holo. †)
C. busseanum Harms in E.J. 33: 156 (1902); L.T.A.: 743 (1930); T.T.C.L.: 99 (1949). Type: Tanganyika, Songea District, Madjanga–Kwa-Bagaya, *Busse* 633 (B, holo. †, BM, BR, EA, K, iso. !)

NOTE. I am in agreement with the opinion expressed by J. Léonard (in F.C.B. 3: 490), that this is an exceedingly variable species and that the synonymy given above is consequently justified.

FIG. 43. *CRYPTOSEPALUM MARAVIENSE*—**A,** habit, × 1; **B, C,** lower surface of leaflets, showing pattern of venation, × 1; **D,** flower-bud, with bract and bracteoles, × 5 **E,** flower, opened out, × 5; **F,** longitudinal section of flower, × 5; **G,** pod, × 1; **H,** seed, × 2. A, B, D–F, from *Heusghem* in *Delevoy* 1296; C, from *Quarré* 3384; G–H, from *Schmitz* 3084. Reproduced by permission of the Director-General of the Institut National pour l'Étude Agronomique du Congo.

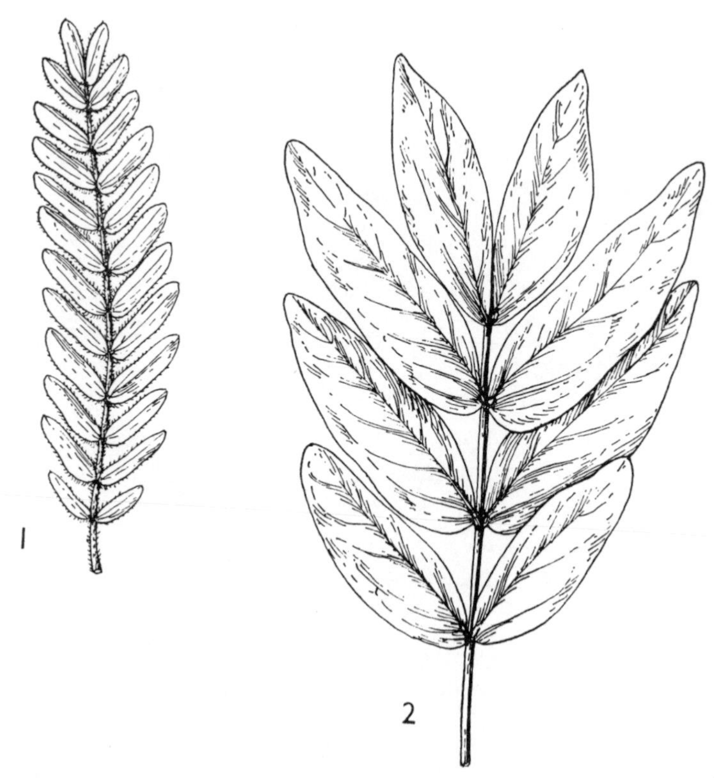

FIG. 44. *CRYPTOSEPALUM MARAVIENSE*—Extremes of variation in leaves, × $\frac{2}{3}$. 1, from *Stolz* 1679; 2, from *Groome* 38.

As Léonard states, the variable characters are fairly constant on each individual plant, and are thus likely to be frequently genetically controlled. This variation is seen in the indumentum, length of stem, number, size and shape of leaflets, length of inflorescence, size of bracteoles, and to some extent in number of sepals, petals and stamens.

I am at present of the opinion that *Cryptosepalum maraviense*, taken in a very wide sense, is a taxonomic group which evolutionarily is in full cry, so to speak. While the range of variation is, as I have implied, enormous, and there are some correlations between different characters and with geography, they appear to be still very imperfect, and generally cannot be used as a basis for clearly defining subspecies within the Flora area. Given sufficient material, however, I strongly suspect that they might be proved statistically.

I have had the recent privilege of examining the splendid range of material brought together by Professor P. Duvigneaud (for whose unstinted help and advice I am most grateful) at the Université Libre at Brussels, and I have attempted to analyse in a general way the variation occurring in the Flora area, using that material as a basis, together with that at Kew, the East African Herbarium and the British Museum (Natural History).

The variations, discussed more fully in K.B. 20: 12–23 (1966), may be grouped in the following way:—

Annual shoots (measured to end of inflorescence) very commonly more than 15 cm. high; basal scales on shoots mostly congested at base, sometimes a few spaced out towards lower leaves, but upper ones usually not much elongated; number of leaflets variable:

Number of pairs of leaflets not more than 8 (occurring in **T**4, 6, 7, 8)*:

* *C. boehmii* belongs to this group, but without authentic material it is hard to be sure exactly where.

Leaflets up to 13 mm. or less wide (" variant A ", partly):
Stems with short straight spreading hairs (e.g. *Hoyle* 1090, Songea District, etc.; typical *C. maraviense* also comes here)
Stems with longer straight spreading hairs (e.g. *Hoyle* 1050, Kigoma District)
Stems with short crisped or curled ± appressed hairs (e.g. *Schlieben* 1507, Ulanga District)
Leaflets, at least some of them, up to 15 mm. or more wide, see fig. 44/2 (" variant C "):
Stems with short straight spreading hairs (e.g. *C. H. N. Jackson* 107, Kigoma District, *Milne-Redhead & Taylor* 9510, Songea District, etc.)
Stems with longer straight spreading hairs (e.g. *C. H. N. Jackson* 106, ? Kigoma District, etc.; *C. busseanum* also comes here)
Stems with short crisped or curled ± appressed hairs (e.g. *Semsei* 155 in *F.H.* 2586, Mpanda District, etc.)
Stems glabrous or almost so (e.g. *Wigg* in *F.H.* 1495, Kigoma District, *Eggeling* 6371, Tunduru District, etc.)
Number of pairs of leaflets, of some leaves at least, 10–13; leaflets not more than 13 mm. wide (" variant A ", partly, occurring in **T7**, only):
Stems with rather long straight spreading hairs (e.g. *Stolz* 435, Rungwe District, etc.; *C. pulchellum* also comes here)
Stems glabrous or nearly so (e.g. *Newbould & Jefford* 2378, 2792, Mpanda District, *Stolz* 1763, Rungwe District, etc.)
Annual shoots short, not more than 15 cm. high (measured to end of inflorescence); basal scales on stem congested at base and also spaced out between base and lowest leaves, the upper scales usually considerably longer and bigger than the lower; some leaves always with 9 or more pairs of leaflets, see fig. 44/1 (occurring in **T7** only):
Annual shoots usually very short, usually 4·5 cm. or less from base of shoot to base of inflorescence (" variant F ", e.g. *B. D. Burtt* 5852, Tanganyika, Iringa District, between Sao Hill and Iheme)
Annual shoots usually 5 cm. or more from base of shoot to base of inflorescence (" variant E "):
Stems ± densely spreading-hairy (e.g. *Stolz* 1679, Rungwe District, etc.; *C. dasycladum* apparently also comes here)
Stems glabrous (e.g. *Greenway* 3620, Mbeya District)

I must emphasize the fact that this key is, at least in its primary divisions, attempting the impossible by separating imperfectly separated tendencies and inconstant correlations, although I suspect that the two primary divisions represent a natural divergence.

The greatest range of variation in *C. maraviense*, taken in its widest sense, is found in Zambia, the Congo and Angola. As one might therefore expect, in Tanganyika the Province **T7** shows the most diversity, the variation decreasing both northwards and eastwards from Lake Nyasa.

In **T4** forms with few pairs of leaflets occur exclusively (range 3–7 pairs, 22 gatherings), and the same applies in **T6** and 8 (range 4–8 pairs, 9 gatherings). In **T7** the range is 4–13 pairs, with 11 gatherings. I suspect that the few pairs found to the N. and E. represent a gradual dwindling of genetic potential within the species rather than any clearly defined entity.

It is of interest to note that in spite of the rather narrow range in number of pairs of leaflets found in **T4**, and also that the leaflets here have a distinct tendency to be large, there is a full range of stem indumentum in this province, from glabrous to quite densely though shortly pubescent.

The other variations confined to **T7** are sufficiently indicated in the key above.

It may seem invidious to give so much space here to the variation of a single species. *C. maraviense* in its widest sense is, however, certainly one of the most protean species in all tropical Africa. It would also be a splendid subject for investigation by modern taxonomic techniques—statistical study of populations, cytology and experimental cultivation.

33. GIGASIPHON

Drake, Hist. Pl. Madag. 1 : 88 (1902); De Wit in Reinwardtia 3 : 418 (1956); Torre & Hillcoat in C.F.A. 2 : 163 (1956)

Small to large evergreen trees, without tendrils (or rarely—*G. gossweileri*—a climbing shrub with simple tendrils). Leaves simple, ± acuminate and not at all bilobed at apex. Flowers large to very large, hermaphrodite, in short terminal or axillary racemes sometimes (*G. gossweileri*) aggregated into panicles. Hypanthium elongate, narrowly cylindric, comprising most of the apparent " pedicel ". Sepals 5, elongate, rather narrow, free or sometimes irregularly joined above. Petals 5. Stamens 10, usually all fertile; filaments glabrous or pubescent below. Ovary long-stipitate; style elongate; stigma small (*fide* De Wit). Pods large (small in *G. gossweileri*), irregularly elliptic-oblong, up to ± 10-seeded, ± woody, dehiscent or indehiscent. Seeds large, with a narrowly U-shaped line extending round most of their circumference; endosperm absent.

Six species have been placed in the genus (two from Tropical Africa*, one from Madagascar, and one each from Timor, the Philippines and New Guinea).

G. macrosiphon (*Harms*) *Brenan* in K.B. 17 : 214 (1963). Types: Tanganyika, Usambara Mts., near Amani, *Braun* 1033 & *Zimmermann* 3088 (B, syn. †) & near Kihuhwi [Kiuhui], *Grote* 3763 (B, syn. †, EA, isosyn., BM, drawing!)

Tree 6–20 m. high. Bark whitish- or pinkish-grey. Branchlets rusty-puberulous, soon glabrescent. Leaves papery, broadly ovate to suborbicular-ovate or ± obovate, 8–17 cm. long, 6–15·5 cm. wide, ± cordate or subcordate at base, subglabrous. Racemes erect (*Gardner* 1), short, terminal, 4–10 cm. long (*fide* Harms). Hypanthium 8–13 cm. long, rusty-pubescent outside. Sepals ± 5–8 cm. long. Petals " magnolia-like, pure white except for yellow splash within on one petal " (*Eggeling* 6408) or " white flushed creamy-pink " (*Greenway* 4915), obovate-elliptic, 9–13 cm. long, 4–6 cm. wide. Stamens all fertile; filaments slightly pubescent. Pods up to ± 6-seeded, flattened, indehiscent, breaking irregularly, ± 30 cm. long and 6–6·8 cm. wide. Seeds suborbicular-compressed, ± 1·8–3 cm. in diameter, 1–1·6 cm. thick, dull purplish-brown. Fig. 45.

KENYA. Kwale District: Mrima Hill, 16 July 1960, *T. A. M. Gardner* 1! & 31 Aug. 1959, *Verdcourt* 2404! & about halfway Msambweni–Lungalunga, 15 Jan. 1964, *Verdcourt* 3935!

TANGANYIKA. Lushoto District: Amani, cultivated, 23 Feb. 1937, *Greenway* 4915!; Lindi District: Rondo Plateau, Mchinjiri, Nov. 1951, *Eggeling* 6408!

DISTR. **K**7; **T**3, 8; not known elsewhere

HAB. Lowland rain-forest; 120–910 m.

SYN. *Bauhinia macrosiphon* Harms in E.J. 53 : 467 (1915); L.T.A.: 658 (1930); T.T.C.L.: 88 (1949)
[*Gigasiphon humblotianum* sensu K.T.S.: 105 (1961), *non* (Baill.) Drake]

NOTE. The large and beautiful flowers of this tree would make it a worthwhile introduction for ornament in other parts of the tropics.

The suggested identity of *Bauhinia macrosiphon* and *Gigasiphon humblotianum* (Baill.) Drake, from Madagascar (see T.T.C.L.: 88 (1949)) has not been confirmed. *G. humblotianum* is, according to Drake and Baillon, a scrambling shrub with leaves 20 cm. long, racemes more than 50 cm. long, a hypanthium 22–25 cm. long, and partly sterile stamens. Examination of the type-material shows no difference in the length of the raceme from *G. macrosiphon*, the 50 cm. mentioned above apparently including a length of the branch supporting the raceme; the leaves are about 18 cm. long. The hypanthium is, however, 20–25 cm. long, and the indumentum on the outside of the sepals is only a short sparse puberulence. The leaves are also glabrous beneath.

* One of the tropical African species, *G. gossweileri* (Bak. f.) Torre & Hillcoat, is aberrant in being a climber with tendrils, in the tendency of the racemes to become paniculate, in the much shorter hypanthium than usual, and in the small pods. Its generic position needs reconsideration.

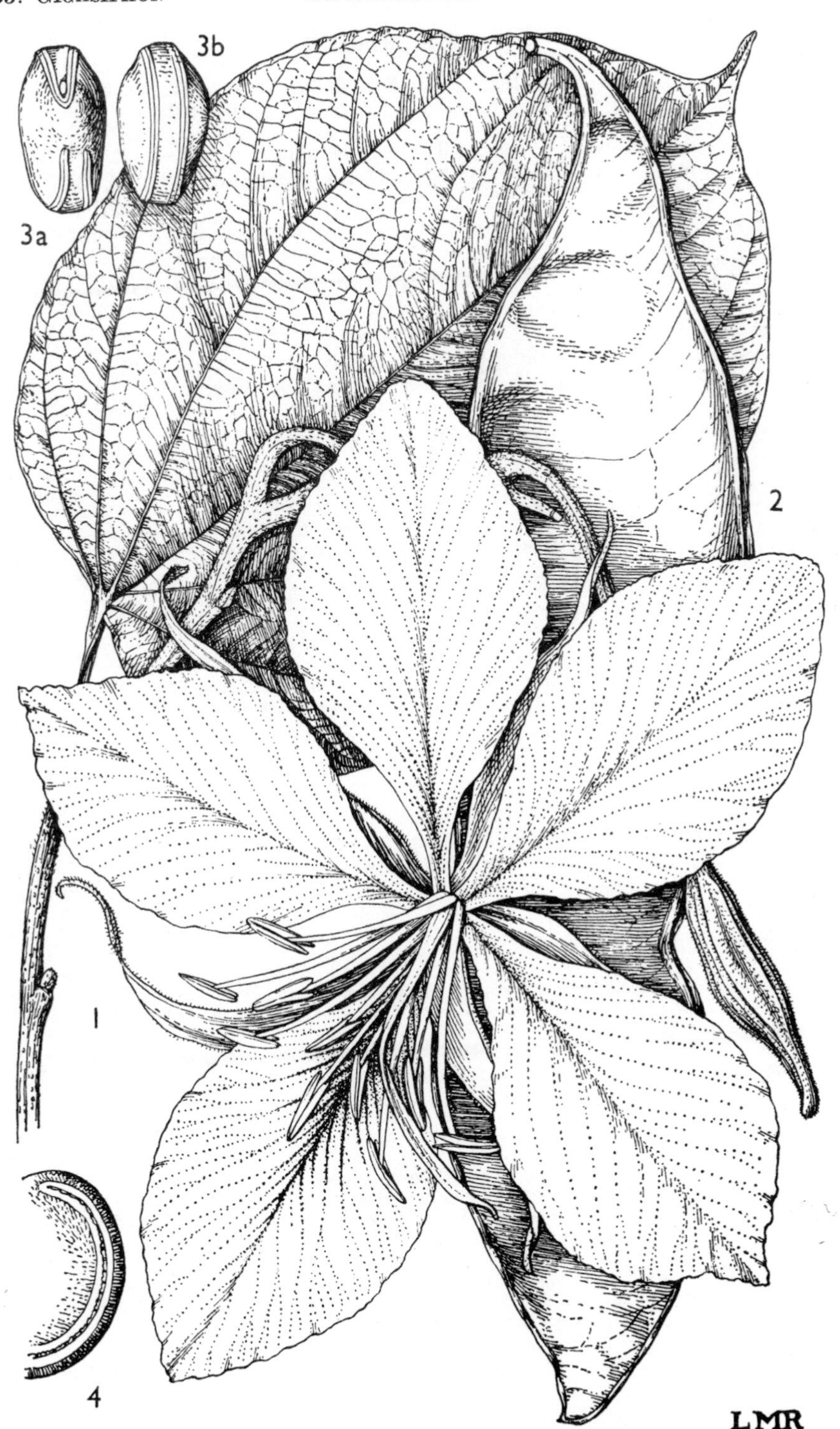

FIG. 45. *GIGASIPHON MACROSIPHON*—**1,** part of flowering branch, showing inflorescence; **2,** young pod; **3,** seed, (a) from funicular end, (b) from opposite end; **4,** part of seed, side view, showing persistent end portion of one funicular outgrowth. All × ⅔. 1, from *Greenway* 4915; 2–4, from *Verdcourt* 2404.

34. PILIOSTIGMA

Hochst. in Flora 29: 598 (1846); Milne-Redh. in Hook., Ic. Pl. 35, t. 3460 (1947); De Wit in Reinwardtia 3: 530 (1956)

Small deciduous trees, occasionally shrubby, not climbing. Tendrils absent. Leaves simple, conspicuously bilobed. Flowers medium to small, normally dioecious and unisexual, rarely monoecious, very rarely hermaphrodite (see F.C.B. 3: 278 (1952)), in terminal axillary or leaf-opposed racemes or panicles. Calyx with a turbinate tube and 4–5 short acute lobes. Petals 5. Fertile stamens 10 (in ♂ flowers), reduced to staminodes in ♀ flowers; filaments villous below. Stigma thick, capitate, flattened-globose, sessile on the ovary; funicle several times as long as ovule. Pods linear to oblong, many-seeded, leathery or woody, indehiscent. Seeds irregularly arranged, embedded in pulp, with a U-shaped line on one side; endosperm present.

A genus of three species in tropical Africa, Asia and Australia.

P. thonningii (*Schumach.*) *Milne-Redh.* in Hook., Ic. Pl. 35: 2, t. 3460 (1947); I.T.U., ed. 2: 67, fig. 16 (1952); Torre & Hillcoat in C.F.A. 2: 199 (1956); Roti-Michelozzi in Webbia 13: 174 (1957); F.W.T.A., ed. 2, 1: 444 (1958); K.T.S.: 107, fig. 20 (1961); F.F.N.R.: 126, fig. 20/N (1962). Type: Ghana, Aquapim, *Thonning* (C, holo.!)

Tree (2–)3–10 m. high, occasionally shrubby. Bark rough, dark brown to grey or black. Branchlets rusty-tomentellous or shortly rusty-tomentose when young. Leaves mostly 5–17 cm. long and 6–19 cm. wide, bilobed about one-eighth to one-third way down, densely reticulate and rusty-puberulous or -pubescent beneath. Panicles usually alternately leaf-opposed and axillary along branches, the ♂ ones 5–19 cm. long, the ♀ 2–7 cm. long. Flowers white to pinkish. Calyx rusty-tomentose or -tomentellous, 1·3–2·3 cm. long. Petals obovate, 1·4–2·6 cm. long, rugose or bullate, hairy on basal claw and outside of limb. Pods stipitate, woody, oblong or linear-oblong, mostly 13–26(–37) cm. long and 3–6·3 cm. wide. Seeds obovoid to ellipsoid, somewhat compressed, dark brown to blackish, ± 7–9 mm. long, 5–7 mm. wide and 3–4 mm. thick. Fig. 46.

UGANDA. W. Nile District: Koboko, May 1938, *Hazel* 481!; Teso District: Serere, May–June 1932, *Chandler* 699!; Mengo District: Bugerere, Busana, May 1932, *Eggeling* 422 in *F.H.* 656!

KENYA. Elgon, SE. slopes, 4 May 1953, *Padwa* 29!; Uasin Gishu District: Kipkarren, Mar. 1932, *Brodhurst Hill* 718!; S. Nyeri/Embu Districts: Sagana–Embu, June 1941, *Bally* 1487!

TANGANYIKA. Moshi, May 1928, *Haarer* 1436!; Ufipa District: Kipili, 31 Jan. 1950, *Bullock* 2366!; Morogoro District: Uluguru Mts., Matombo, 11 Apr. 1935, *E. M. Bruce* 1017!

ZANZIBAR. Pemba I., Miulani, 14 Feb. 1929, *Greenway* 1444!

DISTR. **U**1–4; **K** ?1, 2–7; **T**1–8; **P**; widespread in tropical Africa from Senegal to the Sudan Republic and southwards to South West Africa and the Transvaal

HAB. Woodland, wooded grassland, bushland; near sea-level to 1830 m.

SYN. *Bauhinia thonningii* Schumach. in Schumach. & Thonn., Beskr. Guin. Pl.: 203 (1827); L.T.A.: 657 (1930); T.S.K.: 62 (1936); I.T.U.: 30, fig. 11 (1940); T.T.C.L.: 88 (1949); U.O.P.Z.: 142 (1949); Wilczek in F.C.B. 3: 275, t. 22 (1952)

[*B. reticulata* sensu P.O.A. C: 200 (1895), *non* DC.]

35. BAUHINIA

L., Sp. Pl.: 374 (1753) & Gen. Pl., ed. 5: 177 (1754)

Shrubs or small trees or rarely (and not in Flora area) climbers. Tendrils absent. Leaves simple*, conspicuously bilobed, rarely divided as far as the base. Flowers usually large and showy, ⚥, arranged in short usually few-flowered racemes or solitary. Calyx spathaceous (the sepals ± cohering after the calyx has opened). Petals 5. Fertile stamens 1–10, sometimes accompanied by staminodes; filaments ± hairy below in native species, often glabrous in the introduced ones. Style elongate; stigma capitate or small, sometimes ± unilateral; funicle of ovule short, at top often with 2 short outgrowths appressed to the seed, one of which may be ± suppressed. Pods oblong to linear, few- to many-seeded, ± woody, dehiscent or rarely (not in East Africa) indehiscent.

A very wide concept has usually been given to the genus *Bauhinia* in the past. Here it is taken in a narrower sense which is explained by De Wit in his revision of the Malaysian Bauhinieae in Reinwardtia 3: 390 (1956). In its emended form the genus is distributed through tropical Africa, Asia and America, but the number of species is uncertain without revision of the American Bauhinias. Other East African species formerly in *Bauhinia* may be sought in the genera *Piliostigma* (p. 206), *Gigasiphon* (p. 204), and *Tylosema* (p. 213).

The beauty of their flowers has caused a number of species to be frequently cultivated in East Africa. A brief key to these follows:

Key to exotic species

Flowers small, white, in ± elongate slender racemes; petals less than 1 cm. long; fertile stamens 10, pubescent; pods indehiscent . . . *B. racemosa* Lam. (Native of India)

Flowers much larger, white or coloured, if in racemes then latter not both elongate and slender; petals 3–6·5 cm. long; fertile stamens 1–10; pods dehiscent:

  Fertile stamen 1; petals long-clawed, white to pink with purple speckles inside, upper petal deeper . . . . *B. monandra* Kurz (Native of ? America)

  Fertile stamens 3–10:

    Petals crimson to brick-red, with conspicuous claw as long as the lamina . . . *B. galpinii* N.E. Br. (*B. punctata* Bolle, *non* Jacq.) (Native of southern Africa)

    Petals white, pink or purple, not clawed, or with a very short claw much shorter than the lamina:

      Fertile stamens 10; flower-buds acuminate above, the apex ending in 5 free linear tips; flowers white . . . . *B. acuminata* L. (Native of tropical Asia)

* Although the leaves of *Bauhinia* (and of the related genera Nos. 33–36) appear to be and are described here as simple, they are in origin more complicated. R. E. Fries, in Arkiv för Botanik 8, No. 10: 1–16 (1909), considered them to have been derived through lateral fusion of the leaflets from a pinnate leaf with a single pair of leaflets. Goebel, Organographie Pfl., ed. 2, 3: 1354 (1923) emended Fries's theory by maintaining that there is no evidence of any ontogenetic fusion, but rather that each one of the pair of leaflets has failed to separate along one side from its partner, like a Siamese twin.

Fertile stamens 3–5; buds without free tips; flowers pink, purple, or sometimes white:
Buds winged or sharply ridged above; fertile stamens 3 . . . . . *B. purpurea* L. (Native of Asia)
Buds not winged or ridged; fertile stamens 5:
Petals blotched or striped with purple . *B. variegata* L. * var. *variegata* (Native of Asia)
Petals without purple, either white or partly yellow . . . . . *B. variegata* L. var. *candida* Voigt (var. *alboflava* De Wit)** (? Native of Asia)

In addition to those species mentioned above, the yellow-flowered *B. tomentosa* L. is often cultivated; it is also, however, indigenous (see opposite).

KEY TO NATIVE SPECIES

Hypanthium (calyx-tube) short, funnel-shaped, 1–5 mm. long; fertile stamens 10; petals yellow, or some red-blotched, rarely (3, *B. taitensis*) white, not or scarcely clawed at base and without glands on midrib outside; pods 0·6–2 cm. wide; seeds (where known) 7–9 mm. long:
Leaves divided to $\frac{1}{2}$–$\frac{3}{5}$ of the way down, or less; branchlets glabrous or pubescent; flowers one to many together:
Petals yellow, sometimes with a red or purplish blotch at base, ± 1·5–4·6 cm. wide, rounded at base:
Bracteoles small and inconspicuous, not enfolding the flower-buds; pedicels 0·7–2·7 cm. long; racemes solitary or occasionally aggregated. 1. *B. tomentosa*
Bracteoles broad, conspicuous, boat-shaped, enfolding the flower-buds; pedicels short, up to ± 0·7 cm. long; racemes aggregated into lateral and terminal several- to many-flowered panicles . . . . . . 2. *B. mombassae*
Petals white, 0·5–1 cm. wide, cuneate below; flowers and leaves comparatively small . . . 3. *B. taitensis*
Leaves divided to $\frac{4}{5}$ of the way down, or more, or even to base; branchlets glabrous; flowers solitary . 4. *B. kalantha*
Hypanthium (calyx-tube) 2–4 cm. long; fertile stamens 5; petals white or with pink spots, rather narrow but distinctly clawed at base; midrib of petals pubescent

* It is possible, though the evidence is as yet insufficient, that *B. variegata* may become naturalized in East Africa. Thus *Kirrika* 200, from **K4**, ? Kiambu District, Theta road, is said to have occurred in grassland. The combination of five fertile stamens and a stigma scarcely broader than the style is enough to separate it from all the native species.
** De Wit rejected *candida* in favour of his new name *alboflava* on the assumption that the former was first used as a varietal epithet by Corner in 1940 without a Latin description. In fact it had been already used long previously by Voigt (Hort. Suburb. Calcutt.: 253 (1845)). As he based it on *B. candida* Roxb., *non* Ait., a double citation is not allowable.

outside and with small orange probably glandular bodies; pods 2·5–4·5 cm. wide; seeds 1·6–2 cm. long 5. *B. petersiana*

1. **B. tomentosa** *L.*, Sp. Pl.: 375 (1753); L.T.A.: 654 (1930); T.S.K.: 62 (1936); T.T.C.L.: 88 (1949); Wilczek in F.C.B. 3: 271 (1952); De Wit in Reinwardtia 3: 409 (1956); Torre & Hillcoat in C.F.A. 2: 192 (1956); Roti-Michelozzi in Webbia 13: 153, figs. 3, 4 (1957); K.T.S.; 97 (1961); F.F.N.R.: 99 (1962). Type: Burmann, Thesaurus Zeylanicus, t. 18 (1737) (lecto.!, see Roti-Michelozzi, l.c.; typotype at G)*

Shrub or small tree 1–8 m. high. Branchlets glabrous, puberulous or ± pubescent. Leaves variable in size, 1–6·9(–10·2) cm. long, 1·2–6·5(–11) cm. wide, bilobed at apex to one-third way down or less, rarely to half way or a little more; lobes rounded; lower surface glabrous, appressed-puberulous, or ± pubescent. Racemes 1–2(–7)-flowered, lateral or terminal, occasionally aggregated. Bracteoles small and inconspicuous, not enfolding the flower-buds. Pedicels 0·7–2·7 cm. long. Flower-buds: upper part (i.e. sepals) ovate in outline, 1·2–2·6 cm. long before anthesis, glabrous or pubescent outside; hypanthium 2–6 mm. long; smooth or slightly sulcate. Petals subcircular to obovate or elliptic, (2·4–)3–5·5 cm. long, (1·5–)2–4·6 cm. wide, not or scarcely clawed, sulphur-yellow, 1–3 of them often (not always) blotched at base with dark brown or purplish**. Fertile stamens 10. Stigma 2·5–3 mm. across, rather variable in appearance, terminal, peltate or with one side produced downwards; style gradually enlarged towards the stigma. Pods thinly woody, dehiscent, 6·5–13 cm. long, (1–)1·3–2 cm. wide. Seeds blackish or blackish-brown, ± elliptic, 7–9 mm. long, 5–6 mm. wide; funicle short.

KENYA. Northern Frontier Province: Mathews Range, 10 June 1959, *Kerfoot* 1059!; S. Nyeri District: Tana R., about 1·5 km. above Sagana road suspension bridge, 5 Feb. 1933, *C. G. Rogers* 403!; Teita District: Wusi–Mwatate road, 18 Sept. 1953, *Drummond & Hemsley* 4407!

TANGANYIKA. Moshi, 9 Dec. 1925, *Durham*!; Dodoma District: 22 km. N. of Great Ruaha R. on the Dodoma road, *Wigg* 982!; Uzaramo District: Msua–Bagala, 5 Nov. 1925, *Peter* 31866!

DISTR. **K**1, 3, 4, ? 6, 7; **T**2, 3, 6, 8; also in Ethiopia, French Somaliland, Somali Republic (S.), the Congo Republic, and extending southwards to Angola, Zambia, Rhodesia, the Transvaal and Natal; also in Asia

HAB. Near dry lowland and riverine forest, in wooded grassland and deciduous bushland; 0–1520 (?–2130) m.

SYN. *B. tomentosa* L. var. *glabrata* Hook. f. in Bot. Mag., t. 5560 (1866), as " *glabra* "; Chiov., Racc. Bot. Miss. Consol. Kenya: 39 (1935). Type: cultivated at Kew, origin Angola

B. volkensii Taub. in P.O.A. C: 200 (1895); L.T.A. 654 (1930); T.T.C.L.: 88 (1949). Type: ? Kenya, Teita District, NE. of Lake Chala, *Volkens* 1765a (B, holo. †; *Volkens* 1765 (sic) at BM & K!, from the same locality, may well be isotypes)

B. wituensis Harms in E.J. 26: 275 (1899); L.T.A.: 654 (1930); T.T.C.L.: 89 (1949). Type: Kenya, Lamu District, Witu, *F. Thomas* 132 (B, holo. †, BM, K, iso.!)

NOTE ON SYNONYMY. *B. tomentosa* L. " var. *parvifolia-hirtella* Oliv. in Trans. Linn. Soc., ser. 2, 2: 332 " (1887) is quoted in L.T.A.: 654 (1930) as if it were a varietal name. I do not think that either it or " var. *glabra* ", both given by Oliver, were intended as formal new varieties but as brief descriptive phrases of two variants collected by Johnston on Kilimanjaro.

* In view of the fact that syntypes of *B. tomentosa* are available (as demonstrated by Roti-Michelozzi), the absence of authentic specimens in the Linnaean Herbaria cannot be sufficient reason for choosing a neotype, as De Wit proposed (Reinwardtia 3: 411 (1956)).

** De Wit (in Reinwardtia 3: 410–11 (1956)) calls the blotched form forma *tomentosa*, and that without blotches forma *concolor* De Wit.

VARIATION. *B. tomentosa* shows a considerable range of variation—in indumentum from glabrous to strongly pubescent, in leaf-size, in the shape, size and blotching of the petals, in the degree to which the tip of the flower-bud is acuminate and in the shape of the stigma. No clear pattern of variation is apparent, however, and I therefore recognize no named varieties.

One character, however, perhaps shows in its variation a link with geography: in East Africa the racemes are generally simple and single, but especially in Mozambique and Rhodesia there is a decided tendency for them to become aggregated into little corymbose panicles with up to 20 or more flowers. *Gillman* 1128, from Tanganyika, Lindi District, Sudi, 12 Dec. 1942, and *Schlieben* 5568, also from Lindi District, Lake Lutamba, 30 Oct. 1934, show this tendency.

NOTE. *B. tomentosa* is frequently cultivated (see, for example, U.O.P.Z.: 143, where it is said that the form grown in Zanzibar has unblotched flowers).

2. **B. mombassae** *Vatke* in Oesterr. Bot. Zeitschr. 30: 279 (1880); L.T.A.: 655 (1930); K.T.S.: 96 (1961). Type: Kenya, Mombasa, *Hildebrandt* 2006 (? B, holo. †, BM, K, iso. !)

Shrub (*Kelly* 1230). Young branchlets ± densely brownish-pubescent. Leaves ± 3·5–12·5 cm. long, 4–10 cm. wide, mostly larger than in *B. tomentosa*, bilobed at apex to about one-third to half way down with lobes ± tapering to an obtuse or rounded point and usually ± divergent, cordate at base, densely and softly pubescent beneath. Flowers in lateral and terminal several- to many-flowered panicles, whose lateral branches are sometimes few and short. Bracteoles broad and conspicuous, enfolding the flower-buds. Pedicels short, up to ± 7 mm. long. Flower-buds: upper part (i.e. sepals) lanceolate in outline, ± 1·9–2·2 cm. long before anthesis, pubescent outside; hypanthium ± 3–4 mm. long, slightly sulcate. Petals apparently obovate to suborbicular, ± 3–5 cm. long, not clawed, 4 of them yellow with bright orange bases, the fifth with a deep crimson patch at base. Fertile stamens 10. Stigma apparently peltate and terminal. Pods thinly woody, 7–12 cm. long, 1·4–2 cm. wide. Seeds unknown.

KENYA. Without precise locality, comm. May 1880, *Wakefield*!; Mombasa, June 1876, *Hildebrandt* 2006!; Kwale/Kilifi Districts: banks of Saponi R. near Mazeras, *Kelly* 1230! & in *C.M.* 13911!
DISTR. **K7**; not so far known with certainty from elsewhere, but see note below
HAB. Unknown

NOTE. Much allowance should be made for imperfections and errors in the description of this species, of which I have only seen the four gatherings cited above, none of them good. Although clearly related to *B. tomentosa*, *B. mombassae* in my opinion is beyond question a good species, distinguished by its rather large leaves with cordate bases and divergent tapering lobes, by the paniculate inflorescences, short pedicels, and especially by the broad boat-shaped convex bracteoles enfolding the unopened flower-buds.

A fruiting gathering similar to *B. mombassae* but with rounded-subtruncate, not at all cordate leaf-bases was made in Tanganyika, Morogoro District, Nguru Mts., Koruhamba, Nov. 1953, *Paulo* 216! More material is needed of this, as it is of typical *B. mombassae*. A flowering isotype of *B. loeseneriana* Harms in E.J. 33: 158 (1902); L.T.A.: 655 (1930); T.T.C.L.: 88 (1949) [Type: SE. Tanganyika, Kwa-Mkopo on the Ruvuma R., *Busse* 1027 (B, holo. †, iso. (fragment)!, EA, iso.!)] exactly matches *Paulo* 216, as far as they can be compared. At present I would consider the non-cordate leaf-bases and the (in *B. loeseneriana*) 5–6 mm. long points to the sepals of little taxonomic importance and *B. loeseneriana* to be thus only a minor variation of *B. mombassae*.

3. **B. taitensis** *Taub.* in P.O.A. C: 200 (1895); L.T.A.: 654 (1930); K.T.S.: 97 (1961). Type: Kenya, Teita District, between Ndi and Tsavo R., *Hildebrandt* 2603 (B, holo. †)

Shrub up to ± 2 m. high. Young branchlets densely spreading-pubescent. Leaves small, 1–2·6 cm. long, 1·5–3·3 cm. wide, bilobed to about half to three-fifths way down, pubescent on both surfaces; lobes rounded. Flowers solitary, often produced while the young foliage is emerging. Pedicels 5–8

(–20) mm. long. Flower-buds: upper part (i.e. sepals) lanceolate to ovate in outline, 0·8–1·3 cm. long before anthesis, pubescent outside. Petals white, fading to pink, oblanceolate to narrowly obovate, 1·3–2·4 cm. long, 0·5–1 cm. wide, cuneate below but not clawed. Fertile stamens 10. Stigma ± 1–2 mm. in diameter, terminal, peltate, abruptly enlarged from the style. Pods ± 2–3·8 cm. long (? immature), (0·6–)1–1·3 cm. wide.

Kenya. Northern Frontier Province: Koia [? Melka Koja], 29 Apr. 1946, *J. Adamson* in *Bally* 5876! & in *C.M.* 20590!; Machakos District: 221 km. on Mombasa–Nairobi road, Kenani area, 30 Aug. 1959, *Verdcourt* 2389!; Teita District: between Voi and Tsavo, 6 Apr. 1953, *Bally* 8852!

Distr. **K**1, 4, 7; not known elsewhere

Hab. Deciduous bushland, ? dry scrub with trees; 330–610 m.

Note. Possibly most closely related to *B. tomentosa* L., but easily separated by the normally smaller and more deeply divided leaves, by the much smaller flowers, and especially by the cuneate-based white petals.

4. **B. kalantha** *Harms* in E.J. 28: 398 (1900); L.T.A.: 655 (1930); T.T.C.L.: 88 (1949). Type: S. Tanganyika, Ruaha R., *Goetze* 470 (B, holo. †, K, iso. !)

Shrub up to ± 3 m. high, glabrous except for stamens and ovary. Leaves mostly 1–4·5 cm. long and 1·5–4 cm. wide, bilobed to four-fifths way down or more, sometimes even to base; each lobe rounded at apex; at bottom of sinus a flattened or boat-shaped scale-like projection 2–4 mm. long. Flowers solitary. Flower-buds: upper part (i.e. sepals) narrowly ovate to lanceolate in outline, 1·3–1·7 cm. long before anthesis; hypanthium 3–4 mm. long, smooth. Petals yellow, obovate, 2·5–3·7 cm. long, 1·2–2·1 cm. wide, neither clawed at base nor crisped on margins. Fertile stamens 10. Stigma ± 3 mm. in diameter, terminal, rounded, peltate, abruptly enlarged. Pods thinly woody, dehiscent, length uncertain, 1·3–1·5 cm. wide. Seeds deep brown, ± irregularly elliptic to obovate-elliptic, 7–8 mm. long, 6–7 mm. wide; funicle short.

Tanganyika. Dodoma District: Fufu stream, 72 km. S. of Dodoma, 16 Jan. 1932, *Wigg* 981!; Mpwapwa District: Gulwe, 7 Dec. 1925, *Peter* 32853!; Iringa District: Pawaga area, Jan. 1937, *Ward* P7!

Distr. **T**5, 7; not known elsewhere

Hab. Deciduous thickets; 600–960 m.

Note. The unusually deeply divided leaves distinguish this from all other species of the genus occurring in East Africa.

5. **B. petersiana** *Bolle* in Peters, Reise Mossamb., Bot. 1: 24 (1861); Oliver, F.T.A. 2: 288 (1871); L.T.A.: 656 (1930); T.T.C.L.: 88 (1949); F.F.N.R.: 99 (1962). Type: Mozambique, near Sena, *Peters* (B, holo. †, K, iso. !)

Shrub or tree 2–8 m. high. Young branchlets ± densely brown-pubescent or -puberulous. Leaves often wider than long, 2–8 cm. long, 2–10 cm. wide, bilobed to about one-third to two-thirds way down, minutely appressed-puberulous on lower surface; lobes rounded. Racemes short, often aggregated terminally. Flower-buds: upper part (i.e. sepals) linear to linear-lanceolate in outline, 3–5 cm. long before anthesis; hypanthium 2–4 cm. long, finely longitudinally sulcate. Petals white or marked with pink spots, narrowly elliptic to oblanceolate-elliptic, 4–7·5 cm. long, 1–2·7 cm. wide, with very crisped margins and the midrib pubescent outside and with many small orange ? glands which are also present in the indumentum of the young vegetative parts. Fertile stamens 5. Stigma 2·5–3 mm. in diameter, terminal, peltate, abruptly enlarged from the style. Pods woody, dehiscent, linear-oblong or oblanceolate-oblong, 10–24 cm. long, (2·5–)3–4·7 cm. wide. Seeds deep chestnut-purple, irregularly oblong to obovate or subcircular, 1·6–3 cm. long, 1·1–1·8 cm. wide; funicle very short and thick. Fig. 47, p. 212.

FIG. 47. *BAUHINIA PETERSIANA*—**1,** part of flowering branch, × $\frac{2}{3}$; **2,** lower surface of leaflet, showing appressed indumentum, × 4; **3,** part of petal, showing orange ? glands, × 4; **4,** stamen, × 2; **5,** pod, × $\frac{2}{3}$; 6, part of suture of pod, showing attached funicle with outgrowths, × 2; **7,** seed, × $\frac{2}{3}$; **8,** hilum of seed, showing the two unequal linear marks made by the outgrowths from the funicle, × 2. 1–3, from *Milne-Redhead & Taylor* 8524; 4 from *Milne-Redhead & Taylor* 7560; 5–8, from *Verdcourt* 2779.

TANGANYIKA. Ufipa District: Edith Bay on coast of Lake Tanganyika, 14 Apr. 1936, *B. D. Burtt* 6305!; Morogoro District, 15 Feb. 1932, *Wallace* 371!; Iringa, 18 Jan. 1952, *Wigg* 998!; Songea District: about 1·5 km. S. of Gumbiro, 24 Jan. 1956, *Milne-Redhead & Taylor* 8524!

DISTR. **T4**, 5, 6–8; Congo Republic, Mozambique, Zambia and Rhodesia

HAB. Woodland and wooded grassland; 150–1830 m.

NOTE. *B. petersiana* is a rather uniform species in Tanganyika. A few gatherings from the south-western part of the territory have the petals broader than usual (about 1·5–2·5 cm.), e.g. *Richards* 7418, *Bullock* 2368, *Geilinger* 3073. In this character there is some approach to the very closely related *B. macrantha* Oliv., which does not occur north of Zambia, and differs from *B. petersiana* in the pubescence of curved but non-appressed hairs on the lower surface of the leaf, the scattered not aggregated inflorescences, in the generally smaller leaves up to about 4·7 cm. long and bilobed to more than half-way, in the often wider (to 4·2 cm.) petals, and in the narrower (1·7–3 cm.) pods.

36. TYLOSEMA

(Schweinf.) Torre & Hillc. in Bol. Soc. Brot., sér. 2, 29: 38 (1955) & in C.F.A. 2: 198 (1956)

Bauhinia L. sect. *Tylosema* Schweinf., Reliq. Kotsch.: 17 (1868)

Stems trailing or climbing, herbaceous or woody below, arising from a large or very large underground tuber; tendrils usually present, forked (absent in *T. humifusa*). Leaves simple, bilobed at apex or sometimes divided almost to base, which is very cordate. Flowers medium to rather small, yellow, in lateral short to elongate racemes. Hypanthium short. Sepals 5, the two upper usually completely or partly fused, the others free. Petals 5; upper one smaller than the rest and bicallose at base. Stamens: 2 fertile; remainder ((7–)8) staminodial, unequal, some ± flattened. Ovary long-stipitate; style elongate; stigma very small, not wider than top of style. Pods woody, dehiscent or indehiscent, (? 1–)2-seeded. Seeds large, with a U-shaped line extending only for a short distance from the hilum; funicle short, with a bifid projection at apex; a thin layer of endosperm present.

About four species in eastern and central tropical Africa, from the Sudan southwards to the Transvaal and Swaziland.

Although Roti-Michelozzi (in Webbia 13: 171 (1957)) implies that heterostyly is a feature specific to *T. humifusa*, I find it to occur also in *T. fassoglensis* and *T. argentea*, and in fact it seems to be characteristic of the genus. The two fertile stamens have short filaments in the dolichostylous flowers, and long filaments exceeding the style in the brachystylous ones.

Tendrils present:
- Leaves 5–20 cm. long, 5·5–23 cm. wide, bilobed to $\frac{1}{10}$–$\frac{1}{3}$ (very rarely to $\frac{1}{2}$) distance from lobe-ends to junction with petiole; four larger petals 2·2–3·8 cm. long . . . . 1. *T. fassoglensis*
- Leaves up to 2 cm. long and wide, bilobed to $\frac{1}{2}$–$\frac{2}{3}$ distance from lobe-ends to junction with petiole; four larger petals 1·5–2 cm. long . . . . 2. *T. argentea*

Tendrils absent; leaves up to ± 3·5 cm. long, bilobed to ± $\frac{2}{3}$–$\frac{5}{6}$ distance from lobe-ends to junction with petiole; four large petals 1–1·8 cm. long. . . 3. *T. humifusa*

1. **T. fassoglensis** (*Schweinf.*) *Torre & Hillc.* in Bol. Soc. Brot., sér. 2, 29: 38 (1955) & in C.F.A. 2: 198 (1956). Types: Sudan Republic, Fazoghli, *Boriani* 131 (W, syn.) & *Cienkowski* 92 (? LE or W, syn.) & Metemma, Gallabat, *Schweinfurth* 2250, 2252, 2253 (B, syn. †, isosyn. of 2250 at BM!, isosyn. of 2252 at BM! and K!)

Stems prostrate and trailing, or climbing up to 6 m. or more, herbaceous, or woody below. Young parts ± rusty-tomentose or rusty-pubescent; indumentum becoming greyish, or ± disappearing. Tendrils present, forked. Leaves: petiole 1·5–22 cm. long; blade 5–11·5(–20) cm. long, 5·5–12(–23) cm. wide, usually ± rusty-pubescent beneath especially on nerves, sometimes subglabrous or densely tomentose, shallowly bilobed at apex to about one-tenth to one-third (very rarely to half) the length of the leaf from lobe-ends to junction with petiole; lobes rounded. Racemes: peduncle (2–)5–17·5 cm. long; axis (2–)5–23(–42) cm. long; pedicels (1·5–)2–6 cm. long. Hypanthium 3–8 mm. long. Sepals 1–1·7 cm. long, ± conspicuously keeled along back, the two upper ones fused, the other three free. Petals yellow, fading to pink; the four larger ones obovate to obovate-suborbicular, crinkled-bullate, 2·2–4 cm. long, 1·3–3·3 cm. wide, tapering into a basal claw. Pods 7–12 cm. long, 4·5–7·3 cm. wide. Seeds suborbicular or ellipsoid, 1·7–2·8 cm. long, 1·5–2 cm. wide, chestnut-brown to blackish. Fig. 48.

UGANDA. W. Nile District: Terego, Apr. 1938, *Hazel* 485!; Mbale District: near Busia, W. Bugwe Reserve, 6 Sept. 1950, *Dawkins* 631!; Mengo District: Bugerere, Bale, 28 June 1956, *Langdale-Brown* 2114!

KENYA. Northern Frontier Province: Moyale, 28 Apr. 1952, *Gillett* 12962!; S. Kavirondo, Nov. 1941, *Opiko* in *Bally* 1793!; Teita District: Voi, 6 Feb. 1953, *Bally* 8591!

TANGANYIKA. Lushoto District: SW. Umba steppe, Bombo, 3 Jan. 1930, *Greenway* 2032!; Mpwapwa, 4 Jan. 1929, *M. G. Hornby* 54!; Lindi District: 1·5 km. S. of R. Mbemkuru, 6 Dec. 1955, *Milne-Redhead & Taylor* 7474!

DISTR. **U**1–4; **K**1–7; **T**1, 3–8; distribution as for the genus

HAB. Wooded grassland, grassland, deciduous bushland; 100–1830 m.

SYN. *Bauhinia fassoglensis* Schweinf., Reliq. Kotsch.: 14, t. 12, 13 (1868); L.T.A.: 659 (1930); T.S.K.: 62 (1936); W.F.K.: 41 (1948); T.T.C.L.: 87 (1949); Wilczek in F.C.B. 3: 272 (1952); Roti-Michelozzi in Webbia 13: 163 (1957); F.F.N.R.: 99 (1962)

B. cissoïdes Oliv., F.T.A. 2: 287 (1871). Type: Angola, Ambaca, *Welwitsch* 552 (LISU, holo., BM, K, iso.!)

B. welwitschii Oliv., F.T.A. 2: 287 (1871). Type: Angola, Pungo Andongo, Tunda Quilombo, *Welwitsch* 554 (LISU, holo., BM, K, iso.!)

B. kirkii Oliv., F.T.A. 2: 288 (1871). Type: Zambia, Batoka highlands, *Kirk* (K, holo.!)

B. bainesii Schinz in Mém. Herb. Boiss. 1: 121 (1900). Types: South West Africa, *Schinz* 2061 (Z, syn.) & Rhodesia, ? Gwelo District, *Baines* (K, syn.!)

B. fassoglensis Schweinf. forma *cissoïdes* (Oliv.) Bak. f. in J.B. 66, suppl., Polypet.: 139 (1928)

VARIATION. This species shows a considerable range of variation in indumentum, leaf-size, length of peduncle and inflorescence-axis, and size of flower. This variation has led to the separation of several alleged species here considered to be synonymous. *B. fassoglensis* forma *cissoïdes* has been used for an extreme with the leaves densely tomentose beneath, but there are too many intermediates to make it worth separation.

NOTE. An unusual character shown by *T. fassoglensis*, and shared with *T. argentea*, is the frequent presence of some pubescence on the connectives of the fertile anthers.

2. **T. argentea** (*Chiov.*) *Brenan* in K.B. 17: 214 (1963). Types: Somali Republic (S.), Matagassile on the R. Giuba [Juba], *Paoli* 832 & between Dorianle and Oneiatta, *Paoli* 909 (both FI, syn.!)

Herbaceous climber, with stems up to 90 cm. or more long. Young parts with short appressed grey pubescence. Tendrils present, forked. Leaves: petiole 0·5–2 cm. long; blade up to 2·0 cm. long and wide (*fide* Roti-Michelozzi), pubescent on nerves beneath, bilobed at apex to about half to two-thirds the length of the leaf from lobe-ends to junction with petiole; lobes rounded. Racemes: peduncle 0·4–2 cm. long (but adnate to stem for ± 1–3 cm. above nodes in addition); axis 0·7–5 cm. long; pedicels 0·7–3·5 cm. long. Hypanthium 3–5 mm. long. Sepals ± 6–9 mm. long, somewhat keeled along back in upper part, the upper ones free to near base. Petals bright

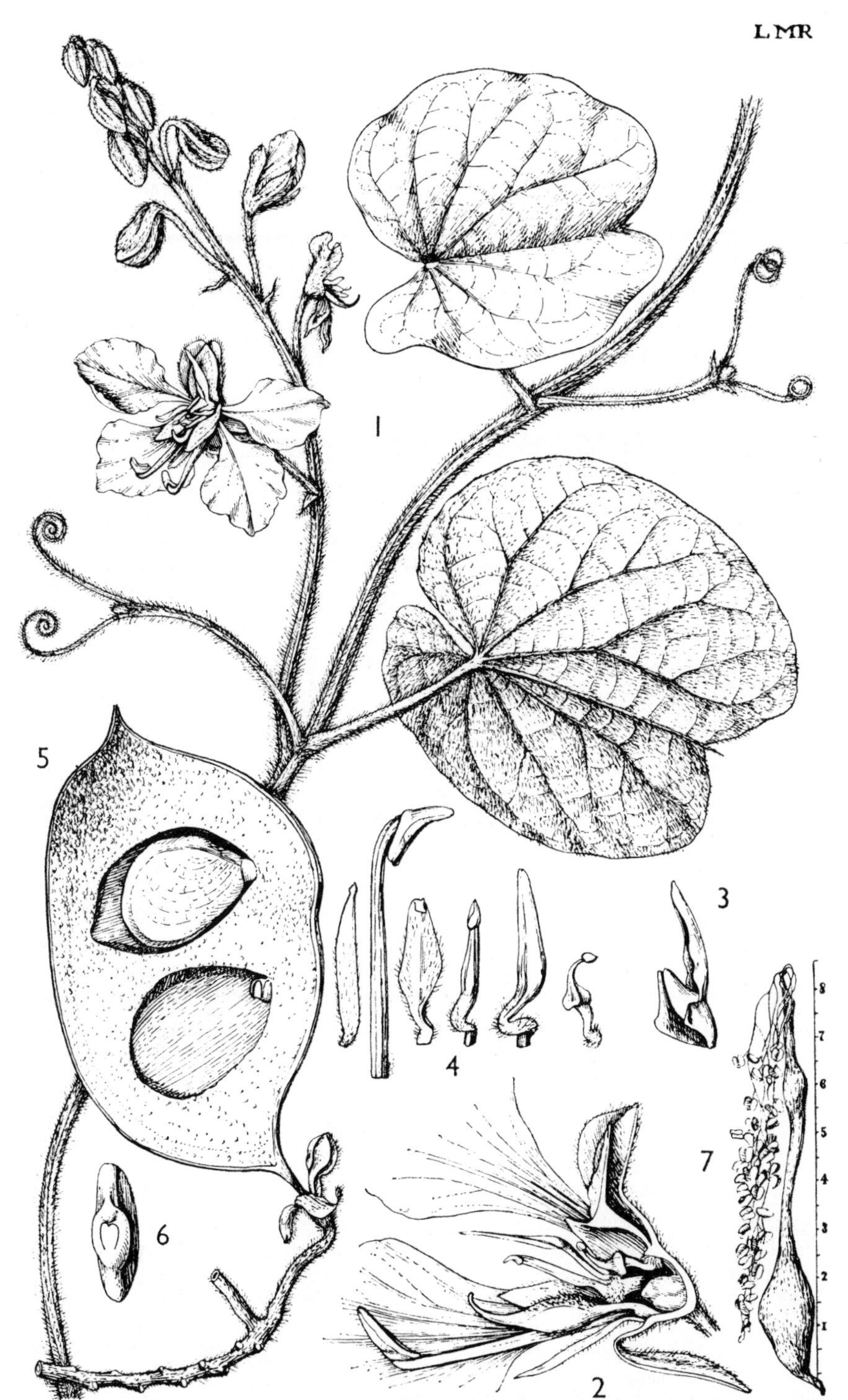

FIG. 48. *TYLOSEMA FASSOGLENSIS*—**1,** part of flowering stem, × ⅔; **2,** longitudinal section of flower, showing upper petal cut longitudinally, four unequal staminodes, one fertile stamen and the ovary, × 2; **3,** upper petal, seen from one side, × 2; **4,** one fertile stamen and five staminodes, showing inequality of latter, × 2; **5,** dehisced pod, inner side of one valve, × ⅔; **6,** funicular end of seed, showing hilum, × 1; **7,** tuberous root, with scale in feet. 1–4, from *Harley* 9410; 5, from *Rodin* 4341; 6, from *Chandler* 1126; 7, from a photograph by J. H. Hopkins, at Kew.

yellow, the four larger ones rotundate-obovate, 1·5–2 cm. long, 1–1·3 cm. wide, narrowed into a basal claw. Pods and seeds unknown.

KENYA. Northern Frontier Province: 48 km. NE. of Wajir, 27 May 1952, *Gillett* 13360!
DISTR. **K1**; Somali Republic (S.)
HAB. *Acacia-Commiphora* open scrub on edge of red sand; 300 m.

SYN. *Bauhinia argentea* Chiov. in Ann. Bot., Roma 13: 388 (1915); Roti-Michelozzi in Webbia 13: 171 (1957)

3. **T. humifusa** (*Pichi-Serm.* & *Roti-Michelozzi*) *Brenan* in K.B. 17: 214 (1963). Type: Somali Republic (N.), Derinderr on S. border, *Glover & Gilliland* 1036 (K, holo.!, BM, EA, iso.!)

Stems apparently herbaceous, creeping along the ground for at least 45 cm., spreading-pubescent especially when young. Tendrils absent. Leaves: petiole very short, 1–5 mm. long (see note below); blade 0·6–3 cm. long, 0·8–3 cm. wide, pubescent on nerves beneath, bilobed at apex to ± two-thirds to five-sixths length of the leaf from lobe-ends to junction with petiole; lobes rounded. Racemes: peduncle 1·5–6 cm. long; axis 4–10 cm. long; pedicels 1·3–2·7 cm. long. Hypanthium 3–4 mm. long. Sepals ± 7–9 mm. long, at most somewhat ridged but scarcely keeled on back, the upper ones ± fused or free to near base. Petals yellow, the four larger ones 1–1·8 cm. long, ± 0·5–1 cm. wide, obovate-suborbicular, tapering into a basal claw. Pods (not fully mature) rhombic, ± 4 cm. long and 2·5 cm. wide. Seeds unknown.

KENYA. Northern Frontier Province: Yabichu near Ramu, 23 May 1952, *Gillett* 13289!
DISTR. **K1**; Somali Republic (N.)
HAB. *Acacia-Commiphora* open scrub on pale limestone soils; ± 360 m.

SYN. *Bauhinia humifusa* Pichi-Serm. & Roti-Michelozzi in Webbia 13: 165 (1957)

NOTE. Certain plants have been collected in northern Kenya which are closely similar to *T. humifusa* except for having longer petioles 0·8–2·8 cm. long. They also have leaves with lamina up to 3·5 cm. long and wide; pedicels up to 3 cm. long; sepals 8–11 mm. long, the upper ones connate, or free to half-way down; petals 1–1·2 cm. wide. Their pods and seeds are unknown. They are the following: Northern Frontier Province: Yaka, 25 May 1945, *J. Adamson* 82 in *Bally* 4383! & Merti, 24 Apr. 1946, *J. Adamson* in *Bally* 5870! & Melca Murri, 26 Apr. 1958, *Everard* in *E.A.H.* 11441!; Meru District: Meru Game Reserve, Thaichu, 20 June 1963, *Mathenge* 183!

Bally 4383 is a poor specimen without foliage, doubtfully placed here. The petioles of *Bally* 5870 are 1–2·8 cm. long, while those of *Everard* 11441 are 0·8–1 cm. long. The latter specimen thus more or less bridges the gap in petiole-length between *Bally* 5870 and typical *T. humifusa*. The material is insufficient to decide whether these specimens represent a closely related but distinct species or (as seems much more likely) a variety or perhaps a mere form of *T. humifusa*.

37. BAPHIOPSIS

Bak. in Oliv., F.T.A. 2: 256 (1871)

Unarmed evergreen tree or shrub. Leaves alternate, unifoliolate; stipules very small and soon falling off; blade without pellucid dots. Flowers ⚥, in short racemes or almost fasciculate, each flower subtended by 2 small nervose caducous bracteoles. Calyx ellipsoid and entire before dehiscence, opening by a single slit and reflexing, or becoming divided into 2–3 lobes; no disc or cupular hypanthium. Petals 6, almost equal, imbricate, the upper one with its margins overlapped. Stamens 13–41, arranged round the base of the ovary, free or almost so; anthers basifixed, dehiscing by longitudinal slits; connective not glandular. Ovary sessile or nearly so, 2–5-ovuled, tapering upwards into a subulate style which is bent in its upper part and bears a minute capitate stigma. Pods thick, beaked 1-(or, *fide* F.C.B., sometimes

Fig. 49. *BAPHIOPSIS PARVIFLORA*—**1,** part of flowering branch, × $\frac{3}{8}$; **2,** part of leaf-surface, showing venation, × 6; **3,** part of young inflorescence, × 2; **4,** diagrammatic cross-section of bud, showing aestivation of petals; **5,** part of more mature inflorescence, showing one open flower, × 2; **6,** flower, × 4; **7,** ovary, longitudinal section, × 6; **8,** pod, × 1; **9,** seed, × 1. 1–5, from *G. H. S. Wood* 322; 6–7, from *Purseglove* 3416; 8, from *Benedicto* 4; 9, from *Dawe* 972.

2-) seeded. Seeds large, thin-walled, apparently not arillate, without endosperm; embryo curved.

A single species in tropical Africa.

B. parviflora *Bak.* in Oliv., F.T.A. 2: 256 (1871); L.T.A.: 590 (1930); J. Léon. in Bull. Soc. Roy. Bot. Belge 84: 56, fig. 1 (1951); Gilbert & Boutique in F.C.B. 3: 554 (1952); I.T.U., ed. 2: 297 (1952); F.W.T.A., ed. 2, 1: 446 (1958). Type: Cameroun Republic, Ambas Bay, *Mann* 715 (K, holo. !)

Shrub or small tree 2·5–15 m. high, often leaning, sprawling or horizontal; bark smooth, reddish-brown (I.T.U.: 298). Young branchlets (in East Africa) glabrous or thinly and inconspicuously appressed-grey-puberulous, or in some Congo specimens with dense rusty pubescence; branchlets soon glabrescent. Leaves on petioles 0·5–4 cm. long which are glabrous or clothed when young like the young branchlets, as is also at first the lower side of the midrib; blade elliptic or sometimes obovate-elliptic or ovate-elliptic or lanceolate, (3–)4·5–16·5(–25) cm. long, (1·2–)1·5–7(–11) cm. wide, acuminate, rounded at base, glabrous when mature; lateral nerves and reticulate venation prominent on both surfaces. Racemes single or clustered, or themselves racemose on a very short axis, axillary, among or below the leaves, (in East Africa) up to 3·5 cm. long, but their axis often very short or almost absent, grey-puberulous or rusty-pubescent in some Congo specimens; pedicels 0·6–2 cm. long. Buds 5–6 mm. long. Flowers white except for the yellow anthers. Petals 6–10 mm. long, 2·5–3 mm. wide, oblanceolate to narrowly obovate or elliptic, scarcely clawed. Stamens 13–18(–41 in the Congo). Ovary glabrous to thinly grey-pubescent or (in Congo) densely rusty-pubescent. Pods 3–4(–6, *fide* F.C.B.) cm. long, 1·5–2·5 cm. wide, black. Seeds 2·5–4 cm. long and ± 1·5 cm. wide. Fig. 49, p. 217.

UGANDA. Kigezi District: Ishasha Gorge, May 1950, *Purseglove* 3416 !; Busoga District: 16 km. N. of Jinja, 1·5 km. N. of Mutai Crown Forest Reserve, Bunya, 6 Aug. 1952, *G. H. S. Wood* 322 !; Mengo District: Mawokota, Mpanga Forest, 1 Nov. 1953, *Byabainazi* 57 !

TANGANYIKA. Bukoba District: Minziro Forest Reserve, July 1951, *Eggeling* 6255 ! & 13 Feb. 1954, *Benedicto* 4 ! & Sept. 1957, *Procter* 690 !

DISTR. **U2–4**; **T1**; Cameroun Republic, Gabon, Congo Republic and Angola

HAB. Swamp-forest and lowland rain-forest; 1110–1310 m.

SYN. *Baphiopsis stuhlmannii* Taub. in P.O.A. C: 203 (1895); L.T.A.: 591 (1930); T.T.C.L.: 410 (1949). Type: Tanganyika, Bukoba, *Stuhlmann* 1045 (B, holo. †, BM, fragments of holotype and drawing !)

Baphia radcliffei Bak. f. in J.L.S. 37: 147 (1905); Lester-Garland in J.L.S. 45: 228 (1921). Type: Uganda, Masaka District, Musozi, *Bagshawe* 74 (BM, holo. !)

VARIATION. Certain specimens from the Congo Republic diverge from the general range of variation of *B. parviflora* in having more stamens (up to 41) than the normal 13–18. With this trend there appear to be correlated others—a tendency to dense rusty pubescence on the young parts and especially on the inflorescence, a dense indumentum on the ovary, and longer inflorescences (sometimes to 5 cm.).

I have allowed for these Congo specimens in the specific description above, but indicated where I was doing so. Léonard (in Bull. Soc. Roy. Bot. Belge, 84: 56–8 (1951)) has carefully examined the range of variation in the Congo and concluded that it should all be included in a single species. In view, however, of the comparative uniformity of the species elsewhere, I feel it possible that these aberrant specimens may in the future be found worthy of recognition as a variety or subspecies.

38. SWARTZIA

Schreb., Gen. Pl. 2: 518 (1791), *nom. conserv.*

Unarmed trees or rarely shrubs. Leaves alternate, imparipinnate or (but not in East Africa) pinnately trifoliolate or unifoliolate; stipules mostly small or very small; leaflets opposite or more rarely alternate, without pellucid dots. Flowers ⚥, in lateral racemes, or sometimes in panicles or fascicles. Bracteoles extremely inconspicuous or (in East Africa) absent. Calyx globose or ellipsoid and entire before dehiscence, becoming variously lobed or torn on opening; disc 0. Petals normally 1, rarely (and not in East Africa) with 2 small additional lateral ones, or entirely absent. Stamens numerous, (more than 30) arranged in several rows at base of calyx around the gynophore, free or almost so; anthers affixed near base, dehiscing by longitudinal slits; connective not glandular. Ovary long-stipitate, several- to many-ovuled; stigma very small, rarely capitate. Pods stipitate, coriaceous or woody, turgid or cylindrical, rarely merely compressed, shortly boat-shaped to cylindrical or torulose, indehiscent or dehiscing into 2 valves, 1–several-seeded. Seeds not areolate, arillate or not, with or without endosperm; radicle of embryo curved, or straight.*

About 100–120 species, two in tropical Africa, the rest in tropical America.

S. madagascariensis *Desv.* in Ann. Sci. Nat. Paris, sér. 1, 9: 424 (1826); L.T.A.: 605 (1929); T.T.C.L.: 444 (1949); Gilbert & Boutique in F.C.B. 3: 551 (1952); Torre & Hillcoat in C.F.A. 2: 167 (1956); F.F.N.R.: 128, fig. 21/K (1962). Type: locality doubtful, *Herb. Desvaux* (P, holo.)

Tree or occasionally a shrub 3–15 m. high, with rough longitudinally fissured or reticulate bark. Young branchlets densely pubescent to tomentose; indumentum when dry greyish or more usually fawn or rusty. Leaflets alternate, rarely opposite, (3–)7–11(–13), elliptic or obovate-elliptic, rarely oblong-elliptic, 2–10 cm. long, 1·2–5(–5·7) cm. wide, rarely more, rounded at both ends and often ± emarginate at apex, much paler beneath than above, and ± densely appressed-hairy or tomentose beneath, rarely subglabrous. Racemes 2–10-flowered or more, axillary, solitary or up to 3 together; axis to 5(–8) cm. long, sometimes very short or absent so that the flowers appear fascicled; pedicels 1·2–5(–7) cm. long, usually shortly tomentose like the axis and outside of calyx. Calyx at first globose and ± 7 mm. in diameter, then rupturing irregularly into 2–5 lobes and reflexing. Petal 1, white, crinkled, densely rusty-pilose outside, glabrous inside, clawed, 2–3·6 cm. long, (1·8–)2·2–3 cm. wide. Stamens orange-yellow. Pods hard, sausage-shaped, deep chestnut to black, indehiscent, (6–)8–30 cm. long, 1–2·3 cm. in diameter. Seeds 7–8 mm. long, 5–7 mm. wide, 3 mm. thick, pale brown, without aril or endosperm. Fig. 50, p. 220.

TANGANYIKA. Shinyanga, *Koritschoner* 1991!; Dodoma District: Kazikazi, 17 Nov. 1931, *B. D. Burtt* 3408!; Ulanga District: Ifakara, Machipi, 24 Sept. 1958, *Haerdi* 69/0!

DISTR. T1, 4–8; Gambia to the Cameroun Republic, Congo Republic and Tanganyika, southwards to Mozambique, Rhodesia and South West Africa**

HAB. Deciduous woodland (*Julbernardia-Brachystegia*) and wooded grassland; 450–1280 m.

* Bentham in Gen. Pl. 1: 561 (1865) described the radicle in *Swartzia* as curved; Gilbert & Boutique in F.C.B. 3: 550 (1952) stated that the embryo is straight. A straight radicle is also described for *Swartzia pinnata* by Corner in Phytomorphology 1: 141 (1951). That of *S. madagascariensis* is certainly curved. The matter is of more than casual significance, as in *Papilionoïdeae* the radicle is curved and in *Caesalpinioïdeae* normally straight. The presence of both sorts within *Swartzia* emphasizes the borderline position (on other features) that it occupies between *Caesalpinioïdeae* and *Papilionoïdeae*, and supports their treatment as no more than subfamilies.

** In spite of the epithet of the species, there is no evidence for its occurrence in Madagascar or the Mascarenes beyond Desvaux's original statement, which is almost certainly erroneous. I am indebted to M. P. Hallé for help over this problem.

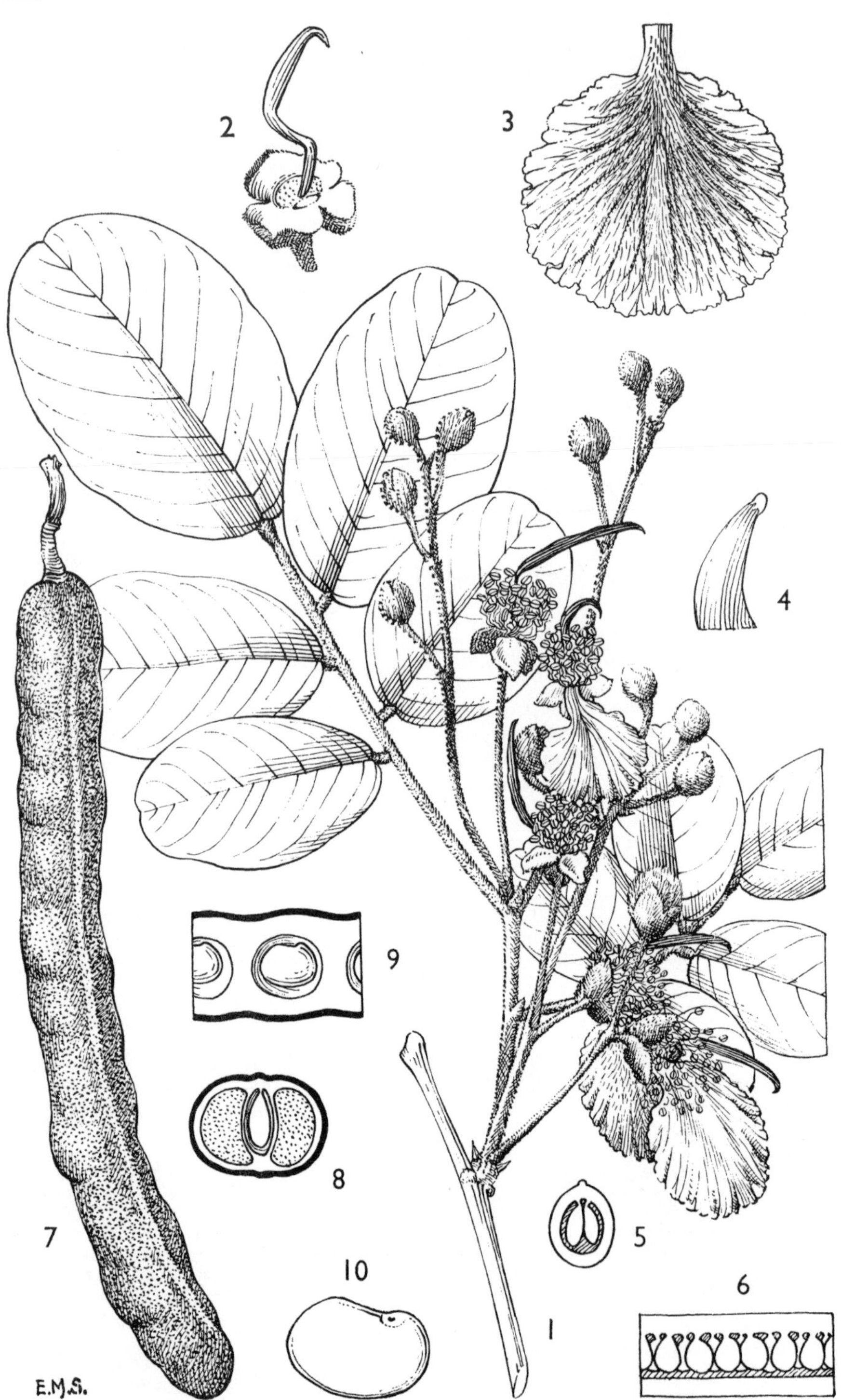

FIG. 50. *SWARTZIA MADAGASCARIENSIS*—**1**, part of flowering branch, × 1; **2**, flower, with petal and stamens removed, × 1½; **3**, petal, under-surface, × 1½; **4**, apex of style, and stigma, × 6; **5**, ovary, cross-section, diagrammatic; **6**, ovary, longitudinal section, diagrammatic; **7**, pod, × $\frac{2}{3}$; **8**, pod, cross-section, diagrammatic; **9**, pod, longitudinal section, diagrammatic; **10**, seed, × 3. 1–6, from *B. D. Burtt* **3417**; 7–10 from *B. D. Burtt* 3382.

39. CORDYLA

Lour., Fl. Cochinch.: 411 (1790); Milne-Redh. in F.R. 41 : 227–235 (1937)

Unarmed deciduous trees, rarely shrubby. Leaves alternate, imparipinnate; stipules small, soon falling off; leaflets petiolulate, alternate or rarely subopposite, with numerous pellucid dots or streaks. Flowers ⚥ or ♂, in racemes which are axillary or clustered at nodes or sometimes terminal. Calyx with a subglobose limb entire before dehiscence, splitting into 3–5 lobes on opening. Receptacle (" calyx-tube ") campanulate; a definite disc (i.e. with a margin) not present, the staminal tube merging evenly with the receptacle. Stamens numerous (± 23–126), usually crowded into several series round the top of the receptacle; filaments very shortly connate at base; anthers dorsifixed, dehiscing by longitudinal slits; connective glandular at top. Ovary (in ⚥ flowers) long-stipitate, several-ovuled, tapering into a subulate style; stigma small. Fruits stipitate, ellipsoid to subglobose, beaked or rounded, indehiscent, with 1–6 seeds embedded in pulp. Seeds large, thin-walled, not arillate, without endosperm; radicle of embryo straight.

A genus of 5 (? 6) species, all tropical African (though *C. africana* extends beyond the tropics into South Africa), except for *C. madagascariensis* R. Vig. from Madagascar, whose description does not separate it clearly from *C. africana*.

Leaflets minutely appressed-puberulous beneath; racemes usually borne on shoots of current year below the leaves; stamens 23–45, orange-yellow . . . 1. *C. africana*
Leaflets with crisped non-appressed hairs beneath, sometimes almost glabrous; racemes borne on woody usually leafless twigs of previous season's growth; stamens 50–126, white or greenish-white:
Pedicels 6–16 mm. long; racemes borne singly or fascicled, usually not densely crowded, 2–7(–10) cm. long, densely pubescent or tomentose; stamens 80–126 . 2. *C. richardii*
Pedicels 4–6(–8) mm. long; racemes usually densely clustered and crowded, 0·7–2(–3·5) cm. long, glabrous to densely pubescent; stamens 50–60 . . . 3. *C. densiflora*

1. **C. africana** *Lour.*, Fl. Cochinch.: 412 (1790); L.T.A.: 606 (1929), pro parte; Milne-Redh. in F.R. 41 : 230 (1937); T.T.C.L.: 410 (1949); F.F.N.R.: 121, fig. 21/F (1962). Type: East African coast, *Loureiro* (P, holo.)*

Tree 9–24(?–40) m. high; bark brown or grey, rough, much fissured; crown bushy. Leaves: petiole with rhachis (5–)9–24 cm. long; leaflets 11–28, oblong, oblong-elliptic or ovate-oblong, (1–)2–5 cm. long, (0·7–)1·2–2·4 cm. wide, rounded at apex and sometimes slightly emarginate, rounded at base, minutely appressed-puberulous beneath. Racemes usually borne on shoots of current season below the leaves, 1·5–11 cm. long; pedicels (and outside of receptacle and calyx) subglabrous to shortly and finely pubescent, 4–9 mm. long. Flowers (including stamens and ovary) 1·5–2·2(–3) cm. long. Receptacle and calyx-lobes green, latter with tuft of yellowish pubescence at tip. Stamens 23–45, orange-yellow. Fruits ellipsoid, oblong or spherical, ± oblique, 4·5–8 cm. long, 3–6 cm. wide, yellow, 1–3-seeded. Fig. 51, p. 222.

Kenya. Teita District: Taveta, 22 Jan. 1936, *Greenway* 4479 !; Kwale District: Malenge Forest, Sept. 1937, *Dale* 3852 !
Tanganyika. Moshi District: Moshi–Arusha road, 19 Nov. 1951, *McCoy-Hill* 19 !; S. Kilosa District, Nov. 1925, *Simmance* in *F.H.* 199 !; Lindi District: Ruponda, 20 Sept. 1947, *Semsei* in *F.H.* 2104 !

* A fragment of a specimen collected by Loureiro, ' 'Ex Herb. Mus. Paris ", is at BM ! This may be a portion of the holotype.

FIG. 51. *CORDYLA AFRICANA*—**1**, part of flowering branch, × 1; **2**, part of leaflet-surface, showing venation and gland-dots, × 6; **3**, longitudinal section of flower, showing attachment of stipe of ovary, × 1½; **4**, ovary, longitudinal section, × 6; **5**, fruit, × 1; **6**, fruit, longitudinal section, × 1. 1, from *McCoy-Hill* 19; 2–4, from *Lewis* 38; 5, 6, from *Wild* 2408 in *G.H.* 19005.

ZANZIBAR. Zanzibar I., Kufile cave-well, 22 Sept. 1935, *Vaughan* 2284!
DISTR. **K7**; **T2, 3, 6, 8**; **Z**; Mozambique, Malawi, Zambia and Rhodesia, the Transvaal and Natal (Zululand)
HAB. Riverine and ground-water forest; ? also in lowland dry evergreen forest; 10–900 m.

2. **C. richardii** *Milne-Redh.* in F.R. 41: 232 (1937); I.T.U., ed. 2: 298 (1952). Type: Sudan Republic, Mongalla Province, S. of Gondokoro, *Grant* (K, holo.!)

Tree, rarely a shrub, 3–12 m. high; bark rough, dark brown on trunk, pale brown to grey-brown on branches (I.T.U.: 298). Leaves: petiole with rhachis 11–31 cm. long; leaflets (11–)17–23(–32), oblong or ovate-oblong or elliptic-oblong, 1·5–4·1 cm. long, 0·8–2 cm. wide, rounded to obtuse and sometimes slightly emarginate at apex, rounded to broadly cuneate at base, glabrous to pubescent above, pubescent beneath with crisped non-appressed hairs varying from dense all over to very sparse except on midrib. Racemes borne singly or fascicled, usually not densely crowded, on woody usually leafless twigs of the previous season, 2–7(–10, *fide* I.T.U.) cm. long; pedicels (and outside of receptacle and calyx) densely pubescent or tomentose, 6–16 mm. long. Stamens 80–126, white; anthers pale yellow. Fruits oblong-globose or ellipsoid-globose, 3–5 cm. long, 2–4 cm. wide, yellow, 2-seeded.

UGANDA. W. Nile District: Midigo, 27 Nov. 1941, *A. S. Thomas* 4075! & W. Madi, Metu, 5 Dec. 1947, *Dawkins* 302!; Acholi District: Chua, Agoro, Feb. 1938, *Eggeling* 3508!
DISTR. **U1**; Sudan Republic
HAB. Deciduous woodland on rocky hillsides; 1070–1220 m.

3. **C. densiflora** *Milne-Redh.* in F.R. 41: 234 (1937); T.T.C.L.: 411 (1949). Type: Tanganyika, Mpwapwa District, Gulwe, *Greenway* 2405 (K, holo.! EA, iso.!)

Tree 4–10 m. high; bark light grey or greyish-brown. Leaves: petiole with rhachis 9–21 cm. long; leaflets 11–19, ovate-oblong to elliptic-oblong, 2–3·5 cm. long, 1·2–2·3 cm. wide, rounded to emarginate at apex, usually cordate but occasionally rounded or truncate at base, glabrous to pubescent above, ± pubescent beneath with crisped non-appressed hairs (which are sometimes almost confined to midrib). Racemes usually densely clustered several together, occasionally single, near ends of woody leafless branches of the previous season, sometimes on short lateral spurs, 0·7–2(–3·5) cm. long; pedicels (and outside of receptacle and calyx) glabrous to densely pubescent, 4–6(–8) mm. long. Stamens ± 50–60, greenish-white; anthers yellowish. Fruits subglobose, sometimes oblique, beaked, ± 2·5–5 cm. in diameter when dry (to 6·5 cm. when fresh), green.

TANGANYIKA. Mpwapwa, 1930, *Hornby* 269! & 18 June 1933, *B. D. Burtt* 4755!; Iringa District: Idodi, June–July 1936, *Ward* I 11! & Nyangolo, 16 July 1956, *Milne-Redhead & Taylor* 11223!
DISTR. **T5, 7**; not known elsewhere
HAB. Deciduous woodland and bushland (*Commiphora*); 850–1220 m.

NOTE. The pedicels and outside of the receptacle and calyx are usually glabrous or nearly so, but in *B. D. Burtt* 4755 they are densely spreading-pubescent. I do not at present consider that this extreme of variation should be made a separate variety until we know how constantly distinct it is.

40. MILDBRAEDIODENDRON

Harms in Z.A.E.: 241, t. 27 (1911)

Unarmed tree. Leaves alternate, imparipinnate; stipules small, soon falling off; leaflets petiolulate, all alternate, or in one and the same leaf the upper opposite and the lower alternate, with numerous pellucid dots. Flowers ⚥ or ♂, in lateral racemes which are simple or occasionally forked

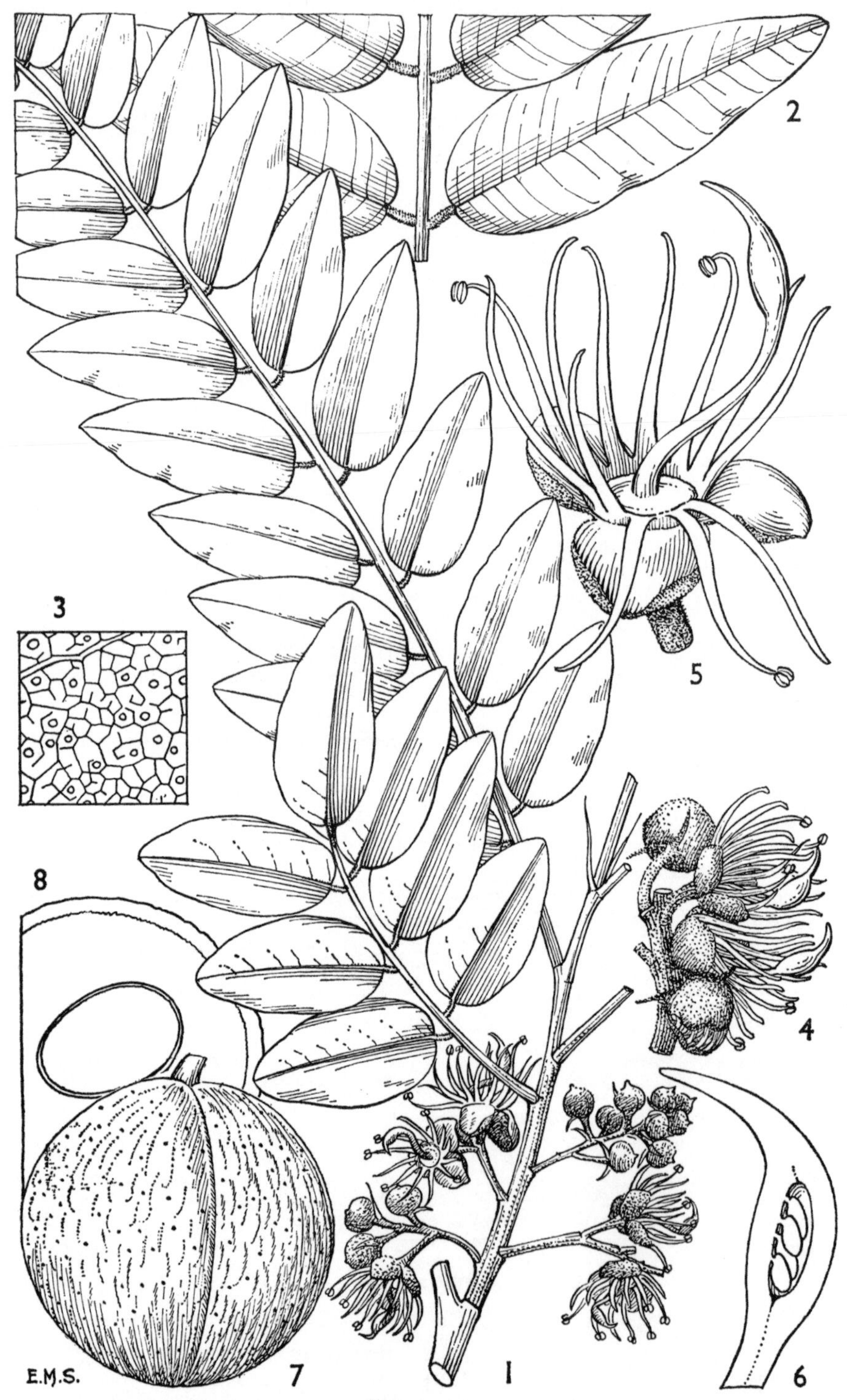

FIG. 52. *MILDBRAEDIODENDRON EXCELSUM*—**1**, part of flowering branch, × 1; **2**, leaflets, × 1; **3**, part of leaflet-surface, showing venation and gland-dots, × 6; **4**, part of inflorescence, × 1½; **5**, flower, × 3; **6**, ovary, longitudinal section, × 6; **7**, fruit, × 1; **8**, fruit, transverse section, × 1. 1, 3, 5,6, from *Eggeling* 1407; 2, from *Dawe* 997; 4, from *Lawton* 63; 7, 8, from *Myers* 6515.

below; each flower subtended by linear quickly falling bracteoles. Calyx subglobose and entire before dehiscence, splitting into 3 (or 2, *fide* Harms) lobes on opening. Petals 0. Stamens 12–18, arranged in a single row round the well-marked margin of the conspicuous flattened disc; filaments slightly connate at base; anthers dorsifixed, dehiscing by longitudinal slits; connective not glandular. Ovary long-stipitate, several-ovuled, tapering into a subulate style; stigma very small. Fruits green, stipitate, spherical, apparently not or scarcely oblique, indehiscent, with 1–3 seeds embedded in pulp. Seeds large, apparently not areolate, not arillate, without endosperm; radicle of embryo straight.

A genus with a single species, in tropical Africa, very closely related indeed to *Cordyla*, differing in the flattened marginate disc and receptacle, fewer stamens, eglandular anther-connective and the always (2–)3-lobed calyx.

M. excelsum *Harms* in Z.A.E.: 241, t. 27 (1911); L.T.A.: 607 (1929); Gilbert & Boutique in F.C.B. 3: 553 (1952); I.T.U., ed. 2: 306 (1952). Type: Congo Republic, between Beni and Ruwenzori, Lumengo, *Mildbraed* 2741 (B, holo. †)

Tall tree 20–50 m. high, buttressed at base; bark grey to brown, splitting into squares or rectangles. Young branchlets sparsely puberulous to glabrous. Leaves finely golden-puberulous when young; petiole and rhachis of mature leaves 25–45 cm. long; leaflets 12–19 pairs, narrowly oblong to lanceolate-oblong, (2·5–)3–8 cm. long, (1–)1·5–2·5 cm. wide, rounded to subcordate at base, obtusely pointed, rarely subacute at apex (apex itself sometimes slightly emarginate), sparsely and very inconspicuously appressed-puberulous beneath and often also above. Flowers yellow-green, in racemes 1–4 cm. long (to 7 cm., *fide* Harms) borne on current growth below the leaves; pedicels 5–7(–10) mm. long. Calyx ± 5–7 mm. in diameter before opening. Fruits 4–5 cm. or more in diameter. Seeds brown, thin-walled, ellipsoid, 4–7 cm. long, 2·5–3·5 cm. wide (mostly *fide* F.C.B.). Fig. 52.

UGANDA. W. Nile District: E. Madi, Zoka Forest, June 1933, *Eggeling* 1241 in *F.H.* 1341!; Bunyoro District: Budongo Forest, 13 Mar. & 20 May 1910, *Dawe* 997! & Budongo Forest, Waisoke R., Sept. 1933, *Eggeling* 1407 in *F.H.* 1355!

DISTR. U1, 2 and (*fide* I.T.U.) 4; Ghana, Nigeria, Cameroun Republic, Congo Republic and Sudan Republic

HAB. Lowland rain-forest; ± 790–1000 m.

INDEX TO CAESALPINIOIDEAE

GEOGRAPHICAL DIVISIONS OF THE FLORA

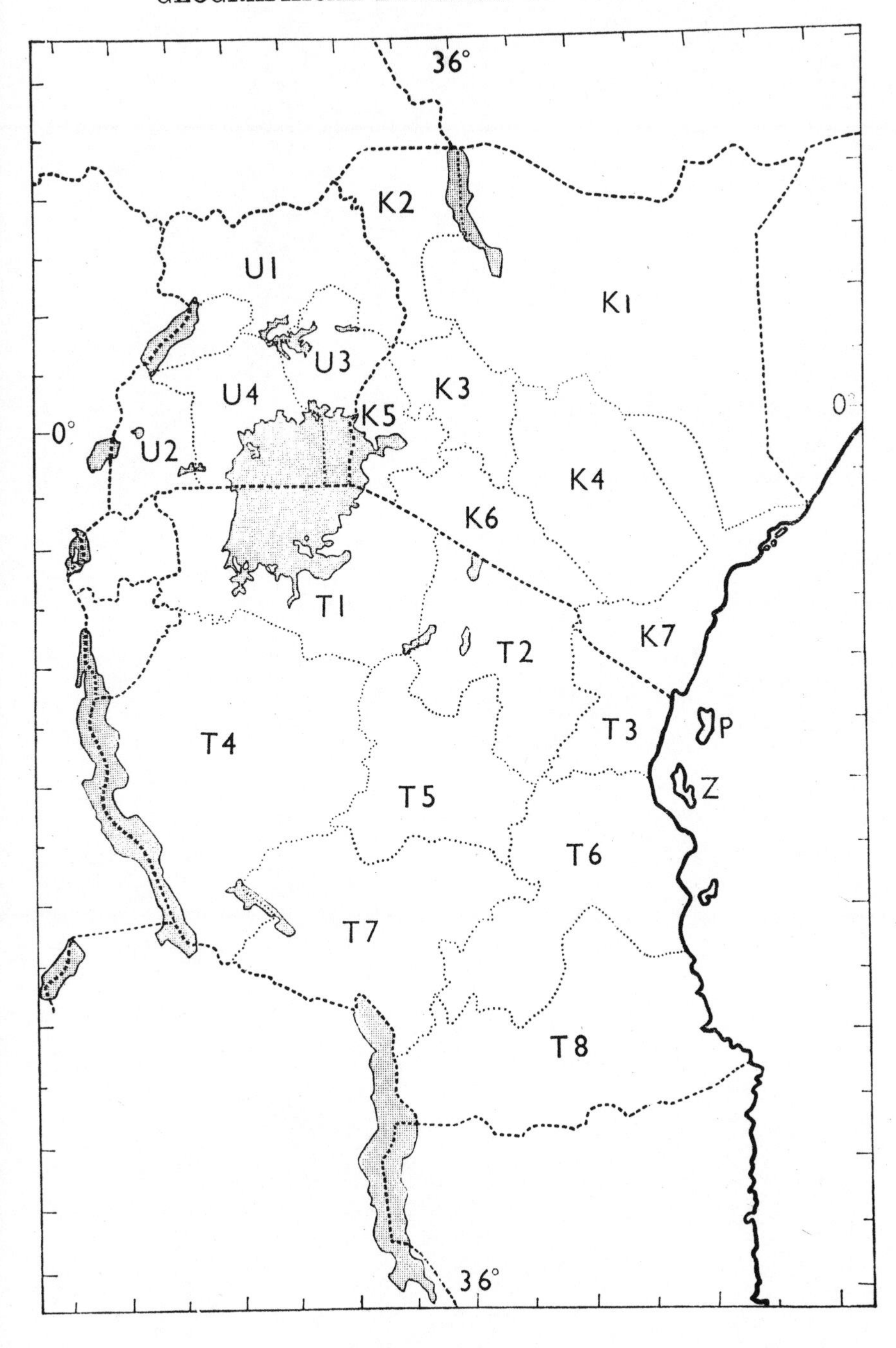